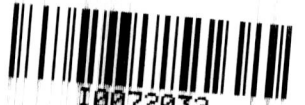

Evolutionary Biology Handbook

Evolutionary Biology Handbook

Edited by **Richard Arber**

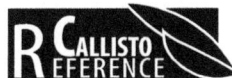

R CALLISTO
REFERENCE

New York

Published by Callisto Reference,
106 Park Avenue, Suite 200,
New York, NY 10016, USA
www.callistoreference.com

Evolutionary Biology Handbook
Edited by Richard Arber

International Standard Book Number: 978-1-63239-334-0 (Hardback)

Printed in the United States of America.

Contents

Preface

Does today's man look the same as say, 2 million years ago? Did you know that Pug and Great Dane have the same ancestor? Why children take on some physical and behavioral traits from their parents? All these questions and many more are answered by evolutionary biology (EB). A subfield of biology, EB studies the evolutionary processes that produced the diversity of life on earth as well as the descent and origin of species.

To understand the purpose of EB, we must know a bit of its history. Ancient Romans, Greeks, Chinese, and medieval Islamic science believed that species alter over a period of time. However, the theory was first formally proposed by Jean-Baptiste Lamarck in early 19th century. Further ahead, theories of natural selection and evolution were propagated by Charles Darwin. Though the acceptance of the theory was slow with some exceptions like Germany, etc.; eventually, evolution was agreed upon. That paved way for modern evolutionary synthesis, which involves evidence from various branches of biology like systematics, cytology, morphology, paleontology, etc.

Eventually, EB has reached its current status in the world of science and evolution studies. And the credit goes to evolutionary biologists who have worked hard and made the concept and different theories of evolution comprehensible. Evolutionary biologists like Theodosius Dobzhansky, R. A. Fisher, J. B. S. Haldane, George Gaylord Simpson are very few to name who have immensely contributed to this field. Moreover, it's heartening to see an expanding literature specifically dedicated to the subject.

Editor

Physiological and Biochemical Responses of *Ulva prolifera* and *Ulva linza* to Cadmium Stress

He-ping Jiang,[1,2] Bing-bing Gao,[1] Wen-hui Li,[1] Ming Zhu,[1,3] Chun-fang Zheng,[2] Qing-song Zheng,[1,2] and Chang-hai Wang[1]

[1] *College of Natural Resources and Environmental Science, Key Laboratory of Marine Biology, Nanjing Agricultural University, Nanjing, Jiangsu 210095, China*
[2] *Zhejiang Mariculture Research Institute, Zhejiang Key Laboratory of Exploitation and Preservation of Coastal Bio-Resource, Wenzhou, Zhejiang 325005, China*
[3] *College of Oceanography, Huaihai Institute of Technology, Lianyungang, Jiangsu 222005, China*

Correspondence should be addressed to Qing-song Zheng; qszheng@njau.edu.cn

Academic Editors: J. Huang and Z. Wang

Responses of *Ulva prolifera* and *Ulva linza* to Cd^{2+} stress were studied. We found that the relative growth rate (RGR), Fv/Fm, and actual photochemical efficiency of PSII (Yield) of two *Ulva* species were decreased under Cd^{2+} treatments, and these reductions were greater in *U. prolifera* than in *U. linza*. *U. prolifera* accumulated more cadmium than *U. linza* under Cd^{2+} stress. While *U. linza* showed positive osmotic adjustment ability (OAA) at a wider Cd^{2+} range than *U. prolifera*. *U. linza* had greater contents of N, P, Na^+, K^+, and amino acids than *U. prolifera*. A range of parameters (concentrations of cadmium, Ca^{2+}, N, P, K^+, Cl^-, free amino acids (FAAs), proline, organic acids and soluble protein, Fv/Fm, Yield, OAA, and K^+/Na^+) could be used to evaluate cadmium resistance in *Ulva* by correlation analysis. In accordance with the order of the absolute values of correlation coefficient, contents of Cd^{2+} and K^+, Yield, proline content, Fv/Fm, FAA content, and OAA value of *Ulva* were more highly related to their adaptation to Cd^{2+} than the other eight indices. Thus, *U. linza* has a better adaptation to Cd^{2+} than *U. prolifera*, which was due mainly to higher nutrient content and stronger OAA and photosynthesis in *U. linza*.

1. Introduction

Heavy metal contamination is an environmental problem in the margin sea [1]. As the economy in Asian countries continues to grow, the release of heavy metals and other contaminants has increased noticeably [2, 3]. Due to their acute toxicity, cadmium (Cd), lead, and mercury are among the most hazardous metals to the environment and living things [4].

Cd, an oxophilic and sulfophilic element, forms complexes with various organic particles and thereby triggers a wide range of reactions that collectively put the aquatic ecosystems at risk. Cadmium also poses a serious threat to human health due to its accumulation in the food chain [5, 6]. It has been classified as group (I) a human carcinogen by the International Agency for Research on Cancer (IARC) [7]. Cadmium toxicity may be characterized by a variety of syndromes and effects, including renal dysfunction, hypertension, hepatic injury, lung damage, and teratogenic effects [8]. To remove Cd pollutants, various treatment technologies, such as precipitation, ion exchange, adsorption, and biosorption, have been employed [9]. Biosorption is one of the promising techniques for removal of heavy metals. Biosorption utilizes the ability of biological materials to accumulate heavy metals from waste streams by either metabolically mediated or purely physicochemical pathways of uptake [10]. Among the biological materials investigated for heavy metal removal, marine macroalgae have high uptake capacities for a number of heavy metal ions [11, 12].

Green algae species of Ulvaceae, especially the members of the green algal genus *Ulva*, have been considered as monitors of heavy metals in estuaries [13–15]. Numerous studies have shown that green macroalgae such as *Ulva lactuca* are able to absorb Cd. These studies mainly focused on metabolism-independent Cd accumulation [6], synthetic surfactants exerting impact on uptake of Cd [12], effect of pH, contact time, biomass dosage and temperature on the Cd uptake kinetics [2], and induced oxidative stress by Cd [7]. However, little information is available regarding physiological responses of different *Ulva* species to increased Cd^{2+} concentrations.

In this study, *Ulva prolifera* and *Ulva linza* were studied for their responses to different Cd^{2+} concentrations. Their growth, chlorophyll fluorescence parameters, osmotic adjustment ability, and accumulation of inorganic ions and organic solutes were investigated in indoor seawater culture systems. The specific objective of this study was to determine if there was species variation in Cd^{2+} adaptation, and what were the major physiological parameters involved in the adaptation.

2. Materials and Methods

2.1. The Seaweed Collection, Cultivation, and Cd^{2+} Treatment.
Green algae were collected from the sea in Dafeng (*Ulva prolifera*) and Lianyungang (*Ulva linza*), Jiangsu province, China. Upon arrival in the laboratory, the seaweeds were washed with distilled water and then cultured in 250 mL flasks containing 200 mL of sterilized artificial seawater (33.33 psu, pH 8.0) enriched with VSE medium [16] for 5 d. The composition of artificial seawater was $(g\,L^{-1})$ HCO_3^- 0.25, SO_4^{2-} 3.84, Cl^- 17.45, Ca^{2+} 0.76, Mg^{2+} 1.00, K^+ 0.57, and Na^+ 9.46. The composition of VSE nutrient solution was $(mg\,L^{-1})$ $NaNO_3$ 42.50, $Na_2HPO_4 \cdot 12H_2O$ 10.75, $FeSO_4 \cdot 7H_2O$ 0.28, $MnCl_2 \cdot 4H_2O$ 0.02, $Na_2EDTA \cdot 2H_2O$ 3.72, vitamin B_1 0.20, Biotin 0.001, and vitamin B_{12} 0.001. After 5 d acclimation, healthy samples (0.5 g fresh weight) were cultured in 250 mL flasks with 200 mL medium as described earlier. $CdCl_2$ was added to each flask at the following concentrations: 0, 5, 10, 20, 40, 80, or 120 $\mu mol\,L^{-1}$. After 7 d treatment, *U. prolifera* and *U. linza* were harvested and analyzed for selected parameters as described later. All experiments were performed in three replicates. During the preculture and the treatment, seaweeds were grown in a GXZ intelligent light incubator at temperature of 20 ± 1°C, light intensity of 50 $\mu mol\,m^{-2}\,s^{-1}$, and photoperiod of 12/12 h. The culture medium was altered every other day.

2.2. Measurement of Relative Growth Rate (RGR).
Fresh weight was determined by weighing the algae after blotting by absorbent paper. RGR was calculated according to the formula RGR $(\%\,d^{-1}) = [\ln(M_t/M_0)/t] \times 100\%$, where M_0 and M_t are the fresh weights (g) at days 0 and 7, respectively [17].

2.3. Measurement of Osmotic Adjustment Ability (OAA).
Saturated osmotic potential was measured by the freezing-point depression principle. Seaweeds were placed in double-distilled water for 8 h and then rinsed 5 times with double-distilled water. After blotting dry with absorbent paper, seaweeds were dipped into liquid nitrogen for 20 min. The frozen seaweeds were thawed in a syringe for 50 min, and the

seaweed sap was then collected by pressing the seaweed in the syringe [18]. The π_{100} was measured by using a fully automatic freezing-point osmometer (8P, Shanghai, China). OAA was calculated by the following equation:

$$\Delta\pi_{100} = \pi_{100}^{\mu} - \pi_{100}^{s}, \tag{1}$$

whereby π_{100}^{μ} was the π_{100} of control seaweeds, and π_{100}^{s} was the π_{100} of Cd^{2+}-stressed seaweeds.

2.4. Measurements of Chlorophyll (Chl) and Carotenoid (Car) Contents.
Determination of Chl and Car was carried out by the method of Häder et al. [19]. Weighed 0.1 g fresh seaweeds were cut with scissors and extracted with 95% (v/v) ethanol (10 mL) in the dark for 24 h. The absorbance of pigment extract was measured at wavelengths of 470, 649, and 665 nm with a spectrophotometer. From the measured absorbance, concentrations of Chl a, Chl b, and Car were calculated on a weight basis.

2.5. Determination of Chlorophyll Fluorescence Parameters.
A PHYTO-PAM Phytoplankton Analyzer (PAM 2003, Walz, Effeltrich, Germany) was used to determine *in vivo* chlorophyll fluorescence from chlorophyll in photosystem II (PSII) using different experimental protocols [19]. Before determination, samples were adapted for 15 min in the total darkness to complete reoxidation of PSII electron acceptor molecules. The maximal photochemical efficiency of PSII (Fv/Fm) and the actual photochemical efficiency of PSII in the light (Yield) were then determined.

2.6. Measurement of Nitrogen (N) and Phosphorous (P) Concentrations.
Dried samples were ground in a mortar and pestle. Total N in seaweed tissue was analyzed by an N gas analyzer using an induction furnace and thermal conductivity. Total P in seaweed tissue was quantitatively determined by Inductively Coupled Plasma Atomic Emission Spectrometry (ICP-AES, Optima 2100 DV, PerkinElmer, USA) following nitric acid/hydrogen peroxide microwave digestion. The total amounts of N and P in the seaweed tissue were calculated by multiplying N and P contents in tissue as a proportion of dry weight by the total dry weight of the sample [20].

2.7. Measurement of Inorganic Elements.
After 7 d, seaweeds were harvested, washed, and oven-dried at 65°C for 3 d. A 50 mg sample was ashed in a muffle furnace. The ash was dissolved in 8 mL of $HNO_3 : HClO_4$ (3 : 1, v : v) and diluted to 50 mL with distilled water. The contents of Cd, Na, K, Ca, and Mg were determined by Inductively Coupled Plasma Atomic Emission Spectrometry (ICP-AES, Optima 2100 DV, PerkinElmer, USA) [21]. To determine Cl content, the ash was dissolved in 100 mL distilled water and analyzed by potentiometric titration with silver nitrate ($AgNO_3$) [18]. Total nitrate was measured as described previously [22] with nitrate extracted from the tissue by boiling fresh seaweeds (20 mg) in distilled water (400 μL) for 20 min. The nitrate concentrations in the samples were measured spectrophotometrically at 540 nm.

2.8. *Measurement of Organic Solutes.* Soluble sugars (SS) determination was carried out by the anthrone method [23]. Water extract of fresh seaweeds was added to 0.5 mL of 0.1 mol L^{-1} anthrone-ethyl acetate and 5 mL H_2SO_4. The mixture was heated at 100°C for 1 min, and its absorbance at 620 nm was read after cooling to room temperature. A calibration curve with sucrose was used as a standard. Soluble proteins (SPs) were measured by Coomassie Brilliant Blue G-250 staining [24]. Fresh seaweeds (0.5 g) were homogenized in 1 mL phosphate buffer (pH 7.0). The crude homogenate was centrifuged at 5,000 g for 10 min. An aliquot of 0.5 mL of freshly prepared trichloroacetic acid (TCA) was added and mixture centrifuged at 8.000 g for 15 min. The pellets were dissolved in 1 mL of 0.1 mol L^{-1} NaOH, and 5 mL of Bradford reagent was added. Absorbance was recorded at 595 nm using bovine serum albumin as a standard. Free amino acids (FAAs) were extracted and determined following the method of Zhou and Yu [23]. A total of 0.5 g fresh tissue was homogenized in 5 mL 10% (w/v) acetic acid, extracts were supplemented with 1 mL distilled water and 3 mL ninhydrin reagent, then boiled for 15 min and fast cooled, and the volume was made up to 5 mL with 60% (v/v) ethanol. Absorbance was read at 570 nm. The content of total free amino acids was calculated from a standard curve prepared using leucine. Proline (PRO) concentration was determined spectrophotometrically by adopting the ninhydrin method of Irigoyen et al. [25]. We first homogenized 300 mg fresh leaf samples in sulphosalicylic acid. To the extract, 2 mL each of ninhydrin and glacial acetic acid were added. The samples were heated at 100°C. The mixture was extracted with toluene, and the free toluene was quantified spectrophotometrically at 528 nm using L-proline as a standard. Organic acids (OAs) were extracted with boiling distilled water. The concentration of total OA was determined by 0.01 mmol L^{-1} NaOH titration method, with phenolphthalein as indicator [26].

2.9. *Statistical Analyses.* All experiments were performed in three replicates. The data are presented as the mean ± SD. Data were analyzed using SPSS statistical software. Significant differences between means were determined by Duncan's multiple range test. Unless otherwise stated, differences were considered statistically significant when $P \leq 0.05$. Statistical analysis on two-way variance analysis (ANOVA), and correlation coefficient was performed using Microsoft Excel.

3. Results

3.1. *Effect of Cadmium Stress on RGR and OAA of U. prolifera and U. linza.* Compared to the control, treatments with $5 \mu\text{mol L}^{-1}$ Cd^{2+} for 7 d did not change RGR of *U. linza*, but significantly decreased RGR of *U. prolifera*. The RGR of both *Ulva* species was significantly decreased as Cd^{2+} concentration increased. After 7 d exposure to 10, 20, 40, 80; or $120 \mu\text{mol L}^{-1}$ Cd^{2+}, RGR of *U. linza* decreased by 53, 75, 116, 177, and 277%, respectively; *U. prolifera* decreased by 93, 139, 271, and 357%, respectively. *U. prolifera* died at $120 \mu\text{mol L}^{-1}$ Cd^{2+} on day 7 (Figure 1).

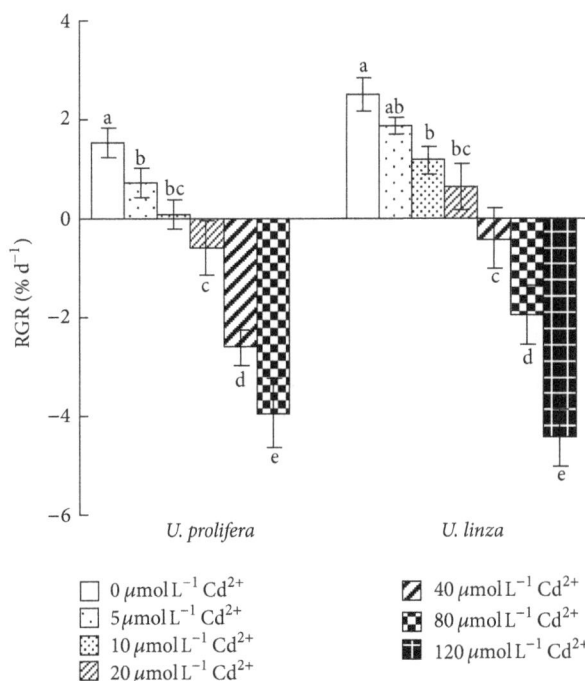

FIGURE 1: Effects of different concentrations of Cd^{2+} (0, 5, 10, 20, 40, 80, and $120 \mu\text{mol L}^{-1}$) on relative growth rate (RGR) in *U. prolifera* and *U. linza*.

The OAA of both species was enhanced by low Cd^{2+} concentration treatments. The enhancement occurred at 5 and $10 \mu\text{mol L}^{-1}$ for *U. prolifera* and 5, 10 and $20 \mu\text{mol L}^{-1}$ for *U. linza* (Figure 2). However, OAA was negative when *U. prolifera* was treated by 20, 40, and $80 \mu\text{mol L}^{-1}$ Cd^{2+}, and *U. linza* treated by 40 and $80 \mu\text{mol L}^{-1}$ Cd^{2+} (Figure 2).

3.2. *Effect of Cadmium Stress on Cadmium Content in U. prolifera and U. linza.* Cadmium contents in *U. prolifera* and *U. linza* increased as Cd^{2+} concentrations increased (Figure 3). At 5, 10, 20, 40, and $80 \mu\text{mol L}^{-1}Cd^{2+}$, Cd contents in *U. prolifera* was 32, 78, 114, 140, and 165 times of the Cd^{2+} = 0 treatment, respectively, and 10, 26, 44, 65, and 79 times of its control treatment in *U. linza*, respectively.

3.3. *Effect of Cadmium Stress on Chl and Car Contents in U. prolifera and U. linza.* Both Chl and Car contents decreased with the increased Cd^{2+} concentration. There was no significant change in Chl and Car when both species were treated by 5 and $10 \mu\text{mol L}^{-1}$ Cd^{2+} for 7 d. However, significant declines in Chl and Car contents were observed when they were exposed to 20, 40, or $80 \mu\text{mol L}^{-1}$ Cd^{2+}. Compared to the control treatment, Chl contents decreased by 18, 25, and 45% at 20, 40, and $80 \mu\text{mol L}^{-1}$ Cd^{2+} in *U. prolifera*, respectively; and the decreases were 16, 20, and 39% in *U. linza*, respectively (Figure 4(a)). The Car content declined by 16, 29 and 54% at 20, 40 and $80 \mu\text{mol L}^{-1}$ Cd^{2+} in *U. prolifera*, respectively; and by 13, 16, and 44% in *U. linza*, respectively (Figure 4(b)).

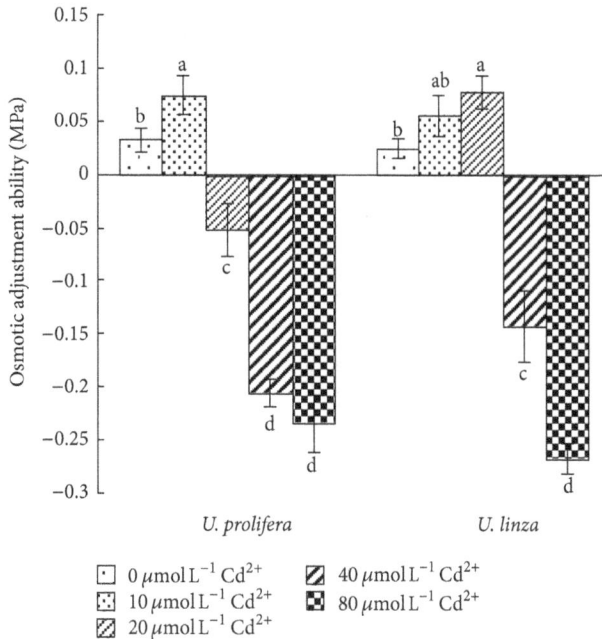

FIGURE 2: Effects of different concentrations of Cd^{2+} (5, 10, 20, 40, and 80 $\mu mol\,L^{-1}$) on osmotic adjustment ability (OAA) of *U. prolifera* and *U. linza*.

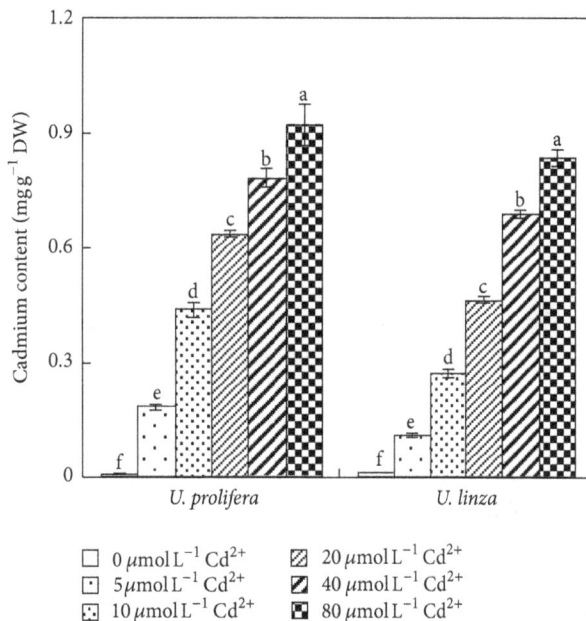

FIGURE 3: Effects of different concentrations of Cd^{2+} (0, 5, 10, 20, 40, and 80 $\mu mol\,L^{-1}$) on cadmium concentration of *U. prolifera* and *U. linza*.

3.4. Effect of Cadmium Stress on Chlorophyll Fluorescence Parameters of U. prolifera and U. linza.

Compared to the control treatment, Fv/Fm of *U. prolifera* and *U. linza* were not significantly affected by the treatments of 5 or 10 $\mu mol\,L^{-1}$ Cd^{2+}. However, Fv/Fm of both *Ulva* species fell significantly

when Cd^{2+} concentrations reached 20 $\mu mol\,L^{-1}$. In comparison with the control, Fv/Fm of *U. prolifera* decreased 17, 22, and 31% at 20, 40, and 80 $\mu mol\,L^{-1}$ Cd^{2+}; whereas Fv/Fm of *U. linza* decreased 9, 10, and 15% after exposure to 20, 40, or 80 $\mu mol\,L^{-1}$ Cd^{2+}, respectively (Figure 5(a)). For actual photochemical efficiency of PSII (Yield) of *U. prolifera*, there was an obvious decrease when Cd^{2+} concentrations rose from 20 to 80 $\mu mol\,L^{-1}$; whereas Yield of *U. linza* showed no significant decline until Cd^{2+} concentration was 80 $\mu mol\,L^{-1}$ (Figure 5(b)).

3.5. Effect of Cadmium Stress on Contents of N and P in U. prolifera and U. linza.

Contents of N and P in both *Ulva* species showed a declining trend after an initial increase. The highest N content was recorded at 10 $\mu mol\,L^{-1}$ Cd^{2+} in *U. prolifera* and at 20 $\mu mol\,L^{-1}$ Cd^{2+} in *U. linza*. N contents in *U. linza* in all Cd^{2+} treatments were higher than those of control; however, in *U. prolifera*, N contents at 20, 40, or 80 $\mu mol\,L^{-1}$ Cd^{2+} were significantly decreased compared to the control (Figure 6(a)).

U. prolifera had the highest P concentration at 5 $\mu mol\,L^{-1}$ Cd^{2+}; but the highest P concentration was observed when *U. linza* was treated by 10 $\mu mol\,L^{-1}$ Cd^{2+}. The P contents decreased 31, 40, and 54% at 20, 40, and 80 $\mu mol\,L^{-1}$ Cd^{2+} in *U. prolifera*, respectively. Compared to the control, the P concentration of *U. linza* at 20 $\mu mol\,L^{-1}$ Cd^{2+} increased significantly, and then decreased by 11 and 27% under 40, and 80 $\mu mol\,L^{-1}$ Cd^{2+}, respectively (Figure 6(b)).

3.6. Effect of Cadmium Stress on Inorganic Elements of U. prolifera and U. linza.

The Na^+ content of *U. prolifera* grown at 5 or 10 $\mu mol\,L^{-1}$ Cd^{2+} was not significantly different from the control, and it increased by 42, 67, and 83% at 20, 40, and 80 $\mu mol\,L^{-1}$ Cd^{2+}, respectively. However, in *U. linza*, 5, 10, 20, and 40 $\mu mol\,L^{-1}$ Cd^{2+} had no significant influence on Na^+ content, and 80 $\mu mol\,L^{-1}$ Cd^{2+} increased Na^+ content by 36% (Table 1). The K^+ content of *U. prolifera* grown at 5 or 10 $\mu mol\,L^{-1}$ Cd^{2+} remained unaffected compared to the control; it decreased significantly by 41, 45, and 62% at 20, 40, and 80 $\mu mol\,L^{-1}$ Cd^{2+}, respectively. In *U. linza*, 5, 10, and 20 $\mu mol\,L^{-1}$ Cd^{2+} had no significant influence on K^+ content, whereas 40 and 80 $\mu mol\,L^{-1}$ Cd^{2+} decreased K^+ content by 34 and 50%, respectively (Table 1). The Ca^{2+} content of *U. prolifera* grown at 5, 10, 20, or 40 $\mu mol\,L^{-1}$ Cd^{2+} remained unaffected, but increased significantly (24%) at 80 $\mu mol\,L^{-1}$ Cd^{2+}. However, in *U. linza*, 5 and 10 $\mu mol\,L^{-1}$ Cd^{2+} had no significant influence on Ca^{2+} contents, whereas 20, 40, and 80 $\mu mol\,L^{-1}$ Cd^{2+} increased Ca^{2+} content by 22, 39, and 50%, respectively (Table 1). The Mg^{2+} content of *U. prolifera* grown at 5, 10, 20, 40 or 80 $\mu mol\,L^{-1}$ Cd^{2+} remained unaffected. With increasing Cd^{2+} concentrations, Mg^{2+} contents of *U. linza* showed an increasing trend after an initial decline (Table 1). The Cl^- contents appeared to have a declining trend with increasing Cd^{2+} concentration similarly to Mg concentrations. However, no obvious difference in Cl^-

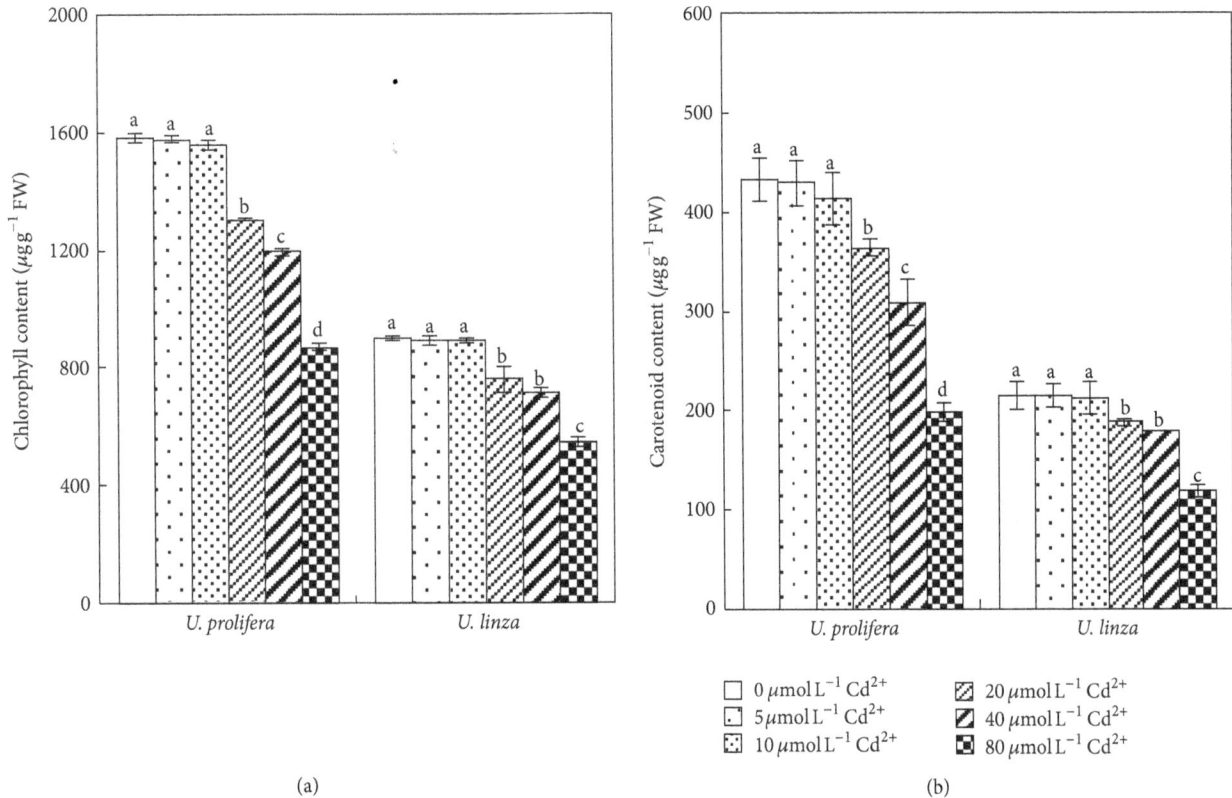

(a)

(b)

☐ 0 μmol L⁻¹ Cd²⁺   ▨ 20 μmol L⁻¹ Cd²⁺
⊡ 5 μmol L⁻¹ Cd²⁺   ◪ 40 μmol L⁻¹ Cd²⁺
⊞ 10 μmol L⁻¹ Cd²⁺   ▩ 80 μmol L⁻¹ Cd²⁺

FIGURE 4: Effects of different concentrations of Cd^{2+} (0, 5, 10, 20, 40, and 80 μmol L^{-1}) on chlorophyll content (a) and carotenoid content (b) in *U. prolifera and U. linza*.

contents among all Cd^{2+} treatments was noted in the two *Ulva* species (Table 1). Nitrate content in *U. prolifera* showed an uptrend with increasing Cd^{2+} concentration; however, with increasing Cd^{2+} concentrations, nitrate content of *U. linza* showed a decline trend after an initial increase. We also found that nitrate contents of *U. linza* were much more than those of *U. prolifera* under all treatments except for 80 μmol L^{-1} Cd^{2+} treatment (Table 1).

The K^+/Na^+ and Ca^{2+}/Na^+ ratios in *U. prolifera* were not influenced by 5 and 10 μmol L^{-1} Cd^{2+}, but they showed declining trends at 20, 40, and 80 μmol L^{-1} Cd^{2+} (Table 1). In *U. linza*, 5 and 10 μmol L^{-1} Cd^{2+} had no significant influence on the K^+/Na^+ ratio, whereas 20, 40, and 80 μmol L^{-1} Cd^{2+} decreased that ratio by 6, 45, and 64%, respectively. However, in *U. prolifera*, 20, 40, and 80 μmol L^{-1} Cd^{2+} decreased the K^+/Na^+ ratio by 55, 65, and 78%. No Cd^{2+} treatment significantly changed the Ca^{2+}/Na^+ ratio in *U. linza*.

3.7. Effect of Cadmium Stress on Organic Solutes in U. prolifera and U. linza. With increasing Cd^{2+} concentration, soluble sugar (SS) content appeared to have an ascending trend after an initial decline in both *Ulva* species. In *U. prolifera*, 40 μmol L^{-1} Cd^{2+} did not change the SS content, and 80 μmol L^{-1} Cd^{2+} increased SS concentration by 27% compared to the control. However, in *U. linza*, 40 and 80 μmol L^{-1} Cd^{2+} increased SS content by 40 and 90%, respectively

(Table 2). In *U. prolifera* and *U. linza*, 5 μmol L^{-1} Cd^{2+} significantly increased free amino acid (FAA) content by 25 and 16%, respectively. However, 10 μmol L^{-1} Cd^{2+} had no obvious change on FAA contents of the two *Ulva* species. Treatments with 20, 40, and 80 μmol L^{-1} Cd^{2+} significantly decreased FAA content by 52, 79, and 87% in *U. prolifera* and by 2, 25, and 43% in *U. linza* (Table 2). Proline (PRO) content was greatly enhanced by Cd^{2+} treatments in both *Ulva* species. At 5, 10, 20, 40, and 80 μmol L^{-1} Cd^{2+}, PRO content was increased 154, 431, 715, 1031, and 1069%, respectively, in *U. prolifera*; and increased 147, 420, 726, 1040, and 1147%, respectively, in *U. linza* (Table 2). Organic acid (OA) content in *U. prolifera* was not affected at 5, 10 and 20 μmol L^{-1} Cd^{2+}, and OA concentration in *U. linza* was not affected at 5, 10, 20, and 40 μmol L^{-1} Cd^{2+}. Treatments with 40 and 80 μmol L^{-1} Cd^{2+} decreased OA content by 29 and 47%, respectively, in *U. prolifera*, whereas in *U. linza* only 80 μmol L^{-1} Cd^{2+} decreased OA content by 27% (Table 2). The soluble protein (SP) content in the two *Ulva* species was not affected at 5, 10 and 20 μmol L^{-1} Cd^{2+} and was decreased at 40 and 80 μmol L^{-1} Cd^{2+}. Treatments with 40 and 80 μmol L^{-1} Cd^{2+} significantly decreased SP content by, respectively, 16 and 42% in *U. prolifera* and by 8 and 25% in *U. linza* (Table 2).

3.8. Correlation Analysis between RGR and Other Physiological and Biochemical Indexes under Cadmium Stress. Correlation

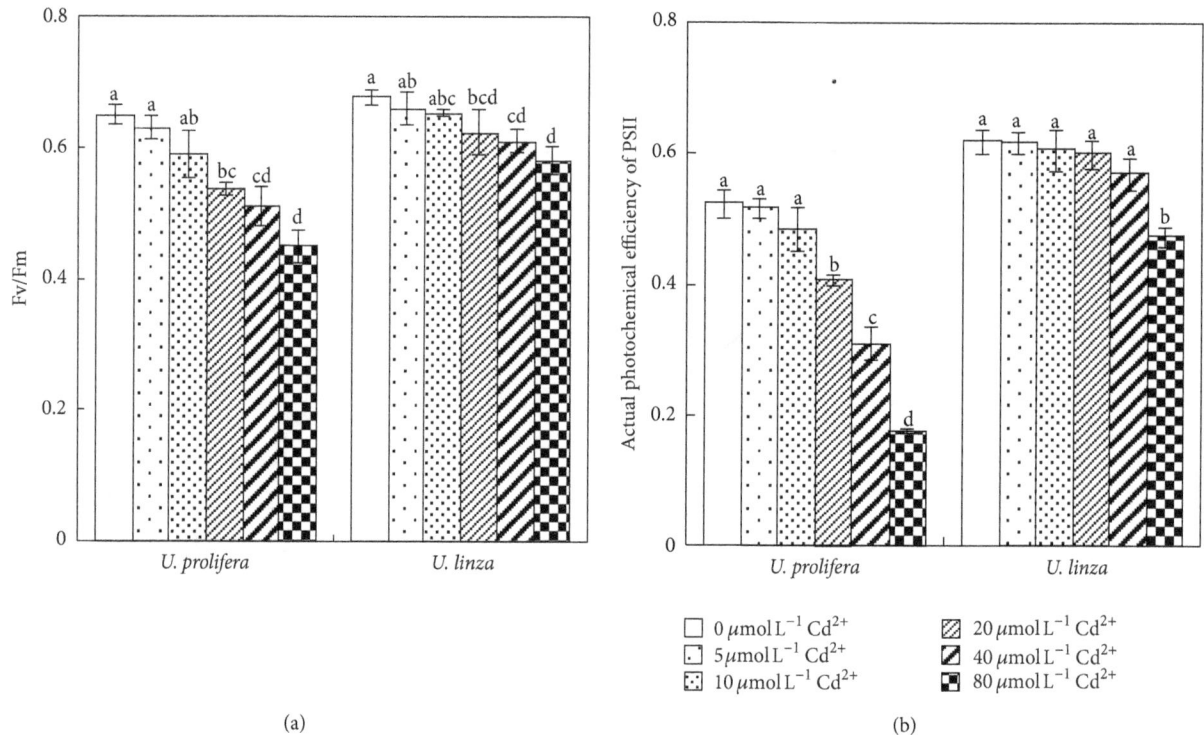

(a)

(b)

FIGURE 5: Effects of different concentrations of Cd^{2+} (0, 5, 10, 30, 40, and 80 μmol L^{-1}) on Fv/Fm (a) and Yield (actual photochemical efficiency of PSII) (b) of *U. prolifera* and *U. linza*.

analysis indicated that RGR of both *Ulva* species was insignificantly related to contents of Chl, Car, Na^+ and Mg^{2+}, and the Ca^{2+}/Na^+ ratio. In contrast, RGR was highly negative correlated with the contents of Cd^{2+}, Ca^{2+}, SS, and PRO, and highly positive correlated with the contents of N, P, K, Cl, FAA, OA and SP, K^+/Na^+ ratio, OAA, Fv/Fm, and Yield (Table 3).

4. Discussion

Plant growth can be suppressed by Cd [7, 17]. It was reported that *Ulva lactuca* was sensitive to cadmium, as obviously shown by growth reduction and lethal effects at 40 μmol L^{-1} Cd^{2+} within 6 days [27]. In the study presented here, *U. prolifera* and *U. linza*, the dominant free-floating *Ulva* species of green tide bloom in the Yellow Sea of China [28], showed sensitivity to Cd^{2+} (reduction in RGR, Fv/Fm, and Yield). Furthermore, this reduction was found to be more pronounced in *U. prolifera* than *U. linza*. After 7 d, *U. prolifera* died at 120 μmol L^{-1} Cd^{2+}, whereas *U. linza* was still alive (Figures 1 and 4). This result indicated that *U. linza* had better adaptation to Cd^{2+} toxicity than *U. prolifera*.

It is known that marine macroalgae can concentrate heavy metals to a large extent [2, 29]. In this study, Cd accumulation in *U. prolifera* and *U. linza* increased significantly in response to increased Cd^{2+} concentrations. However, *U. prolifera* accumulated more Cd than *U. linza* (Figure 3). In general, plant accumulation of a given metal

is a function of uptake capacity and intracellular binding sites [30]. The cell walls of plant cells contain proteins and different carbohydrates that can bind metal ions. After the binding sites in the cell wall become saturated, intracellular Cd accumulation mediated by metabolic processes may lead to cell toxicity [31].

Ulva species are widely distributed in the coastal intertidal zones where had full change on salinity level. Thus, many *Ulva* species have strong OAA to cope with variable and heterogeneous environments. Similarly to a number of other stresses, heavy metal toxicity can decrease cell water content and lower the cell water potential (ψ_w) through increased net concentrations of solutes (osmotic adjustment), which is a common response to water stress and an important mechanism for maintaining cell water content and, thus, turgor [18, 32]. In our experiments, OAA of *U. linza* had positive values in the treatments with 5, 10, or 20 μmol L^{-1} Cd^{2+}, whereas *U. prolifera* had positive OAA only at 5 and 10 μmol L^{-1} Cd^{2+} (Figure 2). When OAA values in *Ulva* were positive, that is, OAA contributed to maintaining turgor, *Ulva* could continue growing, and RGR was positive. However, when OAA in *Ulva* was negative resulting in turgor loss, the growth was stopped, and RGR was negative. Correlation analysis also showed that RGR was positively related to OAA, suggesting that OAA played an important role in maintaining algal growth. Also, good osmotic adjustment enabled plants to maintain high photosynthetic activity (Figure 5).

Cadmium is a nonessential element for plant growth, and it inhibits uptake and transport of many macro- and

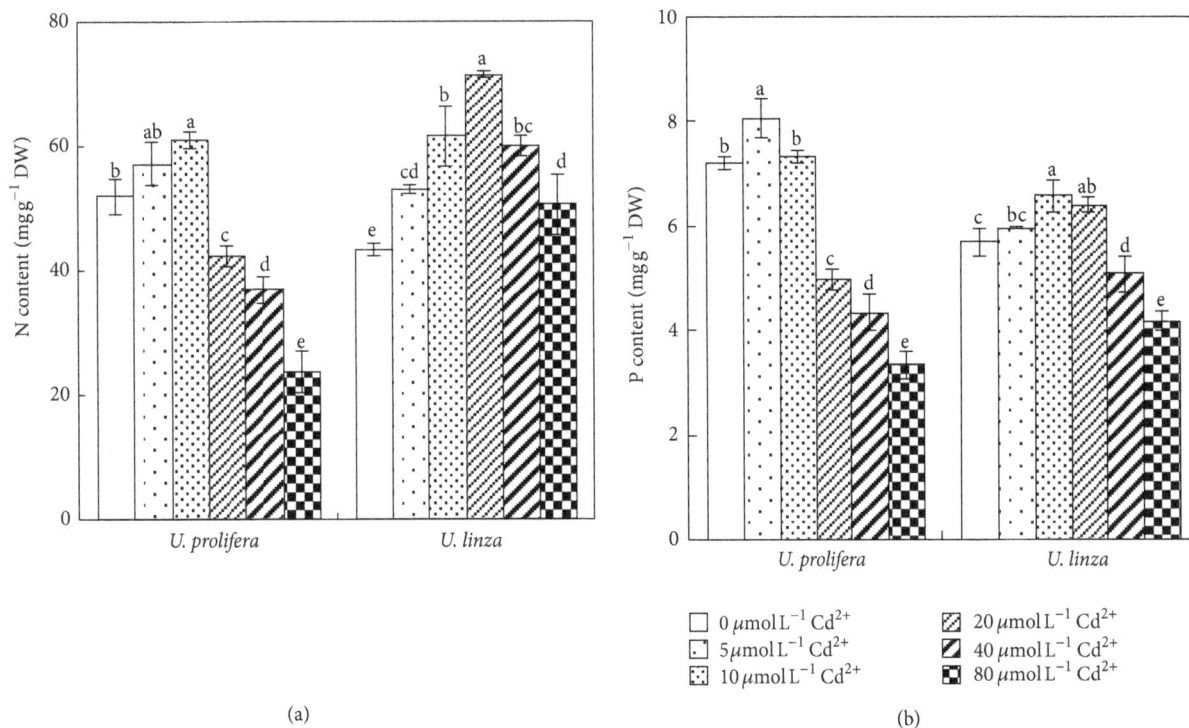

(a)

(b)

FIGURE 6: Effects of different concentrations of Cd^{2+} (0, 5, 10, 20, 40, 80 μmol L^{-1}) on contents of N (a) and P (b) of *U. prolifera* and *U. linza*.

micronutrients, inducing nutrient deficiency [7, 17]. Contradictory data can be found in the literature on the effects exerted by Cd^{2+} on terrestrial plant. Cadmium was reported to reduce uptake of N, P, K, Ca, Mg, Fe, Zn, Cu, Mn, Ni, and Na in many crop plants [33], whereas other authors found reduced K uptake but unchanged P uptake or even an increase in K content of several crop varieties under Cd^{2+} stress [34, 35]. Obata and Umebayashi [36] reported that Cd^{2+} treatment increased Cu content in the roots of pea, rice, and maize, but unchanged Cu content in cucumber and pumpkin plants. With Cd^{2+} stress, Maksimović et al. [37] observed a reduction in the maize root influx and root-shoot transport of Cu, Zn, and Mn, a reduction in the root-shoot transport of Fe, but an increase in Fe influx and Ca and Mg transport. In this study, the response of total N and P concentrations in tissues of the two *Ulva* species to Cd^{2+} treatments was positively related to their Cd resistance. We found that the treatment with low concentration of Cd^{2+} enhanced N and P contents, but high concentrations of Cd^{2+} ($\geq$20 μmol L^{-1}) decreased N and P contents in both *Ulva* species. The maintenance of total N and total P was more pronounced in less Cd-sensitive *U. linza* than Cd-sensitive *U. prolifera* (Figure 6). This suggests that the maintenance of a normal level of total N content upon challenge with Cd is likely to be a feature in relative Cd-resistant marine macroalga, similarly to terrestrial plants [38]. In *Ulva*, we found that the contents of K^+, Ca^{2+}, and Cl^- were related to RGR, especially K^+reduction caused *Ulva* growth reduction significantly (Table 1). Thus, the K^+/Na^+ ratio in both *Ulva* species decreased significantly with increasing Cd^{2+} treatment concentrations, and

Cd^{2+}-sensitive *U. prolifera* showed a greater K^+/Na^+ decline than Cd^{2+}-sensitive *U. linza* (Table 1).

We measured a decline in soluble sugar (SS) concentration at low Cd^{2+} treatment concentrations and an increase at high Cd^{2+} concentrations in both *Ulva* species. Moreover, the SS increase of *U. linza* is more marked than that of *U. prolifera*. In other studies, the decline in SS concentration corresponded with the photosynthetic inhibition or stimulation of respiration rate, affecting carbon metabolism and leading to production of other osmotica [39]. The accumulating soluble sugars in plants growing in presence of Cd^{2+} could provide an adaptive mechanism via maintaining a favorable osmotic potential under adverse conditions of Cd^{2+} toxicity [40].

Soluble protein (SP) content in organisms is an important indicator of metabolic changes and responds to a wide variety of stresses [41]. In this work, SP contents in *U. prolifera* and *U. linza* declined with increasing Cd^{2+} treatment concentrations. Free amino acid (FAA) contents in both *Ulva* species first increased and then declined, with such a decline more pronounced in *U. prolifera* than in *U. linza*. The decreased protein content together with the increased free amino acid content suggest that the protein synthesizing machinery was impaired due to the Cd^{2+} effect [42].

PRO accumulation in plant tissues in response to a number of stresses, including drought, salinity, extreme temperatures, ultraviolet radiation, or heavy metals, is well documented [43]. In this study, even though PRO content was increased in Cd^{2+}-treated *Ulva*, its absolute amount was relatively low. Under assumed localization of inorganic ions

TABLE 1: Effects of different concentrations of Cd^{2+} (0, 5, 10, 30, 40, and 80 μmol L^{-1}) on inorganic ion content (mmol g^{-1} DW), K$^+$/Na$^+$ and Ca^{2+}/Na$^+$ of *U. prolifera* and *U. linza*.

	Cd^{2+} treatment μmol L^{-1}	Na$^+$ mmol g^{-1} DW	K$^+$ mmol g^{-1} DW	Ca^{2+} mmol g^{-1} DW	Mg^{2+} mmol g^{-1} DW	Cl$^-$ mmol g^{-1} DW	NO$_3^-$ mmol g^{-1} DW	K$^+$/Na$^+$	Ca^{2+}/Na$^+$
U. prolifera	0	0.12 ± 0.01 c	0.64 ± 0.04 a	0.20 ± 0.02 b	0.82 ± 0.04 a	0.15 ± 0.01 a	0.34 ×10^{-3} ± 0.03× 10^{-3} c	5.01 ± 0.12 a	1.70 ± 0.08 a
	5	0.13 ± 0.02 c	0.62 ± 0.05 a	0.23 ± 0.01 ab	0.78 ± 0.05 a	0.11 ± 0.01 b	0.49 ×10^{-3} ± 0.06× 10^{-3} c	4.86 ± 0.21 a	1.66 ± 0.07 a
	10	0.12 ± 0.01 c	0.63 ± 0.04 a	0.23 ± 0.02 ab	0.76 ± 0.05 a	0.10 ± 0.01 b	0.78 ×10^{-3} ± 0.06× 10^{-3} b	5.10 ± 0.14 a	1.84 ± 0.12 a
	20	0.17 ± 0.02 b	0.38 ± 0.03 b	0.23 ± 0.02 ab	0.75 ± 0.04 a	0.09 ± 0.01 b	1.41 ×10^{-3} ± 0.08× 10^{-3} a	2.26 ± 0.15 b	1.39 ± 0.10 b
	40	0.20 ± 0.02 ab	0.35 ± 0.03 b	0.25 ± 0.02 ab	0.79 ± 0.04 a	0.10 ± 0.01 b	1.40 ×10^{-3} ± 0.11× 10^{-3} a	1.74 ± 0.11 c	1.24 ± 0.08 bc
	80	0.22 ± 0.01 a	0.24 ± 0.02 c	0.26 ± 0.02 a	0.73 ± 0.05 a	0.10 ± 0.01 b	1.43 ×10^{-3} ± 0.04× 10^{-3} a	1.12 ± 0.08 d	1.14 ± 0.07 c
U. linza	0	0.25 ± 0.02 b	0.74 ± 0.04 a	0.18 ± 0.02 c	0.78 ± 0.04 a	0.16 ± 0.01 a	0.86 ×10^{-3} ± 0.08× 10^{-3} d	3.02 ± 0.15 a	0.72 ± 0.09 a
	5	0.24 ± 0.03 b	0.74 ± 0.04 a	0.17 ± 0.02 c	0.75 ± 0.03 ab	0.12 ± 0.01 b	1.21 ×10^{-3} ± 0.10× 10^{-3} c	3.10 ± 0.23 a	0.73 ± 0.08 a
	10	0.24 ± 0.02 b	0.73 ± 0.04 a	0.18 ± 0.01 c	0.72 ± 0.04 ab	0.10 ± 0.02 b	1.89 ×10^{-3} ± 0.07× 10^{-3} a	3.04 ± 0.12 a	0.75 ± 0.06 a
	20	0.24 ± 0.01 b	0.68 ± 0.03 a	0.22 ± 0.02 b	0.64 ± 0.03 b	0.10 ± 0.01 b	2.07 ×10^{-3} ± 0.12× 10^{-3} a	2.85 ± 0.11 b	0.83 ± 0.07 a
	40	0.26 ± 0.02 b	0.49 ± 0.03 c	0.25 ± 0.02 ab	0.68 ± 0.03 b	0.11 ± 0.01 b	1.65 ×10^{-3} ± 0.05× 10^{-3} b	1.67 ± 0.07 c	0.88 ± 0.08 a
	80	0.34 ± 0.02 a	0.37 ± 0.02 d	0.27 ± 0.02 a	0.72 ± 0.04 ab	0.12 ± 0.02 b	1.12 ×10^{-3} ± 0.11× 10^{-3} c	1.10 ± 0.05 d	0.75 ± 0.06 a

The data in the same column are statistically different if labeled with different letters according to Duncan's multiple range test ($P \leq 0.05$).

TABLE 2: Effects of different concentration of Cd^{2+} (0, 5, 10, 30, 40, and 80 μmol L^{-1}) on organic solute content of *U. prolifera* and *U. linza*.

	Cd^{2+} treatment μmol L^{-1}	SS mmol g^{-1} DW	FAA mmol g^{-1} DW	PRO mmol g^{-1} DW	OA mmol g^{-1} DW	SP mg g^{-1} DW
	0	0.15 ± 0.02 b	1.03 ± 0.05 b	$0.13 \times 10^{-3} \pm 0.02 \times 10^{-3}$ e	0.17 ± 0.01 a	42.15 ± 2.33 a
	5	0.15 ± 0.02 b	1.29 ± 0.12 a	$0.33 \times 10^{-3} \pm 0.02 \times 10^{-3}$ d	0.17 ± 0.01 a	41.38 ± 2.76 a
U. prolifera	10	0.12 ± 0.01 bc	1.10 ± 0.04 ab	$0.69 \times 10^{-3} \pm 0.04 \times 10^{-3}$ c	0.19 ± 0.02 a	40.45 ± 1.86 a
	20	0.10 ± 0.01 c	0.49 ± 0.11 c	$1.06 \times 10^{-3} \pm 0.07 \times 10^{-3}$ b	0.16 ± 0.02 a	38.39 ± 2.75 ab
	40	0.14 ± 0.01 b	0.22 ± 0.05 d	$1.47 \times 10^{-3} \pm 0.09 \times 10^{-3}$ a	0.12 ± 0.01 b	35.53 ± 2.63 b
	80	0.19 ± 0.01 a	0.13 ± 0.06 d	$1.52 \times 10^{-3} \pm 0.12 \times 10^{-3}$ a	0.09 ± 0.01 c	24.35 ± 1.88 c
	0	0.10 ± 0.01 cd	1.23 ± 0.03 b	$0.15 \times 10^{-3} \pm 0.05 \times 10^{-3}$ f	0.11 ± 0.02 a	39.27 ± 1.22 a
	5	0.10 ± 0.01 cd	1.43 ± 0.09 a	$0.37 \times 10^{-3} \pm 0.02 \times 10^{-3}$ e	0.12 ± 0.01 a	38.89 ± 2.37 ab
U. linza	10	0.07 ± 0.01 d	1.21 ± 0.10 b	$0.78 \times 10^{-3} \pm 0.03 \times 10^{-3}$ d	0.13 ± 0.01 a	38.52 ± 2.67 ab
	20	0.10 ± 0.01 c	1.20 ± 0.06 b	$1.24 \times 10^{-3} \pm 0.08 \times 10^{-3}$ c	0.14 ± 0.02 a	37.13 ± 1.89 ab
	40	0.14 ± 0.02 b	0.97 ± 0.06 c	$1.71 \times 10^{-3} \pm 0.07 \times 10^{-3}$ b	0.12 ± 0.01 a	35.95 ± 2.41 b
	80	0.19 ± 0.01 a	0.76 ± 0.08 d	$1.87 \times 10^{-3} \pm 0.15 \times 10^{-3}$ a	0.08 ± 0.01 b	29.34 ± 1.87 c

Different letters in the same column indicate statistical difference according to Duncan's multiple range test ($P \leq 0.05$). "SS, FAA, PRO, OA, and SP" in the table indicate the content of soluble sugar, free amino acid, proline, organic acid, and soluble protein, respectively.

TABLE 3: Correlation coefficients between RGR and other indices for *U. prolifera* and *U. linza*.

Index	Correlation coefficient
Chl content	0.072
Car content	0.198
Fv/Fm	0.830**
Yield	0.858**
Cd^{2+} content	−0.899**
N content	0.561**
P content	0.687**
OAA	0.766**
Na^+ content	−0.138
K^+ content	0.881**
Ca^{2+} content	−0.677**
Mg^{2+} content	0.060
Cl^- content	0.444**
K^+/Na^+	0.627**
Ca^{2+}/Na^+	−0.079
SS content	−0.617**
FAA content	0.828**
PRO content	−0.841**
OA content	0.731**
SP content	0.752**

*Significant at 5% level, **significant at 1% level (two-tailed, $n = 18$).

in the vacuole and organic solutes in the cytoplasm, FAA and PRO may be mainly in the cytoplasm, accounting for about 5%–10% volume in mature cells [44]. A small amount of FAA and PRO accumulating in the cytoplasm can increase concentration significantly and play an important role in balancing vacuolar osmotic potential [44]. It has often been suggested that PRO accumulation may contribute to osmotic adjustment at the cellular level [39]. In addition, PRO as a compatible solute may protect enzymes from dehydration and inactivation [18].

In conclusion, exposing *U. prolifera* and *U. linza* to different concentrations of Cd^{2+} resulted in the changes in growth, pigment content, chlorophyll fluorescence parameters, Cd accumulation, OAA, and concentration of N, P, main inorganic ions, and organic solutes. These changes make *U. linza* better adapted to withstanding Cd^{2+} stress in comparison with *U. prolifera*. Our results highlight the role of osmotic adjustment in *Ulva* during Cd^{2+} stress as an important mechanism enabling *Ulva* to maintain photosynthetic activity and, thus, growth under Cd^{2+} stress.

Authors' Contribution

H. Jiang and B. Gao both contributed equally to this paper.

Acknowledgments

The authors thank professor Zed Rengel, University of Western Australia, for his helpful comments on this study. Also, they thank professor Jianjun Chen, University of Florida, for his good revision on this paper. This work was supported by Open Foundation of Key Laboratory of Exploitation and Preservation of Coastal Bio-resources of Zhejiang Province (2010F30003), the National Key Project of Scientific and Technical Supporting Programs funded by Ministry of Science and Technology of China (no. 2009BADA3 B04-8). The authors also acknowledge members of our laboratory for assistance in this work.

References

[1] L. Zhu, J. Xu, F. Wang, and B. Lee, "An assessment of selected heavy metal contamination in the surface sediments from the South China Sea before 1998," *Journal of Geochemical Exploration*, vol. 108, no. 1, pp. 1–14, 2011.

[2] A. Sari and M. Tuzen, "Biosorption of Pb(II) and Cd(II) from aqueous solution using green alga (*Ulva lactuca*) biomass," *Journal of Hazardous Materials*, vol. 152, no. 1, pp. 302–308, 2008.

[3] B. Xu, X. Yang, Z. Gu, Y. Zhang, Y. Chen, and Y. Lv, "The trend and extent of heavy metal accumulation over last one hundred years in the Liaodong Bay, China," *Chemosphere*, vol. 75, no. 4, pp. 442–446, 2009.

[4] P. Lodeiro, B. Cordero, J. L. Barriada, R. Herrero, and M. E. Sastre De Vicente, "Biosorption of cadmium by biomass of brown marine macroalgae," *Bioresource Technology*, vol. 96, no. 16, pp. 1796–1803, 2005.

[5] B. Volesky and Z. R. Holan, "Biosorption of heavy metals," *Biotechnology Progress*, vol. 11, no. 3, pp. 235–250, 1995.

[6] E. A. Webster, A. J. Murphy, J. A. Chudek, and G. M. Gadd, "Metabolism-independent binding of toxic metals by *Ulva lactuca*: cadmium binds to oxygen-containing groups, as determined by NMR," *BioMetals*, vol. 10, no. 2, pp. 105–117, 1997.

[7] M. Kumar, P. Kumari, V. Gupta, P. A. Anisha, C. R. K. Reddy, and B. Jha, "Differential responses to cadmium induced oxidative stress in marine macroalga *Ulva lactuca* (Ulvales, Chlorophyta)," *BioMetals*, vol. 23, no. 2, pp. 315–325, 2010.

[8] S. Hajialigol, M. A. Taher, and A. Malekpour, "A new method for the selective removal of cadmium and zion ions from aqueous solution by modified clinoptilolite," *Adsorption Science and Technology*, vol. 24, no. 6, pp. 487–496, 2006.

[9] P. X. Sheng, Y. P. Ting, J. P. Chen, and L. Hong, "Sorption of lead, copper, cadmium, zinc, and nickel by marine algal biomass: characterization of biosorptive capacity and investigation of mechanisms," *Journal of Colloid and Interface Science*, vol. 275, no. 1, pp. 131–141, 2004.

[10] A. El-Sikaily, A. E. Nemr, A. Khaled, and O. Abdelwehab, "Removal of toxic chromium from wastewater using green alga *Ulva lactuca* and its activated carbon," *Journal of Hazardous Materials*, vol. 148, no. 1-2, pp. 216–228, 2007.

[11] S. Karthikeyan, R. Balasubramanian, and C. S. P. Iyer, "Evaluation of the marine algae Ulva fasciata and *Sargassum* sp. for the biosorption of Cu(II) from aqueous solutions," *Bioresource Technology*, vol. 98, no. 2, pp. 452–455, 2007.

[12] K. Masakorala, A. Turner, and M. T. Brown, "Influence of synthetic surfactants on the uptake of Pd, Cd and Pb by the marine macroalga, *Ulva lactuca*," *Environmental Pollution*, vol. 156, no. 3, pp. 897–904, 2008.

[13] P. J. Say, I. G. Burrows, and B. A. Whitton, "Enteromorpha as a monitor of heavy metals in estuaries," *Hydrobiologia*, vol. 195, pp. 119–126, 1990.

[14] P. Malea, J. W. Rijstenbil, and S. Haritonidis, "Effects of cadmium, zinc and nitrogen status on non-protein thiols in the macroalgae Enteromorpha spp. from the Scheldt Estuary (SW Netherlands, Belgium) and Thermaikos Gulf (N Aegean Sea, Greece)," *Marine Environmental Research*, vol. 62, no. 1, pp. 45–60, 2006.

[15] A. Lin, S. Shen, G. Wang et al., "Comparison of chlorophyll and photosynthesis parameters of floating and attached *Ulva prolifera*," *Journal of Integrative Plant Biology*, vol. 53, no. 1, pp. 25–34, 2011.

[16] C. Yarish, T. Chopin, R. Wilkes, A. C. Mathieson, X. G. Fei, and S. Lu, "Domestication of Nori for northeast America: the Asian experience," *Bulletin of Aquaculture Association of Canada*, vol. 1, pp. 11–17, 1999.

[17] J. R. Xia, Y. J. Li, J. Lu, and B. Chen, "Effects of copper and cadmium on growth, photosynthesis, and pigment content in *Gracilaria lemaneiformis*," *Bulletin of Environmental Contamination and Toxicology*, vol. 73, no. 6, pp. 979–986, 2004.

[18] Q. Zheng, Z. Liu, G. Chen, Y. Gao, Q. Li, and J. Wang, "Comparison of osmotic regulation in dehydrationand salinity-stressed sunflower seedlings," *Journal of Plant Nutrition*, vol. 33, no. 7, pp. 966–981, 2010.

[19] D. P. Häder, M. Lebert, and E. W. Helbling, "Effects of solar radiation on the Patagonian macroalga *Enteromorpha linza* (L.) J. Agardh—Chlorophyceae," *Journal of Photochemistry and Photobiology B*, vol. 62, no. 1-2, pp. 43–54, 2001.

[20] K. Kamer and P. Fong, "Nitrogen enrichment ameliorates the negative effects of reduced salinity on the green macroalga *Enteromorpha intestinalis*," *Marine Ecology Progress Series*, vol. 218, pp. 87–93, 2001.

[21] F. B. Wu, J. Dong, Q. Q. Qiong, and G. P. Zhang, "Subcellular distribution and chemical form of Cd and Cd-Zn interaction in different barley genotypes," *Chemosphere*, vol. 60, no. 10, pp. 1437–1446, 2005.

[22] M. B. Gilliam, M. P. Sherman, J. M. Griscavage, and L. J. Ignarro, "A spectrophotometric assay for nitrate using NADPH oxidation by *Aspergillus* nitrate reductase," *Analytical Biochemistry*, vol. 212, no. 2, pp. 359–365, 1993.

[23] Q. Zhou and B. Yu, "Changes in content of free, conjugated and bound polyamines and osmotic adjustment in adaptation of vetiver grass to water deficit," *Plant Physiology and Biochemistry*, vol. 48, no. 6, pp. 417–425, 2010.

[24] M. M. Bradford, "A rapid and sensitive method for the quantitation of microgram quantities of protein utilizing the principle of protein dye binding," *Analytical Biochemistry*, vol. 72, no. 1-2, pp. 248–254, 1976.

[25] J. J. Irigoyen, D. W. Emerich, and M. Sanchezdiaz, "Water-stress induced changes in concentrations of proline and total soluble sugars in nodulated alfalfa (*Medicago sativa*) plants," *Plant Physiology*, vol. 84, no. 1, pp. 55–60, 1992.

[26] J. Song, G. Feng, C. Y. Tian, and F. S. Zhang, "Osmotic adjustment traits of *Suaeda physophora*, *Haloxylon ammodendron* and *Haloxylon persicum* in field or controlled conditions," *Plant Science*, vol. 170, no. 1, pp. 113–119, 2006.

[27] J. W. Markham, B. P. Kremer, and K. R. Sperling, "Cadmium effects on growth and physiology of *Ulva lactuca*," *Helgoländer Meeresuntersuchungen*, vol. 33, no. 1-4, pp. 103–110, 1980.

[28] D. Liu, J. K. Keesing, Z. Dong et al., "Recurrence of the world's largest green-tide in 2009 in Yellow Sea, China: *Porphyra yezoensis* aquaculture rafts confirmed as nursery for macroalgal blooms," *Marine Pollution Bulletin*, vol. 60, no. 9, pp. 1423–1432, 2010.

[29] H. A. Baumann, L. Morrison, and D. B. Stengel, "Metal accumulation and toxicity measured by PAM—Chlorophyll fluorescence in seven species of marine macroalgae," *Ecotoxicology and Environmental Safety*, vol. 72, no. 4, pp. 1063–1075, 2009.

[30] M. P. Benavides, S. M. Gallego, and M. L. Tomaro, "Cadmium toxicity in plants," *Brazilian Journal of Plant Physiology*, vol. 17, no. 1, pp. 21–34, 2005.

[31] S. Hu, C. H. Tang, and M. Wu, "Cadmium accumulation by several seaweeds," *Science of the Total Environment*, vol. 187, no. 2, pp. 65–71, 1996.

[32] M. T. Milone, C. Sgherri, H. Clijsters, and F. Navari-Izzo, "Antioxidative responses of wheat treated with realistic concentration of cadmium," *Environmental and Experimental Botany*, vol. 50, no. 3, pp. 265–276, 2003.

[33] L. Sanità Di Toppi and R. Gabbrielli, "Response to cadmium in higher plants," *Environmental and Experimental Botany*, vol. 41, no. 2, pp. 105–130, 1999.

[34] F. F. Nocito, L. Pirovano, M. Cocucci, and G. A. Sacchi, "Cadmium-induced sulfate uptake in maize roots," *Plant Physiology*, vol. 129, no. 4, pp. 1872–1879, 2002.

[35] Z. Ciecko, S. Kalembasa, M. Wyszkowski, and E. Rolka, "The effect of elevated cadmium content in soil on the uptake of nitrogen by plants," *Plant, Soil and Environment*, vol. 50, no. 7, pp. 283–294, 2004.

[36] H. Obata and M. Umebayashi, "Effects of cadmium on mineral nutrient concentrations in plants differing in tolerance for cadmium," *Journal of Plant Nutrition*, vol. 20, no. 1, pp. 97–105, 1997.

[37] I. Maksimović, R. Kastori, L. Krstić, and J. Luković, "Steady presence of cadmium and nickel affects root anatomy, accumulation and distribution of essential ions in maize seedlings," *Biologia Plantarum*, vol. 51, no. 3, pp. 589–592, 2007.

[38] T. Ghnaya, I. Slama, D. Messedi, C. Grignon, M. H. Ghorbel, and C. Abdelly, "Effects of Cd2$^+$ on K$^+$, Ca2$^+$ and N uptake in two halophytes *Sesuvium portulacastrum* and *Mesembryanthemum crystallinum*: consequences on growth," *Chemosphere*, vol. 67, no. 1, pp. 72–79, 2007.

[39] R. John, P. Ahmad, K. Gadgil, and S. Sharma, "Effect of cadmium and lead on growth, biochemical parameters and uptake in *Lemna polyrrhiza* L.," *Plant, Soil and Environment*, vol. 54, no. 6, pp. 262–270, 2008.

[40] S. Verma and R. S. Dubey, "Effect of cadmium on soluble sugars and enzymes of their metabolism in rice," *Biologia Plantarum*, vol. 44, no. 1, pp. 117–123, 2001.

[41] P. K. Singh and R. K. Tewari, "Cadmium toxicity induced changes in plant water relations and oxidative metabolism of *Brassicajuncea* L. plants," *Journal of Environmental Biology*, vol. 24, no. 1, pp. 107–112, 2003.

[42] S. Jana and M. A. Choudhuri, "Synergistic effects of heavy metal pollutants on senescence in submerged aquatic plants," *Water, Air, and Soil Pollution*, vol. 21, no. 1–4, pp. 351–357, 1984.

[43] D. Delmail, P. Labrousse, P. Hourdin, L. Larcher, C. Moesch, and M. Botineau, "Differential responses of *Myriophyllum alterniflorum* DC (Haloragaceae) organs to copper: physiological and developmental approaches," *Hydrobiologia*, vol. 664, no. 1, pp. 95–105, 2011.

[44] R. Munns and M. Tester, "Mechanisms of salinity tolerance," *Annual Review of Plant Biology*, vol. 59, pp. 651–681, 2008.

2

Genomics Approaches for Crop Improvement against Abiotic Stress

Bala Anı Akpınar,[1] Stuart J. Lucas,[1,2] and Hikmet Budak[1,2]

[1] Faculty of Engineering and Natural Sciences, Sabanci University, Orhanlı, Tuzla, 34956 Istanbul, Turkey
[2] Sabanci University Nanotechnology Research and Application Centre (SUNUM), Sabanci University, Orhanlı, Tuzla, 34956 Istanbul, Turkey

Correspondence should be addressed to Hikmet Budak; budak@sabanciuniv.edu

Academic Editors: A. Levine and Z. Wang

As sessile organisms, plants are inevitably exposed to one or a combination of stress factors every now and then throughout their growth and development. Stress responses vary considerably even in the same plant species; stress-susceptible genotypes are at one extreme, and stress-tolerant ones are at the other. Elucidation of the stress responses of crop plants is of extreme relevance, considering the central role of crops in food and biofuel production. Crop improvement has been a traditional issue to increase yields and enhance stress tolerance; however, crop improvement against abiotic stresses has been particularly compelling, given the complex nature of these stresses. As traditional strategies for crop improvement approach their limits, the era of genomics research has arisen with new and promising perspectives in breeding improved varieties against abiotic stresses.

1. Introduction

Abiotic stresses are the most significant causes of yield losses in plants, implicated to reduce yields by as much as 50% [1]. Among abiotic stresses, drought is the most prominent and widespread; consequently the drought stress response has been dissected into its components and extensively studied in order to understand tolerance mechanisms thoroughly [2]. To improve abiotic stress, particularly drought, tolerance of cereals is of extreme importance, as cereals, including wheat and barley, are the main constituents of the world food supply. However, many abiotic stresses are complex in nature, controlled by networks of genetic and environmental factors that hamper breeding strategies [3]. As traditional approaches for crop improvement reach their limits, agriculture has to adopt novel approaches to meet the demands of an ever-growing world population.

Recent technological advances and the aforementioned agricultural challenges have led to the emergence of high-throughput tools to explore and exploit plant genomes for crop improvement. These genomics-based approaches aim to decipher the entire genome, including genic and intergenic regions, to gain insights into plant molecular responses which will in turn provide specific strategies for crop improvement. In this paper, genomics approaches for crop improvement against abiotic stresses will be discussed under three generalized classes, functional, structural, and comparative genomics. However, it should be noted that genomics approaches are highly intermingled, in terms of both the methodologies and the outcome (Figure 1).

2. Functional Genomics

Genomics research is frequently realized by functional studies, which produce perhaps the most readily applicable information for crop improvement. Functional genomics techniques have long been adopted to unravel gene functions and the interactions between genes in regulatory networks, which can be exploited to generate improved varieties. Functional genomics approaches predominantly employ sequence or hybridization based methodologies which are discussed below.

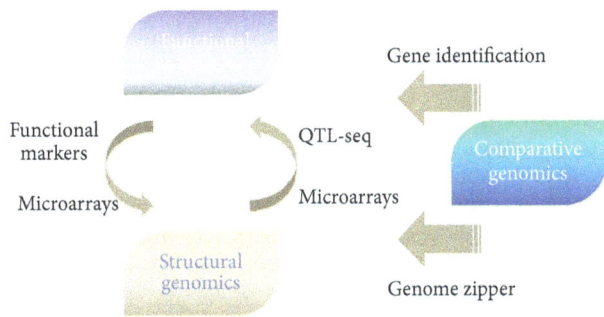

FIGURE 1: Functional, structural, and comparative genomics approaches are highly interrelated. For example, microarrays can be used either to anchor markers to genome maps or to analyze gene expression; functional markers indicate both phenotypes and genetic locations; QTL-seq utilizes a reference genome sequence to isolate QTLs based on phenotypic variation. As more structural genomics information becomes available, comparative genomics tools such as genome zippers can be used both to elucidate the structure of unsequenced genomes and as a shortcut to design targeted functional studies.

2.1. Sequencing-Based Approaches. One way to explore the expressed gene catalogue of a species is to analyze Expressed Sequence Tags (ESTs). ESTs are partial genic sequences that are generated by single-pass sequencing of cDNA clones [4]. Despite the concerns over the quality of ESTs as well as the representation of the parental cDNA [5], ESTs have been shown to identify corresponding genes unambiguously in a rapid and cost-effective fashion [4]; therefore, ESTs have been a major focus on functional studies.

Large-scale EST sequencing has been one of the earliest strategies for gene discovery and genome annotation [5, 6]. Currently, over a million ESTs are deposited in the EST database at National Center for Biotechnological Information (NCBI) for important crops such as maize, soybean, wheat, and rice, along with several thousands of ESTs for other plants (http://www.ncbi.nlm.nih.gov/dbEST/). cDNA libraries from various tissues, developmental stages, or treatments generally serve as the sources for EST sequencing to reveal differentially expressed genes [7]. These approaches can successfully identify tissue or developmental stage-specific and treatment-responsive transcripts. However, such cDNA libraries may underrepresent rare transcripts or transcripts that are not expressed under certain conditions. In addition, ESTs are usually much shorter in length than the cDNAs from which they are obtained. Assembly of overlapping EST sequences into consensus contigs is likely to be more informative on the structure of the parental cDNA, which may reveal polymorphisms. However, assembly and interpretation must be handled cautiously, as paralogous genes may lead to misassemblies of sequences, particularly in polyploid species such as wheat [5]. EST sequencing is utilized extensively in the absence of whole genome sequences, particularly in crops with large and repetitive genomes, although the entire transcriptome is unlikely to be fully represented and resolved. Even so, EST sequencing is still a valid approach, and a recent study has demonstrated its potential in gene discovery via the

comparison of different genotypes under control and stress conditions [8].

An alternative approach, Serial Analysis of Gene Expression (SAGE), has been developed to quantitate the abundance of thousands of transcripts simultaneously. In this approach, short sequence tags from transcripts are concatenated and sequenced, giving an absolute measure of gene expression [9, 10]. The ability of these short tags to identify genes unambiguously depends on the existence of comprehensive EST databases for the respective species [11]. Although SAGE is not widely applied in plants, there are a number of examples, including modifications to the original methodology, such as SuperSAGE and DeepSAGE [12–17]. The first report of SAGE in plants not only identified novel genes but also implied novel functions for known genes in rice seedlings [12]. SAGE has also been used to investigate stress-responsive genes [12, 15]. A similar tag-based approach, Massively Parallel Signature Sequencing (MPSS), where longer sequence tags are ligated to microbeads and sequenced in parallel, enables analysis of millions of transcripts simultaneously [18]. Due to longer tags and high-throughput analysis, MPSS is likely to identify genes with greater specificity and sensitivity. The ability of MPSS to capture rare transcripts is particularly beneficial in species that lack a whole genome sequence [19]. In plants, besides mRNA transcripts, MPSS has been employed in the expression studies of small RNAs [20, 21], which are increasingly implicated in abiotic stress responses [22]. Currently, plant MPSS databases (http://mpss.udel.edu/) contain publicly available MPSS expression data for a number of plant species, including important crops such as rice, maize, and soybean [23]. These MPSS data can be extracted, compiled, and compared with newly generated MPSS data for functional analysis of gene expression, as demonstrated by Jain et al. [24].

2.2. Hybridization-Based Approaches. In contrast to sequence-based approaches, array-based techniques utilize hybridization of the target DNA with cDNA or oligonucleotide probes attached to a surface to assess expression [25, 26]. These array-based methods are targeted; that is, prior knowledge of the transcript to be analyzed, either sequence or clone, is a prerequisite to design probes [27]. Extensive microarray expression data exists for *Arabidopsis thaliana* and rice [28–31], model species with fully sequenced genomes. In addition, microarray studies have been widely employed in crop species such as wheat [32], barley [33], and maize [34], as well as less emphasized but still industrially and agriculturally important plant species, such as cotton [35], cassava [36], and tomato [37] to unravel stress responses.

Besides inherent limitations such as cross-hybridization and background noise, microarray studies investigating stress-responsive genes suffer from technical considerations that may limit their usefulness. Isolation of total RNA from complex tissues that are composed of different types of cells may obscure transcript changes occurring in cell types that are particularly relevant to the stress response. Subtle transcriptional changes may be diluted in the overall stress response of the whole tissue and, thus, remain unnoticed.

Similarly, the choice of tissue or genotype that is sampled in a microarray study is closely related to the relevance of results. Reproductive tissues and stress-tolerant genotypes are most relevant in terms of agricultural gain and stress adaptation mechanisms, respectively [38]. In addition, laboratory-based stress treatments rarely represent field conditions, where multiple stresses usually act together. Interestingly, a comparison of microarray studies carried out using different water deficit stress conditions revealed only a small number of commonly regulated genes [39]. Abiotic stresses are generally complex in nature, eliciting intricate mechanisms of responses in plants. Consequently, slight differences in the experimental application of stress conditions may produce significant differences in stress responses. A further caveat when interpreting microarray studies is that many transcripts are known to undergo posttranscriptional and posttranslational modifications, which results in uncorrelated transcriptomic and proteomic data in some cases.

For species with an available whole genome sequence, a successful expansion of array-based transcript profiling is whole genome tiling arrays [27]. Tiling arrays can identify novel transcriptional units on chromosomes and alternative splice sites and can map transcripts and methylation sites [40, 41]. Tiling arrays have already been applied in model species to investigate abiotic stress responses [42–44].

2.3. Expansions to Functional Genomics Approaches. Genome wide expression profiles are most useful in the detection of candidate genes for desired traits, such as stress tolerance. A fraction of functional studies then adopt inactivation or overexpression of such candidate genes for further characterization and utilization. Of these, Targeting Induced Local Lesions IN Genomes (TILLING) enables high-throughput analysis of large number of mutants [45]. TILLING is applicable to virtually all genes in all species where mutations can be induced and has been reported in several crop species, including hexaploid wheat [46]. TILLING mutants are reported in sorghum [47], maize [48], barley [49], soybean [50], rice [51], and other crops. Although TILLING populations are conventionally screened by phenotypic or genotypic variations, further use of certain TILLING mutants in elucidation of stress responses has been demonstrated. In such a study, TILLING mutants for a specific kinase were used to assess salt stress response in legume species [52].

Importantly, a modified strategy, called EcoTILLING, has been developed to identify natural polymorphisms, analogous to TILLING-assisted identification of induced mutations. Polymorphisms demonstrating natural variation in germplasms are valuable tools in genetic mapping. Furthermore, via the discovery of polymorphisms among individuals, EcoTILLING is able to implicate favorable haplotypes for further analyses, such as sequencing. Similar to TILLING, EcoTILLING is applicable to polyploid species, where it can be utilized to differentiate between alleles of homologous and paralogous genes [53]. In a recent study, EcoTILLING not only provided allelic variants of a number of genes involved in salt stress response but also emphasized the complex nature of salt stress; salt-tolerant genotypes were revealed to harbor

different combinations of favorable alleles indicating the presence of multiple pathways conferring salt stress tolerance [54]. Transcription factors, diversifying stress responses, have also been targeted via EcoTILLING to examine natural rice variants exposed to drought stress [55].

The availability of comprehensive EST databases is central to the success of the above-mentioned approaches to identify genes accurately and unambiguously. Besides their utility in genome annotation and expression profiling, ESTs also provide a source of sequences for designing "functional markers." Functional markers refer to polymorphic sites on genes that are attributed to phenotypic variation of traits among individuals of a species. Functional marker design requires the knowledge of the allelic sequences of functionally characterized genes [56]. In contrast to random DNA markers, functional markers are completely linked to the trait of interest; hence, these markers are also called "perfect markers." The use of random DNA markers in breeding studies necessitates validation and revalidation of linkage between the marker and the trait over generations, since genetic recombination may break the linkage [56, 57]. In addition, functional markers may explore natural variation and biodiversity better, particularly compared to random DNA markers with absence/presence polymorphisms, where allelic variations of a trait exceed that of the linked DNA marker. In the case of such random DNA markers, the locus tested during genotyping will only exhibit biallelic variation, whereas the linked gene may actually have more variants [56]. The importance of functional markers has been highlighted in stress tolerance studies as well [58, 59].

3. Structural Genomics

While functional genomics focus on the functions of genes and gene networks, structural genomics focus on the physical structure of the genome, aiming to identify, locate, and order genomic features along chromosomes. Together, structural genomics and functional genomics can characterize a genome to its full extent.

3.1. Genome Sequencing and Mapping. In the last decade, advances in DNA sequencing technologies have enabled the generation of a wealth of sequence information including whole genome sequences. Next-generation sequencing (NGS) platforms such as Roche 454 GS FLX Titanium (http://www.454.com/) or Illumina Solexa Genome Analyzer (http://www.illumina.com/) can carry out high capacity sequencing at reduced costs and increased rates compared to conventional Sanger sequencing [60]. These advances have paved the way for the exploitation of plant genomics studies for breeding improved varieties. Through NGS technologies, sequencing and resequencing of even large genomes have become feasible. Accordingly, reference or draft genome sequences for a number of species, including the model species *Arabidopsis thaliana* and *Brachypodium distachyon*, along with important crop species such as rice, sorghum, soybean, and maize, have been published [61]. Whole genome sequences provide remarkably detailed information on genomic features including coding and noncoding genes, regulatory sequences, repetitive elements, and GC content which can be exploited in functional studies such as microarray or tiling arrays [41]. A high-quality reference genome sequence is considered pivotal to crop improvement via molecular breeding, particularly for complex traits. Despite their usefulness, producing such reference genomes requires a major investment of resources, and currently they are only available for species with relatively small genomes of low repetitive content [61].

Triticeae genomics, including that of the staple crops barley and wheat, has lagged behind recent advances primarily due to their large and complex genomes (~5 Gb for barley and ~17 Gb for wheat) [62]. As pointed out by Morrell et al. [61], 25x coverage sequencing of *Drosophila* is equivalent to approximately 1x coverage of wheat genome in terms of sequence read counts, demonstrating the challenging genome size of wheat. The high content of repetitive elements is another major challenge, causing ambiguities in sequence assembly. In polyploid species such as wheat, the sequence assembly problem is further exacerbated due to the presence of homoeologous genomes and paralogous loci [61]. For such genomes, construction of a reference sequence has been considered unattainable until recently.

Over the last few years, advances in chromosome sorting technologies have enabled construction of chromosome-specific Bacterial Artificial Chromosome (BAC) libraries to tackle the challenges of complex genomes. Physical mapping of the 1 Gb chromosome 3B of hexaploid wheat has proven the feasibility of a chromosome-by-chromosome approach to explore and exploit complex genomes [63]. Physical maps not only compile genetic mapping data into physical contigs but also serve as scaffolds for sequence assembly into a reference genome. The physical mapping and reference genome sequencing of wheat and barley are ongoing with combined efforts from a number of consortia [62].

In the absence of reference genome sequences, whole genome or BAC-end shotgun sequences provide valuable insights into genome structure and evolution [64–69]. Intriguingly, whole genome shotgun sequences have also been proposed for Quantitative Trait Loci (QTL) detection via a very recently developed methodology named QTL-seq. In this method, extremes of a population exhibiting a normal distribution with respect to a trait of interest are bulked, sequenced, and compared to detect putative QTLs [70].

3.2. Molecular Markers. Genomics applications involving molecular markers are largely dominated by Single Nucleotide Polymorphisms (SNPs) [71] as reflected in the predominance of software related to SNP discovery [60]. The high abundance of SNPs in genomes is particularly beneficial for their use in genomics. SNPs are readily identified by genome or transcript resequencing and by comparison of different genotypes in species where reference genome sequences or extensive transcript databases are available. Transcriptome resequencing not only avoids repetitive sequences of complex genome but also identifies SNPs within transcripts that may serve as functional markers [72]. However, due to low-quality

sequences obtained by most NGS platforms, over-sampling may be required to differentiate SNPs from sequencing errors [71]. In addition, the presence of homoeologous and paralogous loci must be taken into account in SNP identification in polyploid species [72]. Despite the challenges of SNP discovery on the repetitive portion of genomes, efforts are underway to improve SNP identification even in gene-poor regions [73]. In fact, these regions are of functional importance as well; for example, an important vernalization gene *Vrn-D4* has recently been mapped to the centromeric region of chromosome 5D of hexaploid wheat [62, 74].

A recently developed molecular marker type, Insertion Site-Based Polymorphisms (ISBPs), utilizes the insertional polymorphisms observed in the repeat junctions of complex genomes [75]. ISBP markers are readily designed from low coverage shotgun sequences, such as BAC-end sequences [64, 69]. Typically, 50–60% of ISBP markers tested are specific for the locus from which they were designed, and in one study which these ~70% contained SNPs in at least some members of a panel of 14 wheat genotypes [75]. This approach may break the ground for genome saturation particularly for crops with highly repetitive genomes that are impractical to exploit otherwise.

3.3. Applications of Structural Genomics in Crop Improvement. A major impact of NGS-mediated shotgun sequences has been their substantial contribution to the development of molecular markers. These markers indicate diagnostic polymorphisms at the DNA sequence level, and in contrast to morphological markers which once had been the focus of traditional breeding studies, they are not affected by the environment [76]. In general terms, Marker-Assisted Selection (MAS) refers to the utilization of molecular markers in breeding improved varieties with respect to desired traits, such pathogen resistance, abiotic stress tolerance, or high yield [77]. Through MAS, phenotype can be predicted from genotype [71]. For efficient and accurate MAS, the trait of interest should be tightly linked to a molecular marker [78] or more preferably flanked by two close markers. Recombination between both flanking markers and the trait is less likely to occur compared to a single marker, due to the low frequency of double crossovers. In both cases, a genetic distance of less than 5 cM for each marker from the trait is crucial to the success of MAS [77].

Additionally, for efficient MAS, markers should be highly polymorphic in the germplasm used for breeding. MAS can make use of molecular markers at multiple levels. Plant breeding depends on genetic diversity to improve crops [57]. Molecular markers may aid in the exploration of the variation among the germplasm to select the best candidate parental lines. Similarly, molecular markers may identify heterotic groups or ensure genomic purity of cultivars to achieve heterosis. In addition, molecular markers also assist in backcrossing. Plant breeding conventionally involves several backcrossing steps to enable transfer of one or a few traits to an elite cultivar while retaining most of the recurrent genomes. In general, at least six rounds of backcrossing are required to achieve the desired homozygosity, particularly for the selection of traits with low heritability. In contrast, MAS can greatly accelerate this process by utilizing both the flanking markers linked to the trait for selecting the trait and a set of unlinked markers for tracking the recurrent genome. Flanking markers and selection for recombination also reduces "linkage drag," which is the reduction in crop performance due to the cotransfer of undesirable traits that are located in the vicinity of the trait of interest [77]. Typically, a conventional QTL analysis can provide a resolution of approximately 15 cM intervals which may contain hundreds of genes [79]. The availability of a saturated map can potentially reduce this interval to less than 1 cM by backcrossing [78]. Furthermore, MAS enables early selection of traits that are labor and/or cost-intensive to score phenotypically, that are under complex genetic control, or that are manifested late in development. In cases where genotyping by MAS is affordable, this dramatically reduces the number of the plants to be screened in further steps [77, 78].

The major drawbacks of MAS in breeding are high costs of implementation, typically requiring specialized equipment, and the risk of recombination between the marker and the trait that reduces the reliability of MAS to predict phenotype via genotype. The high cost of MAS is particularly relevant in cases where an effective phenotyping method is already established through conventional breeding. Additionally, MAS usually requires the validation of QTLs when applied in different genetic backgrounds. Functional markers, however, may overcome the issue of QTL validation [78]. Despite its drawbacks, MAS has been successfully utilized to improve crops for abiotic stress tolerance, including drought [80], salinity [81], and waterlogging [82] given that the genetic element responsible for the high tolerance is accurately defined and delineated.

Another use of molecular markers is Map-Based Cloning (MBC) where the gene or a QTL linked to a desired trait is isolated via a "mini" physical map. Such a local physical map flanking the gene must be saturated with molecular markers for efficient MBC [6]. Prior to the construction of high-density physical maps, MBC approaches were inefficient, particularly due to the difficulty of finding unique probes in repetitive sequences for chromosome walking. Importantly, repeat contents of barley and wheat genomes, two staple crops, are estimated to exceed 80% of the whole genome [83, 84], potentiating the utility of physical maps. Accordingly, the physical map of chromosome 3B provided sufficient data to enable fine mapping of 16 genes and QTLs in chromosome 3B, none of which had been previously cloned [64].

4. Comparative Genomics

For species with largely unexplored genomes, comparative genomics is a promising tool to gain information by utilizing the conservation between closely related plant species. In fact, plant genomes share extensive similarities even between distantly related species (Figure 2, [85]). Among the plant kingdom, grasses have been the focus of comparative genomics analyses due to their high agronomic importance. The extent of genome conservation first became evident by

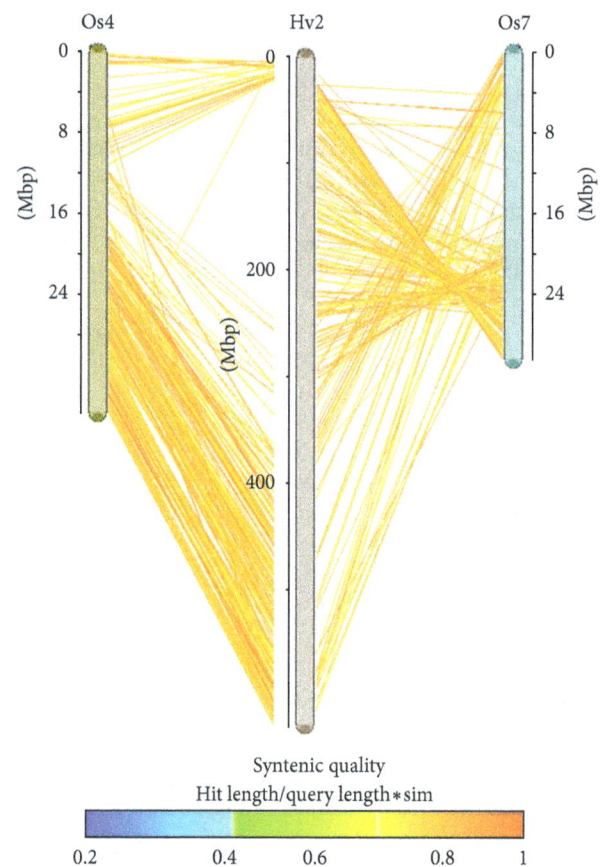

FIGURE 2: Example of colinearity between grass genomes. Analysis of conserved gene sequences between barley (*H. vulgare*) and rice (*O. sativa*) shows that many genes are found in colinear (syntenic) order. In this example, chromosome 2 of barley (Hv2, centre) is compared with chromosomes 4 and 7 of rice (Os4 and Os7). Each colored line represents a gene conserved between the two chromosomes, with the color indicating the strength of the syntenic relationship. It is clear that many genes from both ends of Hv2 are colinear with the ends of Os4, while the centre of the chromosome is largely colinear with Os7, but in the reverse order. Image was generated using the CrowsNest Comparative Map Viewer at MIPS (http://mips.helmholtz-muenchen.de/plant/genomes.jsp).

comparative genome mapping studies, which suggested a colinear order of genes and markers shared by genomes of different species. It is noteworthy that plant genomes differ by several orders of magnitude in size; yet these differences generally correspond to intergenic regions. Although detailed analyses have revealed notable rearrangements such as inversions, deletions, and translocations at the molecular level, large-scale colinearity across grass genomes has been exploited for gene discovery and isolation [86–88].

Comparative genomics has contributed significantly to the emergence of the "genome zipper" concept, which enables the determination of a virtual gene order in a partially sequenced genome. Genome zippers compare the fully sequenced and annotated genomes of *Brachypodium*, sorghum, and rice with various sources of data from less well-studied species, such as genomic survey sequences and genetically mapped markers, to predict the gene order and organization in these species [65, 89]. These genome zippers indicate evolutionary relationships and medium-scale rearrangements, and for the *Triticeae* provide the closest approximation to a reference genome sequence currently available [65]. However, its reliance on synteny means that recently evolved genes and small-scale rearrangements cannot be explored by this approach.

In addition to syntenic genes that are found in colinear blocks of conserved genes, nonsyntenic genes that are found outside their syntenic location in other genomes also provide valuable insight into genome evolution and speciation. Intriguingly, a recent study focused on the nonconserved portion of wheat and barley genome that suggested novel mechanisms, besides transposable element-driven expansion, have driven the evolution and size of these genomes. Many of these non-syntenic genes exhibited pseudogene characteristics, which may have implications for gene content estimate of these large genomes based on survey sequences [90].

Despite the utility of comparative genomics and genome zippers, it is evident that species-specific genomic features can only be accessed through a fully annotated reference genome sequences. Homoeologous genes with different orthologous relationships are examples of such species-specific features [91]. Species-specific rearrangements are also implicated in the formation of gene islands, containing mainly non-syntenic genes, in large crop genomes [92]. Through comparative genomics, Mayer et al. [65] concluded that genomic models can represent the barley genome to a limited extent. Thus, it can be argued that for maximal exploitation of crop genomes, such as wheat and barley, the construction of reference genome sequences scaffolded by highly-saturated physical and genetic maps is indispensable. Accordingly, efforts to accomplish this goal are currently underway (International Wheat Genome Sequencing Consortium for wheat; International Barley Sequencing Consortium for barley).

References

[1] F. Qin, K. Shinozaki, and K. Yamaguchi-Shinozaki, "Achievements and challenges in understanding plant abiotic stress responses and tolerance," *Plant and Cell Physiology*, vol. 52, no. 9, pp. 1569–1582, 2011.

[2] M. Kantar, S. J. Lucas, and H. Budak, "Drought stress: molecular genetics and genomics approaches," in *Advances in Botanical Research*, I. Turkan, Ed., pp. 445–493, Elsevier, Burlington, Mass, USA, 2011.

[3] T. R. Sinclair, "Challenges in breeding for yield increase for drought," *Trends in Plant Science*, vol. 16, no. 6, pp. 289–293, 2011.

[4] D. Bouchez and H. Höfte, "Functional genomics in plants," *Plant Physiology*, vol. 118, no. 3, pp. 725–732, 1998.

[5] S. Rudd, "Expressed sequence tags: alternative or complement to whole genome sequences?" *Trends in Plant Science*, vol. 8, no. 7, pp. 321–329, 2003.

[6] R. K. Varshney, D. A. Hoisington, and A. K. Tyagi, "Advances in cereal genomics and applications in crop breeding," *Trends in Biotechnology*, vol. 24, no. 11, pp. 490–499, 2006.

[7] K. Yamamoto and T. Sasaki, "Large-scale EST sequencing in rice," *Plant Molecular Biology*, vol. 35, no. 1-2, pp. 135–144, 1997.

[8] N. Z. Ergen and H. Budak, "Sequencing over 13 000 expressed sequence tags from six subtractive cDNA libraries of wild and modern wheats following slow drought stress," *Plant, Cell and Environment*, vol. 32, no. 3, pp. 220–236, 2009.

[9] V. E. Velculescu, L. Zhang, B. Vogelstein, and K. W. Kinzler, "Serial analysis of gene expression," *Science*, vol. 270, no. 5235, pp. 484–487, 1995.

[10] M. E. Vega-Sánchez, M. Gowda, and G. L. Wang, "Tag-based approaches for deep transcriptome analysis in plants," *Plant Science*, vol. 173, no. 4, pp. 371–380, 2007.

[11] P. Breyne and M. Zabeau, "Genome-wide expression analysis of plant cell cycle modulated genes," *Current Opinion in Plant Biology*, vol. 4, no. 2, pp. 136–142, 2001.

[12] H. Matsumura, S. Nirasawa, and R. Terauchi, "Transcript profiling in rice (*Oryza sativa* L.) seedlings using serial analysis of gene expression (SAGE)," *Plant Journal*, vol. 20, no. 6, pp. 719–726, 1999.

[13] W. W. Lorenz and J. F. D. Dean, "SAGE profiling and demonstration of differential gene expression along the axial developmental gradient of lignifying xylem in loblolly pine (*Pinus taeda*)," *Tree Physiology*, vol. 22, no. 5, pp. 301–310, 2002.

[14] J. G. Gibbings, B. P. Cook, M. R. Dufault et al., "Global transcript analysis of rice leaf and seed using SAGE technology," *Plant Biotechnology Journal*, vol. 1, no. 4, pp. 271–285, 2003.

[15] J. Y. Lee and D. H. Lee, "Use of serial analysis of gene expression technology to reveal changes in gene expression in Arabidopsis pollen undergoing cold stress," *Plant Physiology*, vol. 132, no. 2, pp. 517–529, 2003.

[16] H. Matsumura, S. Reich, A. Ito et al., "Gene expression analysis of plant host-pathogen interactions by SuperSAGE," *Proceedings of the National Academy of Sciences of the United States of America*, vol. 100, no. 26, pp. 15718–15723, 2003.

[17] K. L. Nielsen, A. L. Høgh, and J. Emmersen, "DeepSAGE—digital transcriptomics with high sensitivity, simple experimental protocol and multiplexing of samples," *Nucleic Acids Research*, vol. 34, no. 19, article e133, 2006.

[18] S. Brenner, M. Johnson, J. Bridgham et al., "Gene expression analysis by massively parallel signature sequencing (MPSS) on microbead arrays," *Nature Biotechnology*, vol. 18, no. 6, pp. 630–634, 2000.

[19] J. Reinartz, E. Bruyns, J. Z. Lin et al., "Massively parallel signature sequencing (MPSS) as a tool for in-depth quantitative gene expression profiling in all organisms," *Briefings in Functional Genomics and Proteomics*, vol. 1, no. 1, pp. 95–104, 2002.

[20] B. C. Meyers, F. F. Souret, C. Lu, and P. J. Green, "Sweating the small stuff: microRNA discovery in plants," *Current Opinion in Biotechnology*, vol. 17, no. 2, pp. 139–146, 2006.

[21] K. Nobuta, R. C. Venu, C. Lu et al., "An expression atlas of rice mRNAs and small RNAs," *Nature Biotechnology*, vol. 25, no. 4, pp. 473–477, 2007.

[22] R. Sunkar, V. Chinnusamy, J. Zhu, and J. K. Zhu, "Small RNAs as big players in plant abiotic stress responses and nutrient deprivation," *Trends in Plant Science*, vol. 12, no. 7, pp. 301–309, 2007.

[23] M. Nakano, K. Nobuta, K. Vemaraju, S. S. Tej, J. W. Skogen, and B. C. Meyers, "Plant MPSS databases: signature-based transcriptional resources for analyses of mRNA and small RNA," *Nucleic Acids Research*, vol. 34, pp. D731–D735, 2006.

[24] M. Jain, A. Nijhawan, R. Arora et al., "F-Box proteins in rice. Genome-wide analysis, classification, temporal and spatial gene expression during panicle and seed development, and regulation by light and abiotic stress," *Plant Physiology*, vol. 143, no. 4, pp. 1467–1483, 2007.

[25] M. Schena, D. Shalon, R. W. Davis, and P. O. Brown, "Quantitative monitoring of gene expression patterns with a complementary DNA microarray," *Science*, vol. 270, no. 5235, pp. 467–470, 1995.

[26] D. J. Lockhart, H. Dong, M. C. Byrne et al., "Expression monitoring by hybridization to high-density oligonucleotide arrays," *Nature Biotechnology*, vol. 14, no. 13, pp. 1675–1680, 1996.

[27] W. A. Rensink and C. R. Buell, "Microarray expression profiling resources for plant genomics," *Trends in Plant Science*, vol. 10, no. 12, pp. 603–609, 2005.

[28] P. Zimmermann, M. Hirsch-Hoffmann, L. Hennig, and W. Gruissem, "GENEVESTIGATOR. Arabidopsis microarray database and analysis toolbox," *Plant Physiology*, vol. 136, no. 1, pp. 2621–2632, 2004.

[29] D. Wang, Y. Pan, X. Zhao, L. Zhu, B. Fu, and Z. Li, "Genome-wide temporal-spatial gene expression profiling of drought responsiveness in rice," *BMC Genomics*, vol. 12, article 149, 2011.

[30] R. Kumar, A. Mustafiz, K. K. Sahoo et al., "Functional screening of cDNA library from a salt tolerant rice genotype Pokkali identifies mannose-1-phosphate guanyl transferase gene (OsMPG1) as a key member of salinity stress response," *Plant Molecular Biology*, vol. 79, no. 6, pp. 555–568, 2012.

[31] A. Singh, A. Pandey, V. Baranwal, S. Kapoor, and G. K. Pandey, "Comprehensive expression analysis of rice phospholipase D gene family during abiotic stresses and development," *Plant Signaling and Behavior*, vol. 7, no. 7, pp. 847–855, 2012.

[32] N. Z. Ergen, J. Thimmapuram, H. J. Bohnert, and H. Budak, "Transcriptome pathways unique to dehydration tolerant relatives of modern wheat," *Functional and Integrative Genomics*, vol. 9, no. 3, pp. 377–396, 2009.

[33] T. J. Close, S. I. Wanamaker, R. A. Caldo et al., "A new resource for cereal genomics: 22K Barley GeneChip comes of age," *Plant Physiology*, vol. 134, no. 3, pp. 960–968, 2004.

[34] M. Luo, J. Liu, R. D. Lee, B. T. Scully, and B. Guo, "Monitoring the expression of maize genes in developing Kernels under drought stress using oligo-microarray," *Journal of Integrative Plant Biology*, vol. 52, no. 12, pp. 1059–1074, 2010.

[35] A. Ranjan, N. Pandey, D. Lakhwani, N. K. Dubey, U. V. Pathre, and S. V. Sawant, "Comparative transcriptomic analysis of roots of contrasting *Gossypium herbaceum* genotypes revealing adaptation to drought," *BMC Genomics*, vol. 13, article 680, 2012.

[36] Y. Utsumi, M. Tanaka, T. Morosawa et al., "Transcriptome analysis using a high-density oligomicroarray under drought stress in various genotypes of cassava: an important tropical crop," *DNA Research*, vol. 19, no. 4, pp. 335–345, 2012.

[37] R. Loukehaich, T. Wang, B. Ouyang et al., "SpUSP, an annexin-interacting universal stress protein, enhances drought tolerance in tomato," *Journal of Experimental Botany*, vol. 63, no. 15, pp. 5593–5606, 2012.

[38] M. K. Deyholos, "Making the most of drought and salinity transcriptomics," *Plant, Cell and Environment*, vol. 33, no. 4, pp. 648–654, 2010.

[39] E. A. Bray, "Genes commonly regulated by water-deficit stress in *Arabidopsis thaliana*," *Journal of Experimental Botany*, vol. 55, no. 407, pp. 2331–2341, 2004.

[40] J. Yazaki, B. D. Gregory, and J. R. Ecker, "Mapping the genome landscape using tiling array technology," *Current Opinion in Plant Biology*, vol. 10, no. 5, pp. 534–542, 2007.

[41] K. Mochida and K. Shinozaki, "Genomics and bioinformatics resources for crop improvement," *Plant and Cell Physiology*, vol. 51, no. 4, pp. 497–523, 2010.

[42] G. Zeller, S. R. Henz, C. K. Widmer et al., "Stress-induced changes in the *Arabidopsis thaliana* transcriptome analyzed using whole-genome tiling arrays," *Plant Journal*, vol. 58, no. 6, pp. 1068–1082, 2009.

[43] A. Matsui, J. Ishida, T. Morosawa et al., "Arabidopsis tiling array analysis to identify the stress-responsive genes," *Methods in Molecular Biology*, vol. 639, pp. 141–155, 2010.

[44] W. Verelst, E. Bertolini, S. de Bodt et al., "Molecular and physiological analysis of growth-limiting drought stress in *Brachypodium distachyon* leaves," *Molecular Plant*, vol. 6, no. 2, pp. 311–322, 2013.

[45] C. M. McCallum, L. Comai, E. A. Greene, and S. Henikoff, "Targeting induced local lesions IN genomes (TILLING) for plant functional genomics," *Plant Physiology*, vol. 123, no. 2, pp. 439–442, 2000.

[46] L. Chen, L. Huang, D. Min et al., "Development and characterization of a new TILLING population of common bread wheat (*Triticum aestivum* L.)," *PLoS ONE*, vol. 7, no. 7, Article ID e41570, 2012.

[47] Z. Xin, M. Li Wang, N. A. Barkley et al., "Applying genotyping (TILLING) and phenotyping analyses to elucidate gene function in a chemically induced sorghum mutant population," *BMC Plant Biology*, vol. 8, article 103, 2008.

[48] B. J. Till, S. H. Reynolds, C. Weil et al., "Discovery of induced point mutations in maize genes by TILLING," *BMC Plant Biology*, vol. 4, article 12, 2004.

[49] D. G. Caldwell, N. McCallum, P. Shaw, G. J. Muehlbauer, D. F. Marshall, and R. Waugh, "A structured mutant population for forward and reverse genetics in Barley (*Hordeum vulgare* L.)," *Plant Journal*, vol. 40, no. 1, pp. 143–150, 2004.

[50] J. L. Cooper, B. J. Till, R. G. Lapor et al., "TILLING to detect induced mutations in soybean," *BMC Plant Biology*, vol. 8, article 9, 2008.

[51] J. L. Cooper, S. Henikoff, L. Comai, and B. J. Till, "TILLING and ecotilling for rice," *Methods in Molecular Biology*, vol. 956, pp. 39–56, 2013.

[52] L. de Lorenzo, F. Merchan, P. Laporte et al., "A novel plant leucine-rich repeat receptor kinase regulates the response of *Medicago truncatula* roots to salt stress," *Plant Cell*, vol. 21, no. 2, pp. 668–680, 2009.

[53] L. Comai, K. Young, B. J. Till et al., "Efficient discovery of DNA polymorphisms in natural populations by Ecotilling," *Plant Journal*, vol. 37, no. 5, pp. 778–786, 2004.

[54] S. Negrão, M. C. Almadanim, I. S. Pires et al., "New allelic variants found in key rice salt-tolerance genes: an association study," *Plant Biotechnology Journal*, vol. 11, no. 1, pp. 87–100, 2013.

[55] S. Yu, F. Liao, F. Wang et al., "Identification of rice transcription factors associated with drought tolerance using the Ecotilling method," *PLoS ONE*, vol. 7, no. 2, Article ID e30765, 2012.

[56] J. R. Andersen and T. Lübberstedt, "Functional markers in plants," *Trends in Plant Science*, vol. 8, no. 11, pp. 554–560, 2003.

[57] R. K. Varshney, A. Graner, and M. E. Sorrells, "Genomics-assisted breeding for crop improvement," *Trends in Plant Science*, vol. 10, no. 12, pp. 621–630, 2005.

[58] M. Bagge, X. Xia, and T. Lubberstedt, "Functional markers in wheat," *Current Opinion in Plant Biology*, vol. 10, no. 2, pp. 211–216, 2007.

[59] B. Garg, C. Lata, and M. Prasad, "A study of the role of gene *TaMYB2* and an associated SNP in dehydration tolerance in common wheat," *Molecular Biology Reports*, vol. 39, no. 12, pp. 10865–10871, 2012.

[60] R. K. Varshney, S. N. Nayak, G. D. May, and S. A. Jackson, "Next-generation sequencing technologies and their implications for crop genetics and breeding," *Trends in Biotechnology*, vol. 27, no. 9, pp. 522–530, 2009.

[61] P. L. Morrell, E. S. Buckler, and J. Ross-Ibarra, "Crop genomics: advances and applications," *Nature Reviews Genetics*, vol. 13, no. 2, pp. 85–96, 2011.

[62] C. Feuillet, N. Stein, L. Rossini et al., "Integrating cereal genomics to support innovation in the Triticeae," *Functional and Integrative Genomics*, vol. 12, no. 4, pp. 573–583, 2012.

[63] E. Paux, P. Sourdille, J. Salse et al., "A physical map of the 1-gigabase bread wheat chromosome 3B," *Science*, vol. 322, no. 5898, pp. 101–104, 2008.

[64] E. Paux, D. Roger, E. Badaeva et al., "Characterizing the composition and evolution of homoeologous genomes in hexaploid wheat through BAC-end sequencing on chromosome 3B," *Plant Journal*, vol. 48, no. 3, pp. 463–474, 2006.

[65] K. F. X. Mayer, M. Martis, P. E. Hedley et al., "Unlocking the barley genome by chromosomal and comparative genomics," *Plant Cell*, vol. 23, no. 4, pp. 1249–1263, 2011.

[66] N. Vitulo, A. Albiero, C. Forcato et al., "First survey of the wheat chromosome 5A composition through a next generation sequencing approach," *PLoS ONE*, vol. 6, no. 10, Article ID e26421, 2011.

[67] S. Fluch, D. Kopecky, K. Burg et al., "Sequence composition and gene content of the short arm of rye (*Secale cereale*) chromosome 1," *PLoS ONE*, vol. 7, no. 2, Article ID e30784, 2012.

[68] P. Hernandez, M. Martis, G. Dorado et al., "Next-generation sequencing and syntenic integration of flow-sorted arms of wheat chromosome 4A exposes the chromosome structure and gene content," *Plant Journal*, vol. 69, no. 3, pp. 377–386, 2012.

[69] S. J. Lucas, H. Šimková, J. Šafář et al., "Functional features of a single chromosome arm in wheat (1AL) determined from its structure," *Functional and Integrative Genomics*, vol. 12, no. 1, pp. 173–182, 2012.

[70] H. Takagi, A. Abe, K. Yoshida et al., "QTL-seq: rapid mapping of quantitative trait loci in rice by whole genome resequencing of DNA from two bulked populations," *Plant Journal*, vol. 74, no. 1, pp. 174–183, 2013.

[71] D. Edwards and J. Batley, "Plant genome sequencing: applications for crop improvement," *Plant Biotechnology Journal*, vol. 8, no. 1, pp. 2–9, 2010.

[72] J. Mammadov, R. Aggarwal, R. Buyyarapu, and S. Kumpatla, "SNP markers and their impact on plant breeding," *International Journal of Plant Genomics*, vol. 2012, Article ID 728398, 11 pages, 2012.

[73] D. F. Simola and J. Kim, "Sniper: improved SNP discovery by multiply mapping deep sequenced reads," *Genome Biology*, vol. 12, no. 6, article 55, 2011.

[74] T. Yoshida, H. Nishida, J. Zhu et al., "Vrn-D4 is a vernalization gene located on the centromeric region of chromosome 5D in hexaploid wheat," *Theoretical and Applied Genetics*, vol. 120, no. 3, pp. 543–552, 2010.

[75] E. Paux, S. Faure, F. Choulet et al., "Insertion site-based polymorphism markers open new perspectives for genome saturation and marker-assisted selection in wheat," *Plant Biotechnology Journal*, vol. 8, no. 2, pp. 196–210, 2010.

[76] M. Mohan, S. Nair, A. Bhagwat et al., "Genome mapping, molecular markers and marker-assisted selection in crop plants," *Molecular Breeding*, vol. 3, no. 2, pp. 87–103, 1997.

[77] B. C. Y. Collard and D. J. Mackill, "Marker-assisted selection: an approach for precision plant breeding in the twenty-first century," *Philosophical Transactions of the Royal Society B*, vol. 363, no. 1491, pp. 557–572, 2008.

[78] G. O. Edmeades, G. S. McMaster, J. W. White, and H. Campos, "Genomics and the physiologist: bridging the gap between genes and crop response," *Field Crops Research*, vol. 90, no. 1, pp. 5–18, 2004.

[79] R. Tuberosa, S. Salvi, M. C. Sanguineti, P. Landi, M. Maccaferri, and S. Conti, "Mapping QTLS regulating morpho-physiological traits and yield: case studies, shortcomings and perspectives in drought-stressed maize," *Annals of Botany*, vol. 89, pp. 941–963, 2002.

[80] M. Ashraf, "Inducing drought tolerance in plants: recent advances," *Biotechnology Advances*, vol. 28, no. 1, pp. 169–183, 2010.

[81] T. Yamaguchi and E. Blumwald, "Developing salt-tolerant crop plants: challenges and opportunities," *Trends in Plant Science*, vol. 10, no. 12, pp. 615–620, 2005.

[82] F. Ahmed, M. Y. Rafii, M. R. Ismail et al., "Waterlogging tolerance of crops: breeding, mechanism of tolerance, molecular approaches, and future prospects," *BioMed Research International*, vol. 2013, Article ID 963525, 10 pages, 2013.

[83] T. Wicker, S. Taudien, A. Houben et al., "A whole-genome snapshot of 454 sequences exposes the composition of the barley genome and provides evidence for parallel evolution of genome size in wheat and barley," *Plant Journal*, vol. 59, no. 5, pp. 712–722, 2009.

[84] F. Choulet, T. Wicker, C. Rustenholz et al., "Megabase level sequencing reveals contrasted organization and evolution patterns of the wheat gene and transposable element spaces," *Plant Cell*, vol. 22, no. 6, pp. 1686–1701, 2010.

[85] R. Guyot, F. Lefebvre-Pautigny, C. Tranchant-Dubreuil et al., "Ancestral synteny shared between distantly-related plant species from the asterid (*Coffea canephora* and *Solanum* Sp.) and rosid (*Vitis vinifera*) clades," *BMC Genomics*, vol. 13, article 103, 2012.

[86] C. Feuillet and B. Keller, "Comparative genomics in the grass family: molecular characterization of grass genome structure and evolution," *Annals of Botany*, vol. 89, no. 1, pp. 3–10, 2002.

[87] B. Keller and C. Feuillet, "Colinearity and gene density in grass genomes," *Trends in Plant Science*, vol. 5, no. 6, pp. 246–251, 2000.

[88] F. Li, C. Ma, Q. Chen et al., "Comparative mapping reveals similar linkage of functional genes to QTL of yield-related traits between *Brassica napus* and *Oryza sativa*," *Journal of Genetics*, vol. 91, no. 2, pp. 163–170, 2012.

[89] K. F. X. Mayer, S. Taudien, M. Martis et al., "Gene content and virtual gene order of barley chromosome 1H," *Plant Physiology*, vol. 151, no. 2, pp. 496–505, 2009.

[90] T. Wicker, K. F. X. Mayer, H. Gundlach et al., "Frequent gene movement and pseudogene evolution is common to the large and complex genomes of wheat, barley, and their relatives," *Plant Cell*, vol. 23, no. 5, pp. 1706–1718, 2011.

[91] C. Rustenholz, P. E. Hedley, J. Morris et al., "Specific patterns of gene space organisation revealed in wheat by using the combination of barley and wheat genomic resources," *BMC Genomics*, vol. 11, no. 1, article 714, 2010.

[92] C. Rustenholz, F. Choulet, C. Laugier et al., "A 3,000-loci transcription map of chromosome 3B unravels the structural and functional features of gene islands in hexaploid wheat," *Plant Physiology*, vol. 157, no. 4, pp. 1596–1608, 2011.

3

Some Problems in Proving the Existence of the Universal Common Ancestor of Life on Earth

Takahiro Yonezawa[1] and Masami Hasegawa[1, 2]

[1] School of Life Sciences, Fudan University, Shanghai 200433, China
[2] Department of Statistical Modeling, Institute of Statistical Mathematics, Tokyo 190-8562, Japan

Correspondence should be addressed to Masami Hasegawa, masamihase@gmail.com

Academic Editor: Yidong Bai

Although overwhelming circumstantial evidence supports the existence of the universal common ancestor of all extant life on Earth, it is still an open question whether the universal common ancestor existed or not. Theobald (Nature 465, 219–222 (2010)) recently challenged this problem with a formal statistical test applied to aligned sequences of conservative proteins sampled from all domains of life and concluded that the universal common ancestor hypothesis holds. However, we point out that there is a fundamental flaw in Theobald's method which used aligned sequences. We show that the alignment gives a strong bias for the common ancestor hypothesis, and we provide an example that Theobald's method supports a common ancestor hypothesis for two apparently unrelated families of protein-encoding sequences (*cytb* and *nd2* of mitochondria). This arouses suspicion about the effectiveness of the "formal" test.

1. Introduction

Data generated by genomic sequencing projects from a wide variety of species now allow for the assembly of combined protein sequence data sets to reconstruct the universal tree of life (e.g., [1]). On the other hand, it is still an open question whether the universal common ancestor (UCA) of all extant life on Earth existed or not. Although molecular phylogenetic methods automatically construct a tree when a sequence data set is provided, the inferred tree does not necessarily guarantee the existence of UCA, because its existence is assumed implicitly from the beginning usually in molecular phylogenetics.

The theory of UCA has enjoyed a compelling list of circumstantial evidence as given by Theobald [2]. However, there had been no attempt to test the UCA hypothesis among three domains (or superkingdoms) of life, that is, eubacteria (Bacteria), archaebacteria (Archaea), and eukaryotes (Eukarya), by using molecular sequences until Theobald [2] challenged this problem with a formal statistical test. By using the sequence data sets compiled by Brown et al. [1] and by using the model selection criterion AIC [3], he

showed that the UCA hypothesis is much superior to any independent origin hypothesis, and he concluded that the UCA theory holds. While the UCA hypothesis postulates that eubacteria, archaebacteria, and eukaryotes descended from a single common ancestor called UCA, the independent origin hypotheses include scenarios such as eubacteria having a different origin from that of archaebacteria/eukaryotes or the three domains have different origins from each other. His attempt is the first step towards the goal of establishing the UCA theory with a solid statistical ground. However, his methodology contains some problems for establishing the UCA theory as discussed by us [4], and, in this communication, we will give further details of our arguments.

The most serious problem of Theobald's analysis is that he used aligned sequences compiled by Brown et al. [1], who were interested in resolving the phylogenetic relationships among archaebacteria, eubacteria, and eukaryotes, including whether each domain of life constitutes a monophyletic clade. So they a priory assumed the existence of UCA. Indeed, alignment is a procedure based on an assumption that the sequences have diverged from a common ancestral sequence. Brown et al. wrote "Individual protein families

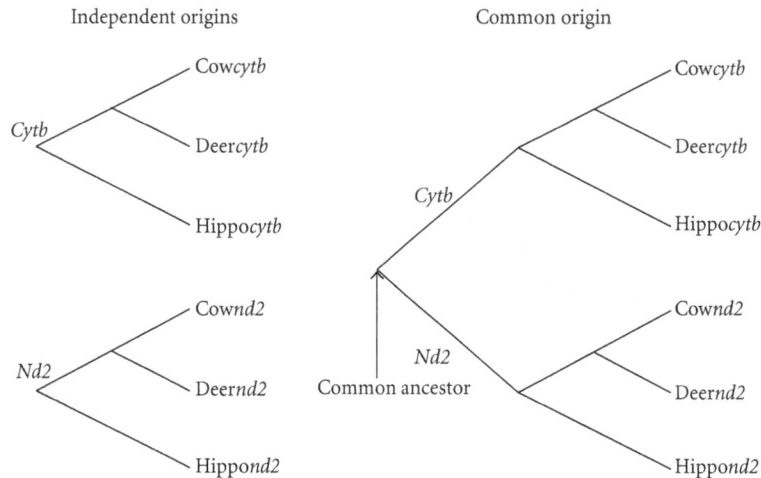

FIGURE 1: Independent origins hypothesis versus common origin hypotheses of *cytb* and *nd2*. No branch exists connecting the two genes in the independent origins hypothesis, while the common ancestor of the two genes exists in the common origin hypothesis.

were first computer aligned and then we manually refined the alignments. We removed poorly conserved regions in individual protein alignments." This procedure clearly assumes the existence of UCA, and this was not a problem for Brown et al., because what they were interested in was the phylogenetic relationship among all species on Earth, and the existence of UCA was supported by circumstantial evidence [2]. However, in proving the existence of UCA, the alignment procedure should not be used, because it gives a strong bias for the UCA hypothesis.

In a previous communication [4], we provided an example from two apparently unrelated families of nucleic acid coding sequences (*cytb* and *nd2* of mitochondria) for which AIC chooses a common origin hypothesis. Since alignment gives a bias for common ancestry, we did not make an alignment between *cytb* and *nd2*, but still the common origin of *cytb* and *nd2* was preferred to the independent origins of these two genes. Probably no one will believe that this result should be regarded as evidence of the ultimate common ancestry of *cytb* and *nd2*. Rather this raises a question mark as to the effectiveness of Theobald's test.

Theobald [5] criticized our analysis by pointing out that our nucleotide substitution model of GTR+Γ is too naïve. We used the same reading frame of the two genes, but, according to Theobald, the constraints of the genetic code are expected to induce correlations between these sequences that are not due to common ancestry. This is a good point, and in this work we will use the amino acid substitution model as well to account of this correlation. We used only the GTR+Γ model of nucleotide substitution in [4] in order to show the most impressive case without alignment, but actually the preference of the common origin model over the independent origin model depends on the assumed substitution model. Therefore, by using several alternative substitution models of nucleotides as well as amino acids, we will study whether default settings of the alignment program, with which the data set of Theobald was made, reject the common origin hypothesis of the two apparently unrelated genes.

2. Materials and Methods

The same sequence data set as used in [4] was provided for the analyses. The 5′-terminal 1,038 bp (excluding the initiation codon) of mitochondrial genes of *cytb* and *nd2* from cow (EU177848), deer (AB210267) and hippopotamus (NC_000889) was analyzed by the maximum likelihood method implemented in PAML [6] assuming the relations of ((cow, deer), hippopotamus) as shown in Figure 1. The independent origin hypothesis shown in left side of Figure 1 is compared with the common origin hypothesis shown in the right with the criterion of AIC [3]. Substitution models used in this work are as follows: JC [7], K80 [8], HKY [9], GTR [10, 11], K80+Γ [8, 12], HKY+Γ [9, 12], and GTR+Γ [10–12] for nucleotide substitutions, and Poisson, JTT [13], mtmam [14], Poisson+Γ [12], JTT+F+Γ [12, 13, 15], mtmam+F+Γ [12, 14, 15] models for amino acid substitutions. CLUSTAL W [16] was used for the alignment with various values for gap open penalty (GOP) and gap extension penalty (GEP). The default values of (GOP, GEP) are (15, 6.66) for nucleotide sequences and (10, 0.1) for amino acid sequences, and the default values for amino acid sequences were used in preparing the data sets used in [1], in which only amino acid sequences were analyzed.

3. Results and Discussion

The result of the analysis in the nucleotide level is given in Table 1. Without alignment, JC, K80+Γ, HKY+Γ, and GTR+Γ models prefer the common origin hypothesis, while K80, HKY, and GTR models prefer the independent origins hypothesis. The best model with respect to AIC is the GTR+Γ model, and it prefers the common origin. Then, sequences aligned with CLUSTAL W with various GOP and GEP values were analyzed. Larger values of GOP and GEP mean stronger penalty for inserting a gap and gap extension, and accordingly the resulting alignment with larger values is closer to the data set without alignment than that produced with smaller values. By changing the GOP and GEP from

TABLE 1: Formal tests of the common ancestry between *cytb* and *nd2* based on the nucleotide sequence data sets aligned with various values of gap penalties (GOP and GEP).

(a)

Model	No alignment (1038 bp)		(GOP, GEP) = (100, 100) (1026 bp)		(GOP, GEP) = (50, 6.66) (1029 bp)	
	Independent	Common	Independent	Common	Independent	Common
JC	11043.8	11005.5[†]	10876.9	10844.5[†]	10935.0	10862.9[†]
K80	10820.8[†]	10821.2	10669.3	10662.2[†]	10727.6	10684.4[†]
HKY	10398.6[†]	10414.7	10255.3[†]	10266.6	10309.7	10294.4[†]
GTR	10307.5[†]	10320.4	10186.5[†]	10192.1	10242.4	10224.3[†]
K80+Γ	10789.5	10723.4[†]	10637.5	10562.7[†]	10695.7	10650.4[†]
HKY+Γ	10329.8	10274.8[†]	10186.4	10119.4[†]	10239.7	10228.4[†]
GTR+Γ	10271.9	<u>10216.4[†]</u>	10129.5	<u>10066.6[†]</u>	10184.1	<u>10168.6[†]</u>
Homology*	0.314		0.317		0.349	

(b)

Model	(GOP, GEP) = (30, 6.66) (1025 bp)		(GOP, GEP) = (15, 6.66) (999 bp)		(GOP, GEP) = (3, 6.66) (974 bp)	
	Independent	Common	Independent	Common	Independent	Common
JC	10890.6	10802.2[†]	**10592.4**	**10409.2[†]**	10262.1	9865.7[†]
K80	10684.6	10623.3[†]	**10395.0**	**10221.3[†]**	10056.9	9613.1[†]
HKY	10271.8	10241.0[†]	**9991.1**	**9875.0[†]**	9645.8	9283.2[†]
GTR	10204.9	10170.3[†]	**9921.1**	**9820.4[†]**	9585.0	9234.3[†]
K80+Γ	10652.5	10577.5[†]	**10363.0**	**10188.2[†]**	10028.1	9595.4[†]
HKY+Γ	10202.4	10162.0[†]	**9920.5**	**9817.6[†]**	9580.9	9249.5[†]
GTR+Γ	10146.3	<u>10099.7[†]</u>	**9863.6**	**<u>9768.5[†]</u>**	9531.1	<u>9201.7[†]</u>
Homology*	0.360		**0.419**		0.504	

AICs of each model comparing the independent and common origin hypotheses were shown. In the comparison between the two hypotheses, the hypothesis with lower AIC was indicated by [†]. The substitution model with the minimal AIC in each data set was indicated by an underline. Default values of GOP and GEP were indicated in bold fonts.
*Homology between *cytb* and *nd2* alignments, which is defined by 1−(average *p*-distance between *cytb* and *nd2*).

TABLE 2: Formal tests of the common ancestry between *cytb* and *nd2* based on the amino acid sequence data sets aligned with various values of gap penalties (GOP and GEP).

Model	No alignment (346 aa)		(GOP, GEP) = (100, 100) (338 aa)		(GOP, GEP) = (15, 6.66) (342 aa)		(GOP, GEP) = (10, 0.1) (330 aa)		(GOP, GEP) = (1, 0.1) (313 aa)	
	Independent	Common	Independent	Common	Independent	Common	Independent	Common	Independent	Common
Poisson	5934.3	5933.5[†]	5748.6	5745.8[†]	5856.9	5838.6[†]	**5664.9**	**5638.0[†]**	5403.1	5288.6[†]
Poisson+Γ	5922.0[†]	5933.5	5735.9[†]	5740.6	5843.9	5832.3[†]	**5651.7**	**5639.0[†]**	5392.7	5288.5[†]
JTT	5591.5	5586.1[†]	5420.3	5414.0[†]	5515.8	5495.6[†]	**5335.5**	**5276.4[†]**	5080.2	4879.8[†]
mtmam	5247.4[†]	5252.5	5083.1[†]	5090.8	5174.7[†]	5176.0	**4995.4**	**4989.9[†]**	4754.3	4688.6[†]
JTT+F+Γ	5304.3[†]	5325.8	5133.7[†]	5152.8	5226.8[†]	5231.7	**5044.8**	**5034.2[†]**	4809.5	4682.4[†]
mtmam+F+Γ	<u>5248.1[†]</u>	5272.3	<u>5082.6[†]</u>	5107.7	<u>5174.6[†]</u>	5185.4	**<u>4995.0[†]</u>**	**4995.6**	4759.7	<u>4678.7[†]</u>
Homology*	0.077		0.083		0.107		**0.123**		0.216	

AICs of each model comparing the independent and common origin hypotheses were shown. In the comparison between the two hypotheses, the hypothesis with lower AIC was indicated by [†]. The substitution model with the minimal AIC in each data set was indicated by an underline. Default values of GOP and GEP were indicated in bold fonts.
*Homology between *cytb* and *nd2* alignments, which is defined by 1−(average *p*-distance between *cytb* and *nd2*).

large to small values, the common origin hypothesis tends to be preferred over the independent origin hypothesis irrespective of the substitution model. Interestingly, such a situation is realized with (GOP, GEP) = (50, 6.66) before the default values of (15, 6.66).

A similar analysis in the amino acid level is given in Table 2. In this case, the common origin hypothesis is preferred only by the Poisson and JTT models without alignment, while the best model of mtmam+F+Γ prefers the independent origins. The aligned sequences with the default setting also give different results depending on the assumed substitution model; while simple models such as the Poisson, JTT, and Poisson+Γ prefer the common origin hypothesis, the best available model with respect to AIC, the mtmam+F+Γ model, prefers the independent origins. Probably, the stronger preference of the common ancestor hypothesis with the nucleotide level analysis is, as Theobald pointed out, due to the constraints of the genetic code which induce correlations between the sequences that are not due to common ancestry. Particularly in the mammalian mitochondrial protein-encoding genes on the heavy strand used in our analysis, second codon positions are biased toward T, whereas third codon positions are biased towards A and biased against G [5]. Therefore, the strong preference of the common origin hypothesis by the nucleotide analysis is probably due to the constraints of the genetic code. However, it is worthwhile to be mentioned that, although the best available substitution model of amino acid analysis without alignment and with alignment of the default setting prefers the independent origin hypothesis, the common origin hypothesis is preferred by some substitution models. This raises a serious problem as to the effectiveness of the formal test. Theobald used a similar data set of amino acid sequences as that of Brown et al. [1], who used the CLUSTALW [16] with default settings to align individual protein data sets. Actually, Theobald [2] used another program called ProbCons [17] instead of CLUSTALW in aligning the sequences, but the difference should not be critically important for our arguments.

Since *cytb* and *nd2* encoded on the heavy strand of mitochondrial DNA have similar amino acid compositions [18], this may induce correlations between these sequences that are not due to common ancestry. This illuminates another flaw in Theobald's analysis; that is, he did not take account of the possibility of convergent evolution as discussed by us [4]. While the examples discussed in [4] were in convergence due to requirement of similar function and to adaptation to similar environment, there is another type of convergence, that is, convergence to similar amino acid composition, which can be achieved by many different ways. A similar amino acid composition between *cytb* and *nd2* may not be bona fide convergence but may only represent constraints due to coexistence of the two genes in the same genome but effectively represents a similar situation of convergent evolution.

As for the bias caused by the alignment, theoretically it can be solved by including the alignment procedure in the framework of maximum likelihood tree estimation [19–21]. Most current alignment programs treat alignment and phylogeny separately, whereas in fact they are interdependent. When a practical method to estimate both alignment and phylogeny simultaneously in the framework of maximum likelihood is developed, we would be able to compare AIC between the UCA and the independent origin hypotheses by taking account of log-likelihood for insertion/deletion process without any bias for the UCA hypothesis. On the other hand, however, it seems not easy to take account of the possibility of convergent evolution, since any currently used maximum likelihood method assumes a stochastic process representing diversifying evolution, and it is difficult to take account of convergent evolution in this framework.

A completely new paradigm might be needed to finally solve the problem which Theobald challenged. Notwithstanding these problems in proving the existence of UCA by statistical testing, it is true that there is strong circumstantial evidence for its existence [2].

Charles Darwin wrote in *On the Origin of Species* [22] as follows: "I should infer from analogy that probably all the organic beings which have ever lived on this earth have descended from someone primordial form, into which life first breathed". Darwin seems to have discarded multiple origins of life on Earth. However, as Theobald [2] correctly noted, the theory of UCA allows for the possibility of multiple independent origins of life [23, 24]. The UCA hypothesis simply states that all extant life on Earth has descended from a single common ancestral species. There must have been a huge amount of extinctions during the course of the history of life, and there is no way to know what kinds of life became extinct during the early evolution of life. Still, it seems likely that a huge amount of trials and errors of different forms occurred during the emergence of life and that UCA if existed was just one of them. Further, as argued by Raup and Valentine [24], the probability of survival of life is low unless there are multiple origins. Even if the UCA hypothesis holds, the survival of the particular form of life does not imply that it was unique or superior.

Acknowledgment

This research was partially supported by Grants-in-Aid for Scientific Research C22570099 to M. Hasegawa from JSPS.

References

[1] J. R. Brown, C. J. Douady, M. J. Italia, W. E. Marshall, and M. J. Stanhope, "Universal trees based on large combined protein sequence data sets," *Nature Genetics*, vol. 28, no. 3, pp. 281–285, 2001.

[2] D. L. Theobald, "A formal test of the theory of universal common ancestry," *Nature*, vol. 465, no. 7295, pp. 219–222, 2010.

[3] H. Akaike, "Information theory and an extension of the maximum likelihood principle," in *Proceedings of the 2nd International Symposium on Information Theory*, B. N. Petrov and F. Csaki, Eds., pp. 267–281, Akademiai Kiado, Budapest Hungary, 1973.

[4] T. Yonezawa and M. Hasegawa, "Was the universal common ancestry proved?" *Nature*, vol. 468, no. 7326, Article ID E9, 2010.

[5] D. L. Theobald, "Theobald reply," *Nature*, vol. 468, no. 7326, Article ID E10, 2010.

[6] Z. Yang, "PAML 4: phylogenetic analysis by maximum likelihood," *Molecular Biology and Evolution*, vol. 24, no. 8, pp. 1586–1591, 2007.

[7] T. H. Jukes and C. R. Cantor, "Evolution of protein molecules," in *Mammalian Protein Metabolism*, H. N. Munro, Ed., pp. 21–123, Academic Press, New York, NY, USA, 1969.

[8] M. Kimura, "A simple method for estimating evolutionary rates of base substitutions through comparative studies of nucleotide sequences," *Journal of Molecular Evolution*, vol. 16, no. 2, pp. 111–120, 1980.

[9] M. Hasegawa, H. Kishino, and T. Yano, "Dating of the human-ape splitting by a molecular clock of mitochondrial DNA,"

Journal of Molecular Evolution, vol. 22, no. 2, pp. 160–174, 1985.

[10] S. Tavare, "Some probabilistic and statistical problems on the analysis of DNA sequences," *Lectures in Mathematics in the Life Sciences*, vol. 17, pp. 57–86, 1986.

[11] Z. Yang, "Estimating the pattern of nucleotide substitution," *Journal of Molecular Evolution*, vol. 39, no. 1, pp. 105–111, 1994.

[12] Z. Yang, "Among-site rate variation and its impact on phylogenetic analyses," *Trends in Ecology and Evolution*, vol. 11, no. 9, pp. 367–372, 1996.

[13] D. T. Jones, W. R. Taylor, and J. M. Thornton, "The rapid generation of mutation data matrices from protein sequences," *Computer Applications in the Biosciences*, vol. 8, no. 3, pp. 275–282, 1992.

[14] Z. Yang, R. Nielsen, and M. Hasegawa, "Models of amino acid substitution and applications to mitochondrial protein evolution," *Molecular Biology and Evolution*, vol. 15, no. 12, pp. 1600–1611, 1998.

[15] Y. Cao, J. Adachi, A. Janke, S. Paabo, and M. Hasegawa, "Phylogenetic relationships among Eutherian orders estimated from inferred sequences of mitochondrial proteins: instability of a tree based on a single gene," *Journal of Molecular Evolution*, vol. 39, no. 5, pp. 519–527, 1994.

[16] J. D. Thompson, D. G. Higgins, and T. J. Gibson, "CLUSTAL W: improving the sensitivity of progressive multiple sequence alignment through sequence weighting, position-specific gap penalties and weight matrix choice," *Nucleic Acids Research*, vol. 22, no. 22, pp. 4673–4680, 1994.

[17] C. B. Do, M. S. Mahabhashyam, M. Brudno, and S. Batzoglou, "ProbCons: probabilistic consistency-based multiple sequence alignment," *Genome Research*, vol. 15, no. 2, pp. 330–340, 2005.

[18] J. Adachi and M. Hasegawa, "Model of amino acid substitution in proteins encoded by mitochondrial DNA," *Journal of Molecular Evolution*, vol. 42, no. 4, pp. 459–468, 1996.

[19] J. L. Thorne, H. Kishino, and J. Felsenstein, "An evolutionary model for maximum likelihood alignment of DNA sequences," *Journal of Molecular Evolution*, vol. 33, no. 2, pp. 114–124, 1991.

[20] J. L. Thorne, H. Kishino, and J. Felsenstein, "Inching toward reality: an improved likelihood model of sequence evolution," *Journal of Molecular Evolution*, vol. 34, no. 1, pp. 3–16, 1992.

[21] G. Lunter, A. J. Drummond, I. Miklos, and J. Hein, "Statistical alignment: recent progress, new applications, and challenges," in *Statistical Methods in Molecular Evolution*, R. Nielsen, Ed., pp. 375–405, Springer, New York, NY, USA, 2005.

[22] C. Darwin, *On the Origin of Species by Means of Natural Selection, or, The Preservation of Favoured Races in the Struggle for Life*, J. Murray, London, UK, 1859.

[23] M. Steel and D. Penny, "Origins of life: common ancestry put to the test," *Nature*, vol. 465, no. 7295, pp. 168–169, 2010.

[24] D. M. Raup and J. W. Valentine, "Multiple origins of life," *Proceedings of the National Academy of Sciences of the United States of America*, vol. 80, no. 10, pp. 2981–2984, 1983.

4

Complex Tasks Force Hand Laterality and Technological Behaviour in Naturalistically Housed Chimpanzees: Inferences in Hominin Evolution

M. Mosquera,[1,2] **N. Geribàs,**[1,2] **A. Bargalló,**[2] **M. Llorente,**[2,3] **and D. Riba**[2,3]

[1] *Universitat Rovira i Virgili (URV), Campus Catalunya, Avinguda de Catalunya 35, 43002 Tarragona, Spain*
[2] *Institut Català de Paleoecologia Humana i Evolució Social (IPHES), Campus Catalunya, Avinguda de Catalunya 35, 43002 Tarragona, Spain*
[3] *Unitat de Recerca i Laboratori d'Etologia, Fundació Mona, Carretera de Cassà 1 km, Riudellots de la Selva, 17457 Girona, Spain*

Correspondence should be addressed to M. Mosquera, marina.mosquera@urv.cat

Academic Editors: L. Kratochvil, A. L. Mayer, and A. V. Peretti

Clear hand laterality patterns in humans are widely accepted. However, humans only elicit a significant hand laterality pattern when performing complementary role differentiation (CRD) tasks. Meanwhile, hand laterality in chimpanzees is weaker and controversial. Here we have reevaluated our results on hand laterality in chimpanzees housed in naturalistic environments at Fundació Mona (Spain) and Chimfunshi Wild Orphanage (Zambia). Our results show that the difference between hand laterality in humans and chimpanzees is not as great as once thought. Furthermore, we found a link between hand laterality and task complexity and also an even more interesting connection: CRD tasks elicited not only the hand laterality but also the use of tools. This paper aims to turn attention to the importance of this threefold connection in human evolution: the link between CRD tasks, hand laterality, and tool use, which has important evolutionary implications that may explain the development of complex behaviour in early hominins.

1. Introduction

Hand laterality is a cognitive factor according to which a group of individuals (populations or species) differentially use one hand (left or right) to perform a task [1] or a group of tasks [2]. From a behavioural point of view, the importance of hand laterality lies in the fact that in humans it is the most developed functional asymmetry. Hand laterality seems to be an indicator of brain hemispheric specialisation, which is not exclusive to humans. It is present in species such as rats (*Rattus norvegicus*) [3], elephants (*Elephas maximus*) [4], humpback whales (*Megaptera novaeangliae*) [5], and crows (*Corvus macrorhynchos*) [6]. Actually, Rogers [7] suggests that all vertebrates share brain hemispheric specialisation. However, brain hemispheric specialisation seems to be also related in humans to linguistic functions. Therefore, its pattern of emergence and development throughout human

evolution can provide insight into the evolution of human cognitive capacities.

In modern humans, 97% of the population is hand lateralised, and between 85% and 90% of individuals are right-handed [8]. However, several studies have found great diversity in the expression of hand laterality [9–17], which appears to be influenced by environmental and cultural factors [18] and by the motor actions involved in performing the task at hand [19]. Despite this variability, research in non-Western societies confirms the universality of hand laterality in the species *Homo sapiens* [20]. Results from three preindustrial cultural groups—the G/wi (Botswana), Himba (Namibia), and Yanomamo (Venezuela)—show right-hand dominance at the population level for all tasks and stronger preferences for conducts involving tools. Even when discounting the strong biases of Western educative influences [21], the pattern of right-handedness in modern humans emerges. This has led to the widely accepted belief that human hand laterality may be conditioned by biological factors [8, 22] with inheritable components [23, 24].

Therefore, most research suggests the existence of a *genetic component* for hand preference, although neither the inherited pattern nor the responsible gene or genes have yet been identified [25–27]. Two main genetic models [22, 28] propose that hand laterality and brain dominance for language depend on a single gene with two alternative alleles. Both models assume that the gene for laterality is unique and exclusive to human beings. However, some studies on chimpanzees contradict this suggestion.

Research on hand laterality in nonhuman primates has been conducted for decades. The aim of these studies is to understand how and when hand laterality was fixed into the evolutionary history of our order. Copious data have been gathered regarding the hand laterality of *Pan*, *Gorilla*, and *Pongo*; however, no clear manual tendencies have been identified. The most abundant data come from studies on chimpanzees, because such animals are more easily accessible and frequently make and use tools both in the wild and in captivity [29, 30].

There are two opposing positions concerning hand laterality in chimpanzees. One position supports right-hand dominance in chimpanzees, given its high incidence (67%) among this species [31]. The other position rejects this manual asymmetry at the population level [2, 32, 33]. These differences are mainly due to different conceptions concerning empirical studies and conflicting viewpoints at the theoretical level [2, 31].

Despite these divergences, some overall tendencies can be observed regarding hand laterality in nonhuman primates. Firstly, nonhuman primates display clear evidence of lateralisation at the individual level. Secondly, they show population asymmetries for some behaviours, particularly complex and structured behaviours. Thirdly, differences between human and nonhuman primates seem to be of more degree than nature, that is, weaker laterality is seen in the latter. Beside interspecies differences, the main disparity of results seems to be related to the living environment of the samples studied (wild or in captivity) and the type of tasks performed (simple or complex).

Therefore, hand laterality in humans has proved to be universal, while hand laterality at the population level in nonhuman primates remains controversial. However, hypotheses on the emergence of hand laterality are based on nonhuman primate studies. Several factors have been suggested as the cause of this emergence, such as body posture, bipedalism, tool use, and task complexity. The primary difference between the hypotheses proposed is the emphasis given to one factor as the key element around which the others turn.

To begin with, the *postural origin hypothesis* [34] stresses the importance of body posture in facilitating right-hand dominance for handling objects from a primate arboreal ancestor. On the other hand, the *bipedalism hypothesis* [35–37] suggests that the emergence of hand dominance in humans developed from bipedal posture, through the improvement of the brain skills needed to keep the body

balanced in this stance. This hypothesis is supported by several studies with nonhuman primates [36, 38–42]. Additionally, the advent of bipedalism may have favoured the development of different tasks performed by the upper limbs, such as gesture communication or the use of tools [43].

Thirdly, the *tool use hypothesis* argues that hand dominance evolved because of the bimanual coordination required in making and using tools. Therefore, the strong manual asymmetry of the genus *Homo* would be the product of the systematic manufacture and use of tools [44–46]. This hypothesis is also supported by several studies with nonhuman primates [29, 47–51].

Finally, the *task complexity hypothesis* [52] considers that hand laterality depends on the nature of the tasks to be performed. Low-level tasks demand low cognitive and motor involvement, so they are poor indicators of hand and brain lateralisation. In contrast, high-level tasks call for precise motor actions and cognitive complexity, so they are good indicators of manual and brain lateralisation. This hypothesis has been empirically supported by several studies with nonhuman primates [51, 53–55]. Actually, it seems that the *task complexity hypothesis* complements both the *tool use hypothesis* and the *bipedalism hypothesis*, since complexity increases both when a vertical position is adopted and when instruments are used.

Uomini [56] has recently published a study that supports the *task complexity hypothesis* for the emergence of hand preference. She proposes that only tasks involving complementary role differentiation (CRD) [57] are indicative of hand laterality. A task of this type requires the action of both hands performing different roles. In contrast, coordinated bimanual tasks are those in which both hands play the same role. CRD tasks are also known as bimanual complementary (see [2] for definition) and bimanual complex tasks [58].

In her study, Uomini [56] conducted two experiments to test handedness in humans. In the first experiment, several people were asked to refit fragments of a flint core. In this task both hands were active, but performing the same role. In the second experiment, the same people were asked to crack nuts, which involved both hands in different roles. As a consequence of this difference, when performing the flint refitting, individuals did not show significant hand laterality, whereas, during the nut-cracking task, hand laterality was evident. The author aimed to demonstrate that, when humans are asked to do the same experiments as chimpanzees, only bimanual CRD tasks, as opposed to coordinated bimanual tasks, are significant indicators of handedness, despite extreme human hand laterality. In our view, this conclusion is extremely important and has implications regarding both hand lateralisation and human evolution that must be further studied.

In light of Uomini's results [56], we have revisited the results of our studies on hand laterality in chimpanzees housed in naturalistic environments. Uomini's research shows that, although hand laterality in humans has been widely proved, it can be as complex and variable as in nonhuman primates. Only CRD tasks appear to express clear hand laterality in humans. In accordance with this assertion, we have reevaluated our results on hand laterality

TABLE 1: Hand preferences and consistency for *simple reaching* and *hose task* at the FM chimpanzees. R: right-hand preference. L: left-hand preference. A: Nonpreferent.

Subject	Hand preference *Simple reaching*	Hand preference *Hose task*	Consistency
Bongo	R	R	Yes
Charly	L	R	No
Julio	L	R	Yes
Juanito	A	R	No
Marco	R	R	Yes
Nico	R	R	Yes
Pancho	R	R	Yes
Romie	L	L	Yes
Sara	R	L	No
Tico	R	L	No
Toni	L	R	No
Toto	R	R	Yes
Victor	R	R	Yes
Waty	A	L	No

in naturalistically housed chimpanzees [49, 51], with special attention to CRD tasks. In this paper, we present a review of these results from an evolutionary perspective. The chimpanzees from our sample appeared to show a link between CRD tasks, hand laterality, and technological behaviour that may provide insight into the development of complex technological behaviour in early hominins.

2. Hand Laterality and Tool Use in Naturalistically Housed Chimpanzees

Research on chimpanzee hand laterality yields contradictory results depending on whether it is conducted in the wild [59, 60] or in captivity [61–63]. It has been argued that these differences are not solely due to the environment but to the different tasks studied as well [64]. Therefore, we performed our research on chimpanzees sheltered in two naturalistic environments—Fundació Mona in Spain and Chimfunshi Wildlife Orphanage in Zambia—and we studied different types of tasks, from unimanual spontaneous tasks to CRD bimanual tasks.

2.1. Fundación Mona. Fundació Mona (FM) (Riudellots de la Selva, Girona, north-eastern Spain) (41° 54′ N, 2° 49′ E) (http://www.fundacionmona.org/) was opened in the year 2000 and is devoted to the rescue, rehabilitation, and sheltering of primates that have been exploited or mistreated. Today, FM shelters a group of chimpanzees (*Pan troglodytes*) made up of 10 males and 3 females, ranging from 6 to 53 years old. (See Table 1 in [49], for additional information about age, classes, sex, and rearing history of each individual.)

The institution consists of a naturalistic outdoor enclosure of $5,640\,m^2$ and two socialisation enclosures of $25\,m^2$

connected to a pavilion measuring $140\,m^2$. The outdoor enclosure has natural ground with Mediterranean and riverside vegetation. Several structures made of wood, rope, and nets, as well as a shallow pond, have been built in this enclosure. Water supply is readily available, and curators provide food four times a day. Juices, fresh fruits, special dehydrated food, fresh vegetables, boiled rice, nuts, and seeds complete the chimpanzees' diet. This food is delivered in special containers or left on the ground. The enclosure is surrounded by a steel fence and a $12\,V$ electrified fence.

Since 2000, three experiments have been performed to evaluate the handedness of FM chimpanzees: spontaneous tasks [65], simple reaching [49], and the hose task [49, 51].

Our first study at FM was an observational study [65]. Ten chimpanzees (8 males and 2 females) were observed while performing daily spontaneous tasks. The aim of this study was to detect hand preference in the chimpanzees at FM. 111 hours of data were recorded over a period of 11 months. The ethological methodology was based on other authors' works. The observational protocol followed the observational rules described by Altmann [66] and Martin and Bateson [67]. The behavioural catalogue was built on the catalogues described for wild chimpanzees [59, 60, 68]. Finally, the recording of the unimanual and bimanual tasks followed the procedures described in McGrew and Marchant [60].

A total of 3,496 bouts were recorded. Results showed that 89.0% of the bouts ($n = 3,110$) corresponded to unimanual tasks (Figure 1) and only 11.0% of the bouts ($n = 386$) corresponded to bimanual tasks. The latter were divided into "coordinated tasks" (96.6%) and to a much lesser extent "complementary tasks" or CRD tasks (3.4%). Three of the ten individuals displayed a statistically significant preference for the left hand, two individuals were on the significance borderline (one for left-hand preference and another for right-hand preference), while the other five individuals did not move significantly away from a chance selection of left or right hand. In terms of manual preferences according to activity, five individuals showed a statistically significant manual preference in some pattern, whereas the remainder showed no significant preference in any task. The one-sample *t*-test concluded that none of the activities studied in this work showed significant differences.

In summary, spontaneous tasks were mainly unimanual, and they did not lead to hand dominance either as a result of the activity or the individual. However, our current analyses with a wider sample are pointing to the existence of low degree of hand laterality at individual level for spontaneous unimanual tasks. In our view, this different result is related to the bigger size of the sample. We understand that much more data than the previously obtained was needed to detect this pattern. Similar results on unimanual and bimanual tasks have been achieved by other authors. Of the actions Marchant and McGrew [59] recorded at Gombe, 86% were unimanual and 14% bimanual. At Mahale, McGrew and Marchant [60] detected 87.4% unimanual actions and 12.6% bimanual actions, of which around 65% were coordinated actions and about 35% complementary tasks. Therefore, in spontaneous tasks bimanual actions are less common than

Complex Tasks Force Hand Laterality and Technological Behaviour in Naturalistically Housed Chimpanzees: Inferences in Hominin Evolution

29

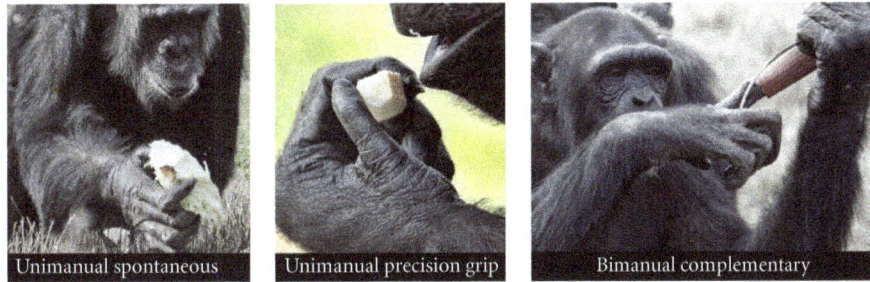

Figure 1: Different tasks performed by some of the chimpanzees at Fundació Mona (Girona, Spain): unimanual spontaneous task, unimanual with precision grip (simple reaching), and bimanual complementary task (hose task) with tool use. (Credit: Miquel Llorente.)

unimanual actions, and bimanual complementary tasks are the least common. These authors also concluded that bimanual actions seem to be more indicative of hand laterality than unimanual actions.

Recently, two experimental tests were used to reevaluate hand preference at FM, this time with 14 individuals (3 females and 11 males). These two experimental tests were the "simple reaching" and "tube task" tests (Figure 1) [49]. Simple reaching involved simple motor actions and consisted of observing the hand responses of individuals undertaking tasks eliciting fine precise manipulation. The tube task was proposed by Hopkins [69] as a measure to test hand preference because it is a bimanual task sensitive to determining hand motor bias [58, 62]. Simple reaching is actually a unimanual task, while the tube task is a bimanual CRD task as defined by Guiard [57] and Uomini [56].

The simple reaching task was observed daily during midday feeding. In order to encourage the use of fine precise manipulation, every item (peanuts, pieces of apple, muesli, bread, etc.) was smaller than 3×3 cm. The procedure consisted of the keeper scattering the food directly on the ground, providing enough food for all the animals in order to prevent any possible dominant-subordinate conflicts. The observation session continued until the subject performed ≥ 100 simple reaching manual events, as proposed in similar studies [70]. For the tube task, a variant called the "hose task" was designed, in which cylindrical rubber hoses were used in place of rigid tubes. Hoses were filled with honey, peanuts, muesli, and seeds, thus preventing extraction with the tongue or by hitting the hose. The subjects had to remove the food with their fingers or with tools such as sticks or branches. All the individuals ($n = 14$) were evaluated for both experimental tasks (simple reaching and hose task). Table 1 offers the results about their hand preferences. The experimental protocol and the methodology for data analyses can be consulted in Llorente and colleagues [49].

The results of the simple reaching task showed that 12 individuals were lateralised and 2 were not: 9 (64.29%) were right-handed, 3 (21.43%) were left-handed, and 2 (14.29%) showed no preference. These results are inconsistent with Fagot and Vauclair's suggestion that simple reaching is a low-level task [52], a type of task from which we would expect to see weak evidence of preference.

The results of the hose task showed that all individuals were lateralised. Ten individuals (71.43%) were right-handed and four (28.57%) were left-handed, supporting Hopkins suggestion that chimpanzees may be preferentially right-handed for this type of bimanual (CRD) task [58, 69]. Also, the strength of hand preference was high in the sample, and it did not vary between groups. This may indicate that the hose task elicited a strong lateralisation in individuals [63]. There were no statistically significant differences between the number of right-handed and left-handed subjects, so there was no population level handedness in our chimpanzee sample. 80.64% of the individuals used their fingers (63.83% index finger), while 19.36% of the individuals used tools to extract the food. We detected no differences between digital and tool techniques regarding hand preference: tools seemed to have no effect on the direction of preference, possibly because both left-handed and right-handed individuals used this technique.

Comparing the simple reaching and hose tasks, our results reveal that chimpanzees are right-handed or on the significance borderline for right-hand preference at the population level. This is the first time that chimpanzees housed at a naturalistic environment have yielded this result. Comparing hand preferences for the hose task and for simple reaching, the bimanual task elicited significantly greater individual asymmetries than the simple reaching task, a low-level task. This may be influenced by tool techniques and by the dominance of the index finger as a method of food extraction in the hose task.

Interestingly, we did not detect handedness at the population level in the earlier study at FM in the performance of spontaneous, low-level tasks [65]. Along with the inconsistency between the hose task and simple reaching, this may suggest that hand laterality is a multidimensional trait, as suggested by other authors [31, 71]. In their opinion, motor and neurological demands and requirements are different for these diverse tasks (spontaneous experimental, unimanual, and bimanual coordinated or complementary) [72].

Another interesting feature is that in the hose task 19.36% of subjects used small sticks as tools to access the food, which means that almost 20% of the individuals took on the complex task assisted by a technological behaviour. In contrast, the use of tools in spontaneous behaviour is only present in around 4% of actions.

In summary, our experiments pointed to two main conclusions. On the one hand, at a methodological level, bimanual CRD tasks are not important by themselves, but as part of the wider group to which they belong: the complex tasks, either unimanual or bimanual. However, complementary bimanual tasks appear to be the most complex tasks, since they entail variables such as precise actions, the number of stages required by the task, the number of elements to be combined, the need for using both hands, the sequence of actions, the use of one hand as subordinate, and a complex control of body balance [73–75]. On the other hand, at an evolutionary level complex tasks, as opposed to spontaneous tasks, force the expression and the emergence of hand laterality and technological behaviour.

2.2. Chimfunshi Wild Orphanage. As a control measure for the FM experiments, we considered the possibility of replicating the hose task at the Chimfunshi Wild Orphanage (CWO) in Zambia [51] with several naturalistically housed chimpanzees, notably less humanised than the FM individuals.

CWO opened 25 years ago and today shelters 120 chimpanzees, 61 of which were born in captivity and reared by their mothers as in the wild. Most of them were confiscated to prevent the smuggling of infant animals to be later sold as pets or were taken from dilapidated zoos and circuses from all over the world. Their ages range between newborn and 33 years old (see Table 1 in [49], for additional information about age, class, sex and rearing history of each group). Chimpanzees at CWO live in groups in different enclosures, including outdoor enclosures and indoor quarters. The average size of the indoor rooms is 6×4 metres. Outdoor enclosures are carved out of the forest and floodplains along the upper Kafue River, with enough thick jungle and fruit groves and open grasslands to allow the chimpanzees to roam almost like in the wild (see [51], for more details).

The aim of our study [51] was to evaluate hand preferences in bimanual complementary actions through observing subjects performing the hose task. We applied the same methodology as used at FM. Out of the 120 individuals in the sample, 100 obtained the minimum number of responses required ($n = 50$) and a minimum of six responses for each test. The experimental protocol and the methodology for data analyses can be consulted in Llorente and colleagues [51].

At CWO the results were similar to those obtained at FM. Overall, a total of 14,854 manual actions were observed: 55.48% ($n = 8,241$) were performed with the right hand and 44.52% ($n = 6,613$) with the left hand. Based on binomial tests, 14% of individuals showed no hand preference, and 86% were lateralised for this task: 48 were right-handed and 38 were left-handed. According to the laterality index of the four tests as a whole (see [51], for details on the analytical method), individuals were not lateralised at the population level, although they were at borderline significance. However, when analysing the four tests individually, two tests showed right-handedness at the population level. When analysing

only the two first experimental tests (test 1 + test 2), the sample was also clearly right-handed at the population level.

In 95.66% of the actions observed, the subjects removed the food with their fingers (mostly the index finger), and in 4.34% of the actions they used tools. According to our results, subjects performing extractions with the index finger preferentially did so with the right hand, which was consistent with other studies on chimpanzees [69] and other primates [54]. It looks like the use of the index finger as an extracting technique encouraged the use of the right hand. On the other hand, subjects performing extractions with their little finger or tools did so with the left hand. Therefore, a relationship was observed between the use of the little finger, tools, and the left hand, although as yet no explanation for this relationship has been proposed. However, it seems that hand laterality is affected by the distal motions of fingers and hands when performing bimanual complementary tasks in which each hand plays a distinct role. According to Brinkman and Kuypers [76], distal movements require frequent use of the contralateral brain hemisphere, what may explain our results. In addition, the index finger is the most sensitive because it has the largest neuronal representation in motor cortex [77], what may explain its higher use.

Finally, the statistical test used to detect different behaviours between human-reared chimpanzees and mother-reared chimpanzees did not reveal significant differences either in the direction or in the degree of preference. Thus, the original environment and context from where these individuals came did not have any effect on their hand preference patterns. This conclusion had previously been reached by other authors in studies with a sample large and varied enough to test this variable [78]. Actually, these results had also come to light in our earlier study [65], where the observation of hand laterality in the FM chimpanzees at spontaneous unimanual tasks yielded similar results to the wild samples. So, these data seem to indicate that environment cannot explain the disparity of results regarding the current pattern of hand preference in nonhuman primates.

3. Technology and Hand Laterality in Human Evolution

Based on the behaviour of great apes [79–83], it is likely that before stone tool manufacture the earliest hominins made use of perishable materials such as sticks and branches and employed materials such as nonmodified bones and stones as tools [84]. It is possible that the first lithic morphotypes were the result of stones being used to crack nuts on anvils, which may have led to accidental flaking, as documented in the Gombe chimpanzees [80] and in Bossou [85]. Some of the flakes with sharp edges may have remained as passive tools until hominins used them to carry out other activities.

As described elsewhere [86], the process of lithic production is derived from objects being used and handled. This adaptive behaviour, which has also been observed in some mammals, birds, and insects, leads to more complex behaviours when the size of the brain increases. Before stone tool production was systematised at African sites, a

Complex Tasks Force Hand Laterality and Technological Behaviour in Naturalistically Housed Chimpanzees: Inferences in
Hominin Evolution

31

background would have been in place that facilitated this leap to exosomatic production. As Toth and Schick state [87, page 299], "a decrease in the size of jaws and teeth over time may be correlated with the rise in exosomatic tool use, with technology creating "synthetic organs" and gradually allowing hominins to move into niches traditionally occupied by other animals, such as the carnivore guild." However, it is not possible in archaeology to identify this basic technological behaviour, or even the manufacture of one simple tool, since such isolated findings are difficult to identify and impossible to classify as intentional. Therefore, it is only possible to identify this process in archaeology when a method of lithic production has been established.

The earliest recorded lithic industry comes from the Ethiopian site of Kada-Gona [88–90], which dates to 2.6 mya. Other sites dated to around 2.4–2.3 mya include Kada-Hadar [91, 92] and Omo-Shungura [93] (both in Ethiopia), Lokalalei (Kenya) [94], and Senga 5A (DR Congo) [95]. The lithic production at these sites was aimed at obtaining flakes with sharp edges, and such artefacts are abundant and diversified, suggesting that the technology was not newly formed [94] but had already been generalised by this time. This means that technology may have originated in Africa some time before this date, perhaps even as early as around 3.5 mya [84, 86, 96]. Recent findings of cut marks on bones at the site of Dikika in Ethiopia [97] confirm this hypothesis.

The archaeo-paleontological scope is rather limited regarding evidence of hand laterality, although not as much as Uomini [56] describes. Actually, hominin hand laterality has been well established for the European *Homo heidelbergensis* of 500,000 years ago [98, 99]. According to our research at Atapuerca (Spain), this hominin species already showed modern-like hand laterality. These results come from two independent sources of evidence: tooth-wear analyses and use-wear traces on tools. Dental microwear analyses have been used to determine hand laterality in hominin species. Since the earliest stages of human evolution, hominins have used their teeth to process their food. Tasks which involved putting the anterior teeth in contact with other materials produced marks and traces on dental surfaces, which are known as dental wear traces of cultural origin. Right-handed individuals and left-handed individuals produce tooth marks oriented in opposing directions. Archaeologically, this tooth wear has been documented in *Homo heidelbergensis* from Sima de los Huesos (Atapuerca, Spain, c. 450 ky) [99], showing the same tendency as in modern humans. On the other hand, use-wear analyses on the edges of the tools made, used, and discarded by the same hominin population (*H. heidelbergensis*) at the site of Galería (Atapuerca, Spain, 400–200 ky) concluded that these tools were used by right-handed individuals [98].

4. Discussion

Our results revealed a certain connection between hand laterality, task complexity, and technology. We believe this same connection may apply to human evolution. To trace it back, we have two different groups of data: present-day primates (both human and nonhuman) and archaeological and paleoenvironmental evidence about extinct hominins.

Two fundamental conclusions can be drawn with regard to present-day primates. Firstly, the more complex the task is, the more hand laterality is expressed in humans and apes, regardless of the differences in their brain capacity. Secondly, modern apes mainly show technological behaviour when performing complementary bimanual tasks (CRD).

Regarding hand laterality and task complexity, we believe there is a gradient of manual motor complexity that influences the expression of hand laterality in apes. The more complex the task, the more hand laterality is expressed. Therefore, according to their increasing complexity, tasks would be ordered as follows: (1) unimanual spontaneous tasks, (2) precision-handling (grip) unimanual tasks (such as simple reaching), and (3) bimanual complementary (CRD) tasks, such as nut-cracking and the hose task. Coordinated bimanual tasks (i.e., Uomini's flint puzzle, [56]) are more complex than unimanual tasks and less than CRD tasks, but they are not indicative of handedness in humans and in apes as yet there is no available data. According to our results, the more complex the tasks, the less common they are in the spontaneous behaviour of an individual. Unimanual tasks with no precision grip are the most common tasks, followed by unimanual tasks with precision grip. Finally, the most seldom performed actions are complementary bimanual tasks (CRD tasks).

Present-day humans appear to be ruled by the same gradient of manual motor complexity. Despite the fact that *Homo sapiens* express manual preference even for unimanual tasks with no precision grip, Uomini's research [56] has shown that some tasks do not elicit the expression of hand laterality, while others clearly do. The former are coordinated bimanual tasks (e.g., the flint puzzle) and the latter complementary bimanual tasks (e.g., nut cracking). Although humans have three times the brain capacity of apes and greater brain organisation complexity and are clearly more lateralised animals, they are as prone as apes to this gradient of manual motor complexity. Therefore, when performing simple tasks, *Homo sapiens* elicit a low degree of significant hand laterality. Meanwhile, hand laterality is much more significant when performing complex tasks, as demonstrated by Uomini [56], hence, the complexity of hand laterality tests for humans. Anyone can hold a glass of water with his or her nondominant hand; however, writing with the nondominant hand is almost impossible.

Concerning task complexity and technological behaviour, our results with FM and CWO chimpanzees showed that CRD tasks not only forced the expression of hand laterality but also seem to be behind a higher use of tools. In fact, CRD tasks, the most complex motor tasks, forced the emergence of technological behaviour. Indeed, the concept of maximum complexity would refer to those tasks in which the body itself does not suffice to complete the task at hand. Hence, the correlation evidenced by Schick and Toth between the reduction in the size of the mandibles and teeth in hominins over time and the increase in tool-assisted strategies, developing what they called "synthetic organs" [100, page 299]. Therefore, the manual or functional complexity of the task forces the expression of hand laterality and the emergence of technological behaviour. In modern

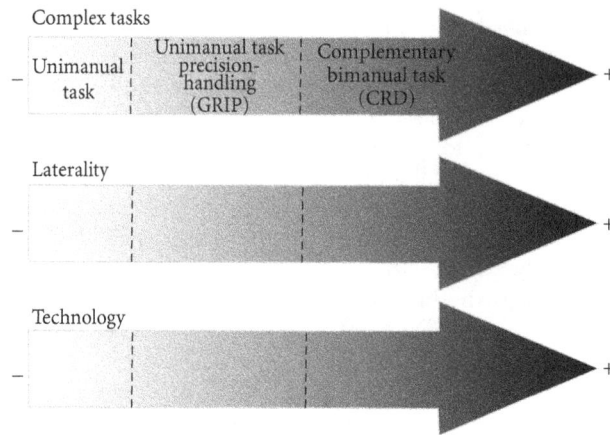

FIGURE 2: Relationships between complex tasks, hand laterality, and technological behaviour.

humans, however, the use of tools is no longer linked to task type. Modern humans use tools to satisfy all tasks, whether simple or complex. Nevertheless, for modern humans and chimpanzees alike, the more complex the task, the more difficult it is to complete without tools.

With our results from FM and CWO chimpanzees and Uomini's results [56] from modern humans, we aim to approach a particular scenario of hominin evolution with the second group of data: archaeological and paleoenvironmental evidence about extinct hominins. We aim to explore the first interactions between complex tasks, hand laterality, and technological behaviour in human evolution (Figure 2).

To begin with, early hominins such as australopithecines had a cranial capacity that exceeded that of present-day chimpanzees (mean values of 400–500 cm^3 versus 300–400 cm^3, resp.) [101]. Therefore, we can assume that they possessed at least the same capacities. It is likely, then, that basic technological behaviours, such as CRD tasks like termite fishing with sticks, would not be unusual among these hominins.

However, unlike chimpanzees, early hominins inhabited an increasingly more arid environment. The close woods that dominated in Africa until ca. 3.5 myr ago were gradually replaced by open forests, savannas, and steppes around 2.5 myr ago. Unlike closed woods, these new landscapes were increasingly unpredictable [96] because resources were more widely dispersed both in terms of space (mosaic) and time (seasonality) [102]. This resource dispersion forced hominins to adopt a generalist diet in order to maximise energy intake ([103], for later species such as *Homo erectus*, [96]).

The adoption of this generalist diet probably involved a diversification of feeding activities, so complex tasks may have become more and more commonplace in this increasingly cruder environment. These conditions may have involved the management of meat, wood, and vegetation, probably requiring cutting actions, which are always CRD tasks. This may be the starting point in human evolution from which cutting tasks become habitual and essential actions. As highlighted by Schick and Toth, cutting tasks are

not usually needed in the world of apes [104]. But, cutting actions, like all other CRD tasks, elicit the most both hand laterality and the use of tools.

Therefore, compared to their ancestors, early hominins more frequently practised complex tasks that forced the expression of hand laterality. While their ancestors may have expressed hand laterality only occasionally, like modern nonhuman primates, early hominins would have displayed this trait so often that it would have become permanent. As pointed out by Teixeira and Okazaki [105], there may be a feedback loop involved. The preferential use of one hand would bestow more skill to that hand, increasing the amount of experience provided to that hand and, thus, reinforcing hand laterality. Actually, as some authors have proved, strong individual laterality is associated with increased efficiency in *Gorilla* [106] and *Pan* [107]. In turn, the increase in hand laterality may have favoured an increase in brain laterality, also in feedback loop fashion. The more hand laterality was reinforced, the more the individual was being lateralised for his or her brain functions.

So, we can assume that early hominins such as *Australopithecus afarensis* and *Australopithecus garhi* were already developing technological behaviour and diversifying their diet, as evidenced by the cut marks on herbivore bones at the site of Dikika [97]. This data indicates that these early hominins had already started manufacturing isolated cutting tools around 3.3 mya, one million years prior to the earliest lithic assemblages known to date. They may have regularly practised complex tasks, especially bimanual tasks and particularly CRD tasks, such as cutting. Therefore, the complexity of these tasks forced the expression of hand laterality in these hominins, probably on a regular basis.

What is more, these hominins moved on to technological production while some of their contemporaries maintained the same technological behaviours. Technically, the difference may have lain in the precision and efficiency of percussion, probably enhanced by better defined hand laterality, and in the incorporation of a particular material: stone. At this point the divergence between early hominins and contemporary primates may have broadened, because task acquisition started becoming more and more complex. As pointed out by Byrne [108] from [109], "hammer and anvil use is much slower to acquire than any other manual skill in any ape species." Actually, according to Schick and Toth [104], apes are poorer stone knappers than early hominins even after years of training. Therefore, Byrne appears to be right when affirming that "the ability to control blows (...) seems to be a crucial adaptation of the human lineage" [108, page 16].

At some point along this process, we find species like *Homo habilis/rudolfensis*. They show a considerable increase in their brain capacity (600–700 cm^3), as well as in their brain reorganisation [101]. The causes for the increase in brain capacity remain controversial; however, it is likely that there was a constant feedback loop involving hand laterality, complex tasks, technological production, and brain increase, enhanced by meat consumption [110] as part of the generalist diet. A constantly developing brain would benefit the enhancement of operative intelligence, with several consequences. Firstly, hominins became capable of performing increasingly complex tasks more frequently. Secondly, technology was indissolubly established in human evolution. Thirdly, complex tasks were performed through standardised technological behaviour. Fourthly, hand laterality was expressed more often and became permanent. Finally, hominins were able to maximise energy from any resource.

Therefore, the technological scenario of early hominins went from basic technological behaviour to the manufacture of isolated tools and eventually to the establishment of systematic methods of technological production. Eventually, around 2.6 mya, hominins (*Homo habilis/H. rudolfensis*) would establish systematic methods of technological production. These systematic methods would include not only the habit of extracting flakes from cores but also the development of production methods with which to do so (centripetal, unipolar, etc.) and even the method of retouching simple flakes to make shaped tools. All these production tasks are necessarily complementary role differentiation (CRD) tasks, as are most of the processes involved in technology. Over the course of this technological development, hand laterality would have become permanent.

The final consequence of this technological development would be the possibility of maximising the energy intake from any resource. This possibility implies better adaptation to any environment (especially if hominins were permanently assisted by technology) and, therefore, the occupation of new and diversified environments and greater biological development and, consequently, a demographic increase. Actually, social learning and cultural transmission would have probably also developed at the earliest stages of tool manufacture or when lithic production methods were established, in order to socialise the innovations into the community [111], which led to the development of populations.

5. Conclusions

Although living environment has been proposed as an important component in explaining the disparity of results regarding hand preferences in chimpanzees, the results of our studies at FM and CWO would reject this hypothesis. The original environment and context from which the animals come do not have any effect on their hand preference patterns.

However, two aspects do seem to be crucial in expressing hand laterality: the type of task being performed and the role performed by the hands during the activity. Our studies confirmed that chimpanzees do not show hand laterality according to activity but may show a low degree according to individual when performing spontaneous *unimanual tasks*, the most common tasks in their daily activities. However, the same individuals displayed higher degree of hand laterality when facing *unimanual tasks that require a precision grip*. Furthermore, *bimanual complementary tasks*, where each hand performed different motions, were infrequent in spontaneous behaviour, but involved the highest degree of hand laterality and the emergence of tool use, as observed during the hose task. Interestingly, although the frequency of tool use varied from FM to CWO chimpanzees, technological behaviour emerged particularly in bimanual complementary tasks (CRD tasks).

Therefore, there appears to be a gradient of task complexity that forces the individual expression of hand laterality and technological behaviour. This gradient would start from spontaneous unimanual tasks, which do not show handedness. Then, we would find the unimanual tasks requiring precision-grip expressing stronger hand laterality. At the extreme of this gradient, there would be the complementary bimanual tasks, such as nut cracking and the tube task.

All processes involved in tool configuration and production are complementary bimanual tasks, as well as most of the subsistence activities carried out by the earliest hominins (and also by modern humans). The need to maximise the supply of energy in an unpredictable landscape forced early hominins to increase the number and complexity of the subsistence activities performed daily. Therefore, previously infrequent complementary bimanual tasks became almost permanent. This, in turn, forced the frequent expression of hand laterality and technological assistance, which had up to then been quite uncommon. As this expression developed, the efficiency of the dominant hand also developed, as well as the efficiency of the tools produced. Hence, this constant loop led to the gradual complexity of the tasks performed, the gradual implementation of hand laterality, and the development of technological support, which in turn favoured the development of the brain motor and associative areas concerned. From this point onwards, brain, technology, and hand laterality were involved in a continuous feedback loop.

Acknowledgments

This research has been funded by the Spanish MICINN (projects CGL2009-12703-C03-01 and HAR2009-07223/HIST), the Generalitat de Catalunya (AGAUR 2009 SGR-00188), and the Universitat Rovira i Virgili (2009AIRE-05). Núria Geribàs's research is funded by the Programa FPU from the MICINN. David Riba's research is funded by Fundación Atapuerca.

References

[1] W. D. Hopkins, "On the other hand: statistical issues in the assessment and interpretation of hand preference data in nonhuman primates," *International Journal of Primatology*, vol. 20, no. 6, pp. 851–866, 1999.

[2] W. C. McGrew and L. F. Marchant, "On the other hand: current issues in and meta-analysis of the behavioral laterality of hand function in nonhuman primates," *Yearbook of Physical Anthropology*, vol. 40, pp. 201–232, 1997.

[3] A. C. Tang and T. Verstynen, "Early life environment modulates "handedness" in rats," *Behavioural Brain Research*, vol. 131, no. 1-2, pp. 1–7, 2002.

[4] F. Martin and C. Niemitz, "'Right-trunkers' and "left-trunkers": side preferences of trunk movements in wild asian elephants (*Elephas maximus*)," *Journal of Comparative Psychology*, vol. 117, no. 4, pp. 371–379, 2003.

[5] P. J. Clapham, E. Leimkuhler, B. K. Gray, and D. K. Mattila, "Do humpback whales exhibit lateralized behaviour?" *Animal Behaviour*, vol. 50, no. 1, pp. 73–82, 1995.

[6] E. I. Izawa, T. Kusayama, and S. Watanabe, "Foot-use laterality in the Japanese jungle crow (*Corvus macrorhynchos*)," *Behavioural Processes*, vol. 69, no. 3, pp. 357–362, 2005.

[7] L. J. Rogers, "Lateralization in vertebrates: its early evolution, general pattern, and development," *Advances in the Study of Behavior*, vol. 31, pp. 107–161, 2002.

[8] M. Annett, *Handedness and Brain Asymmetry: The Right Shift Theory*, Psychology Press, Hove, UK, 2002.

[9] D. Ingram, "Motor asymmetries in young children," *Neuropsychologia*, vol. 13, no. 1, pp. 95–102, 1975.

[10] J. Fagard, "Does manual asymmetry of right-handers change between six and nine years of age?" *Human Movement Science*, vol. 6, no. 4, pp. 321–332, 1987.

[11] G. Hammond, Y. Bolton, Y. Plant, and J. Manning, "Hand asymmetries in interresponse intervals during rapid repetitive finger tapping," *Journal of Motor Behavior*, vol. 20, pp. 67–71, 1988.

[12] S. L. Schmidt, R. M. Oliveira, T. E. Krahe, and C. C. Filgueiras, "The effects of hand preference and gender on finger tapping performance asymmetry by the use of an infra-red light measurement device," *Neuropsychologia*, vol. 38, no. 5, pp. 529–534, 2000.

[13] S. Barthélémy and P. Boulinguez, "Manual reaction time asymmetries in human subjects: the role of movement planning and attention," *Neuroscience Letters*, vol. 315, no. 1-2, pp. 41–44, 2001.

[14] S. Barthélémy and P. Boulinguez, "Manual asymmetries in the directional coding of reaching: further evidence for hemispatial effects and right hemisphere dominance for movement planning," *Experimental Brain Research*, vol. 147, no. 3, pp. 305–312, 2002.

[15] J. A. Agnew, T. A. Zeffiro, and G. F. Eden, "Left hemisphere specialization for the control of voluntary movement rate," *NeuroImage*, vol. 22, no. 1, pp. 289–303, 2004.

[16] M. Hausmann, I. J. Kirk, and M. C. Corballis, "Influence of task complexity on manual asymmetries," *Cortex*, vol. 40, no. 1, pp. 103–110, 2004.

[17] K. Lutz, S. Koeneke, T. Wüstenberg, and L. Jäncke, "Asymmetry of cortical activation during maximum and convenient tapping speed," *Neuroscience Letters*, vol. 373, no. 1, pp. 61–66, 2005.

[18] M. K. Mandal and T. Dutta, "Left handedness: facts and figures across cultures," *Psychology and Developing Societies*, vol. 13, no. 2, pp. 173–191, 2001.

[19] L. A. Teixeira, "Categories of manual asymmetry and their variation with advancing age," *Cortex*, vol. 44, no. 6, pp. 707–716, 2008.

[20] L. F. Marchant, W. C. McGrew, and I. Eibl-Eibesfeldt, "Is human handedness universal? Ethological analyses from three traditional cultures," *Ethology*, vol. 101, pp. 239–258, 1995.

[21] C. Faurie, W. Schiefenhövel, S. Le Bomin, S. Billiard, and M. Raymond, "Variation in the frequency of left-handedness in traditional societies," *Current Anthropology*, vol. 46, no. 1, pp. 142–147, 2005.

[22] I. C. McManus, *Right Hand, Left Hand: The Origins of Asymmetry in Brains, Bodies, Atoms, and Cultures*, Weidenfeld & Nicolson, London, UK, 2002.

[23] S. Coren, "Family patterns in handedness: evidence for indirect inheritance mediated by birth stress," *Behavior Genetics*, vol. 25, no. 6, pp. 517–524, 1995.

[24] K. N. Laland, J. Kumm, J. D. Van Horn, and M. W. Feldman, "A gene-culture model of human handedness," *Behavior Genetics*, vol. 25, no. 5, pp. 433–445, 1995.

[25] C. Francks, S. Maegawa, J. Laurén et al., "LRRTM1 on chromosome 2p12 is a maternally suppressed gene that is associated paternally with handedness and schizophrenia," *Molecular Psychiatry*, vol. 12, no. 12, pp. 1129–1139, 2007.

[26] T. J. Crow, J. P. Close, A. M. Dagnall, and T. H. Priddle, "Where and what is the right shift factor or cerebral dominance gene? A critique of Francks et al. (2007)," *Laterality*, vol. 14, no. 1, pp. 3–10, 2009.

[27] C. Francks, "Understanding the genetics of behavioural and psychiatric traits will only be achieved through a realistic assessment of their complexity," *Laterality*, vol. 14, no. 1, pp. 11–16, 2009.

[28] M. Annett, *Left, Right, Hand and Brain: The Right Shift Theory*, LEA Publishers, London, UK, 1985.

[29] M. Llorente, M. Fabré, and M. Mosquera, "Lateralización cerebral en chimpancés: una aproximación filogenética al estudio del cerebro humano," *Estudios de Psicología*, vol. 29, no. 2, pp. 147–161, 2008.

[30] D. Riba, M. Llorente, M. Mosquera, and O. Feliu, "Hand preference in simple and complex tasks in naturalistically housed chimpanzees at the Mona Foundation (Spain)," *Folia Primatologica*, vol. 80, no. 6, p. 406, 2009.

[31] W. D. Hopkins, "Comparative and familial analysis of handedness in great apes," *Psychological Bulletin*, vol. 132, no. 4, pp. 538–559, 2006.

[32] A. R. Palmer, "Chimpanzee right-handedness reconsidered: evaluating the evidence with funnel plots," *American Journal of Physical Anthropology*, vol. 118, no. 2, pp. 191–199, 2002.

[33] M. Annett, "The distribution of handedness in chimpanzees: estimating right shift in Hopkins' sample," *Laterality*, vol. 11, no. 2, pp. 101–109, 2006.

[34] P. F. MacNeilage, M. G. Studdert, B. Lindblom et al., "Primate handedness reconsidered," *Behavioral and Brain Sciences*, vol. 10, pp. 247–303, 1987.

[35] C. Sanford, K. Guin, and J. P. Ward, "Posture and laterality in the bushbaby (*Galago senegalensis*)," *Brain, Behavior and Evolution*, vol. 25, no. 4, pp. 217–224, 1984.

[36] G. C. Westergaard, H. E. Kuhn, and S. J. Suomi, "Bipedal posture and hand preference in humans and other primates,"

Complex Tasks Force Hand Laterality and Technological Behaviour in Naturalistically Housed Chimpanzees: Inferences in Hominin Evolution

35

Journal of Comparative Psychology, vol. 112, no. 1, pp. 55–64, 1998.

[37] S. Braccini, S. Lambeth, S. Schapiro, and W. T. Fitch, "Bipedal tool use strengthens chimpanzee hand preferences," *Journal of Human Evolution*, vol. 58, no. 3, pp. 234–241, 2010.

[38] D. L. Dodson, D. K. Stafford, C. Forsythe, C. P. Seltzer, and J. P. Ward, "Laterality in quadrapedal and bipedal prosimians: reach and whole-body turn in the mouse lemur (*Microcebus murinus*) and the galago (*Galago moholi*)," *American Journal of Primatology*, vol. 26, pp. 191–202, 1992.

[39] W. D. Hopkins, K. A. Bard, A. Jones, and S. L. Bales, "Chimpanzee hand preference in throwing and infant cradling: implications for the origin of human handedness," *Current Anthropology*, vol. 34, pp. 786–790, 1993.

[40] C. Blois-Heulin, J. S. Guitton, D. Nedellec-Bienvenue, L. Ropars, and E. Vallet, "Hand preference in unimanual and bimanual tasks and postural effect on manual laterality in captive red-capped mangabeys (*Cercocebus torquatus torquatus*)," *American Journal of Primatology*, vol. 68, no. 5, pp. 429–444, 2006.

[41] C. Blois-Heulin, V. Bernard, and P. Bec, "Postural effect on manual laterality in different tasks in captive grey-cheeked mangabey (*Lophocebus albigena*)," *Journal of Comparative Psychology*, vol. 121, no. 2, pp. 205–213, 2007.

[42] D. P. Zhao, W. H. Ji, K. Watanabe, and B. G. Li, "Hand preference during unimanual and bimanual reaching actions in Sichuan snub-nosed monkeys (*Rhinopithecus roxellana*)," *American Journal of Primatology*, vol. 70, no. 5, pp. 500–504, 2008.

[43] J. L. Bradshaw, "Animal asymmetry and human heredity: dextrality, tool use and language in evolution—10 years after Walker (1980)," *The British Journal of Psychology*, vol. 82, pp. 39–59, 1991.

[44] G. T. Frost, "Tool behaviour and the origins of laterality," *Journal of Human Evolution*, vol. 9, pp. 447–459, 1980.

[45] K. A. Provins, "Handedness and Speech: a critical reappraisal of the role of genetic and environmental factors in the cerebral lateralization of function," *Psychological Review*, vol. 104, no. 3, pp. 554–571, 1997.

[46] D. Stout, N. Toth, K. Schick, and T. Chaminade, "Neural correlates of Early Stone Age toolmaking: technology, language and cognition in human evolution," *Philosophical Transactions of the Royal Society B*, vol. 363, no. 1499, pp. 1939–1949, 2008.

[47] G. C. Westergaard and S. J. Suomi, "Hand preference for stone artefact production and tool-use by monkeys: possible implications for the evolution of right-handedness in hominids," *Journal of Human Evolution*, vol. 30, no. 4, pp. 291–298, 1996.

[48] E. V. Lonsdorf and W. D. Hopkins, "Wild chimpanzees show population-level handedness for tool use," *Proceedings of the National Academy of Sciences of the United States of America*, vol. 102, no. 35, pp. 12634–12638, 2005.

[49] M. Llorente, M. Mosquera, and M. Fabré, "Manual laterality for simple reaching and bimanual coordinated task in naturalistic housed chimpanzees (*Pan troglodytes*)," *International Journal of Primatology*, vol. 30, no. 1, pp. 183–197, 2009.

[50] T. Humle and T. Matsuzawa, "Laterality in hand use across four tool-use behaviors among the wild chimpanzees of Bossou, Guinea, West Africa," *American Journal of Primatology*, vol. 71, no. 1, pp. 40–48, 2009.

[51] M. Llorente, D. Riba, L. Palou et al., "Population-level right-handedness for a coordinated bimanual task in naturalistic housed chimpanzees: replication and extension in 114 animals from Zambia and Spain," *American Journal of Primatology*, vol. 73, no. 3, pp. 281–290, 2011.

[52] J. Fagot and J. Vauclair, "Manual laterality in nonhuman primates: a distinction between handedness and manual specialization," *Psychological Bulletin*, vol. 109, no. 1, pp. 76–89, 1991.

[53] A. Chapelain, P. Bec, and C. Blois-Heublin, "Manual laterality in Campbell's Monkeys (*Cercopithecus c. campbelli*) in spontaneous and experimental actions," *Behavioural Brain Research*, vol. 173, no. 2, pp. 237–245, 2006.

[54] C. Schweitzer, P. Bec, and C. Blois-Heulin, "Does the complexity of the task influence manual laterality in De Brazza's monkeys (*Cercopithecus neglectus*)?" *Ethology*, vol. 113, no. 10, pp. 983–994, 2007.

[55] A. L. Lilak and K. A. Phillips, "Consistency of hand preference across low-level and high-level tasks in capuchin monkeys (*Cebus apella*)," *American Journal of Primatology*, vol. 70, no. 3, pp. 254–260, 2008.

[56] N. T. Uomini, "The prehistory of handedness: archaeological data and comparative ethology," *Journal of Human Evolution*, vol. 57, no. 4, pp. 411–419, 2009.

[57] Y. Guiard, "Asymmetric division of labor in human skilled bimanual action: the kinematic chain as a model," *Journal of Motor Behavior*, vol. 19, pp. 486–517, 1987.

[58] W. D. Hopkins, M. J. Wesley, M. K. Izard, M. Hook, and S. J. Schapiro, "Chimpanzees (*Pan troglodytes*) are predominantly right-handed: replication in three populations of apes," *Behavioral Neuroscience*, vol. 118, no. 3, pp. 659–663, 2004.

[59] L. F. Marchant and W. C. McGrew, "Laterality of limb function in wild chimpanzees of Gombe National Park: comprehensive study of spontaneous activities," *Journal of Human Evolution*, vol. 30, no. 5, pp. 427–443, 1996.

[60] W. C. McGrew and L. F. Marchant, "Ethological study of manual laterality in the chimpanzees of the Mahale Mountains, Tanzania," *Behaviour*, vol. 138, no. 3, pp. 329–358, 2001.

[61] M. Colell, M. D. Segarra, and J. Sabater Pi, "Manual laterality in chimpanzees (*Pan troglodytes*) in complex tasks," *Journal of Comparative Psychology*, vol. 109, no. 3, pp. 298–307, 1995.

[62] W. D. Hopkins, M. Hook, S. Braccini, and S. J. Schapiro, "Population-level right handedness for a coordinated bimanual task in chimpanzees: replication and extension in a second colony of apes," *International Journal of Primatology*, vol. 24, no. 3, pp. 677–689, 2003.

[63] W. D. Hopkins, T. S. Stoinski, K. E. Lukas, S. R. Ross, and M. J. Wesley, "Comparative assessment of handedness for a coordinated bimanual task in chimpanzees (*Pan troglodytes*), gorillas (*Gorilla gorilla*), and orangutans (*Pongo pygmaeus*)," *Journal of Comparative Psychology*, vol. 117, no. 3, pp. 302–308, 2003.

[64] W. D. Hopkins and C. Cantalupo, "Individual and setting differences in the hand preferences of chimpanzees (*Pan troglodytes*): a critical analysis and some alternative explanations," *Laterality*, vol. 10, no. 1, pp. 65–80, 2005.

[65] M. Mosquera, M. Llorente, D. Riba et al., "Ethological study of manual laterality in naturalistic housed chimpanzees (*Pan troglodytes*) from the Mona Foundation Sanctuary (Girona, Spain)," *Laterality*, vol. 12, no. 1, pp. 19–30, 2007.

[66] J. Altmann, "Observational study of behavior: sampling methods," *Behaviour*, vol. 49, no. 3-4, pp. 227–267, 1974.

[67] P. Martin and P. Bateson, *Measuring Behaviour: An Introductory Guide*, Cambridge University Press, Cambridge, UK, 1993.

[68] T. Nishida, K. Takayoshi, J. Goodall, W. C. McGrew, and M. Nakamura, "Ethogram and ethnography of Mahale chimpanzees," *Anthropological Science*, vol. 107, no. 2, pp. 141–188, 1999.

[69] W. D. Hopkins, "Hand preferences for a coordinated bimanual task in 110 chimpanzees (*Pan troglodytes*): cross-sectional analysis," *Journal of Comparative Psychology*, vol. 109, no. 3, pp. 291–297, 1995.

[70] J. Vauclair, A. Meguerditchian, and W. D. Hopkins, "Hand preferences for unimanual and coordinated bimanual tasks in baboons (Papio anubis)," *Cognitive Brain Research*, vol. 25, no. 1, pp. 210–216, 2005.

[71] M. J. Wesley, S. Fernandez-Carriba, A. Hostetter, D. Pilcher, S. Poss, and W. D. Hopkins, "Factor analysis of multiple measures of hand use in captive chimpanzees: an alternative approach to the assessment of handedness in nonhuman primates," *International Journal of Primatology*, vol. 23, no. 6, pp. 1155–1168, 2002.

[72] W. D. Hopkins and K. Pearson, "Chimpanzee (*Pan troglodytes*) handedness: variability across multiple measures of hand use," *Journal of Comparative Psychology*, vol. 114, no. 2, pp. 126–135, 2000.

[73] L. F. Marchant and W. C. McGrew, "Laterality of function in apes: a meta-analysis of methods," *Journal of Human Evolution*, vol. 21, no. 6, pp. 425–438, 1991.

[74] R. D. Morris, W. D. Hopkins, and L. Bolser-Gilmore, "Assessment of hand preference in two language-trained chimpanzees (*Pan troglodytes*): a multimethod analysis," *Journal of Clinical and Experimental Neuropsychology*, vol. 15, no. 4, pp. 487–502, 1993.

[75] G. Spinozzi, M. G. Castorina, and V. Truppa, "Hand preferences in unimanual and coordinated-bimanual tasks by tufted capuchin monkeys (*Cebus apella*)," *Journal of Comparative Psychology*, vol. 112, no. 2, pp. 183–191, 1998.

[76] J. Brinkman and H. G. J. M. Kuypers, "Split-brain monkeys: visuomotor coordination after cortical lesions," *Brain Research*, vol. 49, no. 2, p. 507, 1973.

[77] W. W. Sutherling, M. F. Levesque, and C. Baumgartner, "Cortical sensory representation of the human hand: size of finger regions and nonoverlapping digit somatotopy," *Neurology*, vol. 42, no. 5, pp. 1020–1028, 1992.

[78] W. D. Hopkins, "Chimpanzee right-handedness: internal and external validity in the assessment of hand use," *Cortex*, vol. 42, no. 1, pp. 90–93, 2006.

[79] J. Sabater Pi, *El Chimpancé y los Orígenes de la Cultura*, Anthropos, Barcelona, Spain, 1978.

[80] J. Mercader, M. Panger, and C. Boesch, "Excavation of a chimpanzee stone tool site in the African rainforest," *Science*, vol. 296, no. 5572, pp. 1452–1455, 2002.

[81] I. Bila-Isa, "Bonobos dig termite mounds: a field example of tool use by wild bonobos of the Etate, northern sector of the Salonga National Park," in *Proceedings of the Bonobo Workshop: Behaviour, Ecology and Conservation of Wild Bonobos*, Inuyama, Japan, 2003.

[82] C. P. Van Schaik and G. R. Pradhan, "A model for tool-use traditions in primates: implications for the coevolution of culture and cognition," *Journal of Human Evolution*, vol. 44, no. 6, pp. 645–664, 2003.

[83] T. Breuer, M. Ndoundou-Hockemba, and V. Fishlock, "First observation of tool use in wild gorillas," *PLoS Biology*, vol. 3, no. 11, article e380, pp. 2041–2043, 2005.

[84] M. A. Panger, A. S. Brooks, B. G. Richmond, and B. Wood, "Older than the oldowan? Rethinking the emergence of hominin tool use," *Evolutionary Anthropology*, vol. 11, no. 6, pp. 235–245, 2002.

[85] S. Carvalho, D. Biro, W. C. McGrew, and T. Matsuzawa, "Tool-composite reuse in wild chimpanzees (*Pan troglodytes*): archaeologically invisible steps in the technological evolution of early hominins?" *Animal Cognition*, vol. 12, pp. S103–S114, 2009.

[86] E. Carbonell, M. Mosquera, and X. P. Rodríguez, "The emergence of technology: a cultural step or long-term evolution?" *Comptes Rendus—Palevol*, vol. 6, no. 3, pp. 231–233, 2007.

[87] N. Toth and K. Schick, "The oldowan: the tool making of early hominins and chimpanzees compared," *Annual Review of Anthropology*, vol. 38, pp. 289–305, 2009.

[88] S. Semaw, "The world's oldest stone artefacts from Gona, Ethiopia: their implications for understanding stone technology and patterns of human evolution between 2.6-1.5 million years ago," *Journal of Archaeological Science*, vol. 27, no. 12, pp. 1197–1214, 2000.

[89] S. Semaw, P. Renne, J. W. K. Harris et al., "2.5-million-year-old stone tools from Gona, Ethiopia," *Nature*, vol. 385, no. 6614, pp. 333–336, 1997.

[90] S. Semaw, M. J. Rogers, J. Quade et al., "2.6-million-year-old stone tools and associated bones from OGS-6 and OGS-7, Gona, Afar, Ethiopia," *Journal of Human Evolution*, vol. 45, no. 2, pp. 169–177, 2003.

[91] W. H. Kimbel, D. C. Johanson, and Y. Rak, "The first skull and other new discoveries of *Australopithecus afarensis* at Hadar, Ethiopia," *Nature*, vol. 357, no. 6470, pp. 449–451, 1994.

[92] W. H. Kimbel, R. C. Walter, D. C. Johanson et al., "Late pliocene *Homo* and Oldowan tools from the Hadar Formation (Kada Hadar member), Ethiopia," *Journal of Human Evolution*, vol. 31, no. 6, pp. 549–561, 1996.

[93] F. C. Howell, P. Haesaerts, and J. de Heinzelin, "Depositional environments, archeological occurrences and hominids from Members E and F of the Shungura Formation (Omo basin, Ethiopia)," *Journal of Human Evolution*, vol. 16, no. 7-8, pp. 665–700, 1987.

[94] H. Roche, A. Delagnes, J. -P. Brugal et al., "Early hominid stone tool production and technical skill 2.34 Myr ago in West Turkana, Kenya," *Nature*, vol. 399, no. 6731, pp. 57–60, 1999.

[95] J. W. K. Harris, P. G. Williamson, J. Verniers et al., "Late pliocene hominid occupation in Central Africa: the setting, context, and character of the Senga 5A site, Zaire," *Journal of Human Evolution*, vol. 16, no. 7-8, pp. 701–728, 1987.

[96] E. Carbonell, M. Mosquera, X. P. Rodríguez et al., "Eurasian gates: the earliest human dispersals," *Journal of Anthropological Research*, vol. 64, no. 2, pp. 195–228, 2008.

[97] S. P. McPherron, Z. Alemseged, C. W. Marean et al., "Evidence for stone-tool-assisted consumption of animal tissues before 3.39 million years ago at Dikika, Ethiopia," *Nature*, vol. 466, no. 7308, pp. 857–860, 2010.

[98] A. Ollé, *Variabilitat i patrons funcionals en els sistemes tècnics de mode 2. Anàlisi de les deformacions d'ús en els conjunts lítics del Riparo Esterno de Grotta Paglicci (Rignano Garganico, Foggia), Aridos (Arganda, Madrid) i Galeria-TN (Atapuerca,Burgos)*, Ph.D. Dissertation, Universitat Rovira i Virgili, Tarragona, Spain, 2003.

[99] M. Lozano, M. Mosquera, J. M. Bermúdez de Castro, J. L. Arsuaga, and E. Carbonell, "Right handedness of *Homo heidelbergensis* from Sima de los Huesos (Atapuerca, Spain)

Complex Tasks Force Hand Laterality and Technological Behaviour in Naturalistically Housed Chimpanzees: Inferences in Hominin Evolution

37

500,000 years ago," *Evolution and Human Behavior*, vol. 30, no. 5, pp. 369–376, 2009.

[100] K. Schick and N. Toth, *Making Silent Stones Speak: Human Evolution and the Dawn of Technology*, Simon & Schuster, New York, NY, USA, 1993.

[101] R. Holloway, "Evolution of the human brain," in *Handbook of Human Symbolic Evolution*, E. A. Lock and C. R. Peters, Eds., pp. 74–116, Clarendon Press, Oxford, UK, 1996.

[102] C. Finlayson, F. Giles Pacheco, J. Rodríguez-Vidal et al., "Late survival of Neanderthals at the southernmost extreme of Europe," *Nature*, vol. 443, no. 7113, pp. 850–853, 2006.

[103] S. C. Antón, W. R. Leonard, and M. L. Robertson, "An ecomorphological model of the initial hominid dispersal from Africa," *Journal of Human Evolution*, vol. 43, no. 6, pp. 773–785, 2002.

[104] K. Schick and N. Toth, "African origin," in *The Human Past: World Prehistory and the Development d Human Societies*, C. Scarre, Ed., pp. 46–83, Thames & Hudson, Portland, Ore, USA, 2009.

[105] L. A. Teixeira and V. H. A. Okazaki, "Shift of manual preference by lateralized practice generalizes to related motor tasks," *Experimental Brain Research*, vol. 183, no. 3, pp. 417–423, 2007.

[106] R. W. Byrne and J. M. Byrne, "Hand preferences in the skilled gathering tasks of mountain gorillas (*Gorilla gorilla berengei*)," *Cortex*, vol. 27, no. 4, pp. 521–536, 1991.

[107] W. C. McGrew and L. F. Marchant, "Laterality of hand use pays off in foraging success for wild chimpanzees," *Primates*, vol. 40, no. 3, pp. 509–513, 1999.

[108] R. W. Byrne, "The manual skills and cognition that lie behind hominid tool use," in *Evolutionary Origins of Great Ape Intelligence*, A. E. Russon and D. R. Begun, Eds., Cambridge University Press, Cambridge, UK, 2004.

[109] C. Boesch and H. Boesch, "Optimisation of nut-cracking with natural hammers by wild chimpanzees," *Behaviour*, vol. 83, pp. 265–286, 1983.

[110] L. Aiello and P. Wheeler, "The expensive tissue hypothesis: the brain and the digestive system in human and primate evolution," *Current Anthropology*, vol. 36, pp. 199–221, 1995.

[111] E. Carbonell, R. Sala Ramos, X. P. Rodríguez et al., "Early hominid dispersals: a technological hypothesis for 'out of Africa'," *Quaternary International*, vol. 223-224, pp. 36–44, 2010.

5

Drought Tolerance in Modern and Wild Wheat

Hikmet Budak, Melda Kantar, and Kuaybe Yucebilgili Kurtoglu

Biological Sciences and Bioengineering Program, Faculty of Engineering and Natural Sciences, Sabanci University, 34956 Tuzla, Istanbul, Turkey

Correspondence should be addressed to Hikmet Budak; budak@sabanciuniv.edu

Academic Editors: J. Huang, A. Levine, and Z. Wang

The genus *Triticum* includes bread (*Triticum aestivum*) and durum wheat (*Triticum durum*) and constitutes a major source for human food consumption. Drought is currently the leading threat on world's food supply, limiting crop yield, and is complicated since drought tolerance is a quantitative trait with a complex phenotype affected by the plant's developmental stage. Drought tolerance is crucial to stabilize and increase food production since domestication has limited the genetic diversity of crops including wild wheat, leading to cultivated species, adapted to artificial environments, and lost tolerance to drought stress. Improvement for drought tolerance can be achieved by the introduction of drought-grelated genes and QTLs to modern wheat cultivars. Therefore, identification of candidate molecules or loci involved in drought tolerance is necessary, which is undertaken by "omics" studies and QTL mapping. In this sense, wild counterparts of modern varieties, specifically wild emmer wheat (*T. dicoccoides*), which are highly tolerant to drought, hold a great potential. Prior to their introgression to modern wheat cultivars, drought related candidate genes are first characterized at the molecular level, and their function is confirmed via transgenic studies. After integration of the tolerance loci, specific environment targeted field trials are performed coupled with extensive analysis of morphological and physiological characteristics of developed cultivars, to assess their performance under drought conditions and their possible contributions to yield in certain regions. This paper focuses on recent advances on drought related gene/QTL identification, studies on drought related molecular pathways, and current efforts on improvement of wheat cultivars for drought tolerance.

1. Introduction

Current climate change is projected to have a significant impact on temperature and precipitation profiles, increasing the incidence and severity of drought. Drought is the single largest abiotic stress factor leading to reduced crop yields, so high-yielding crops even in environmentally stressful conditions are essential [1, 2]. This is not the first time we face this situation, in which increasing demands on existing resources are not feasible, and higher-yielding crops are required to balance crop production with increasing human food consumption. A similar scenario occurred 50 years ago due to the high rate of population growth, and it was overcome by selective breeding of high grain yielding semidwarf mutants of wheat, a process coined Green Revolution [3]. In relation to current development of cultivars, which are higher yielding even in water-limited environments, one of the major targets is *Triticum* species, being one of the leading human food source, accounting for more than half of total human consumption [2, 4].

The increasing incidence and importance of drought in relation to crop production has rendered it as a major focus of research for several decades. However, studying drought response is challenged by the complex and quantitative nature of the trait. Drought tolerance is complicated with environmental interactions. In the analysis of a plant's drought response, the mode, timing, and severity of the dehydration stress and its occurrence with other abiotic and biotic stress factors are significant [5]. Furthermore different species, subspecies, and cultivars of crops show variation in their drought tolerance under same conditions, emphasizing the importance of genetic diversity as an underlying factor of drought and its significance in drought-related research. Plants exhibiting high drought tolerance are the most suitable targets of drought-related research and are the most promising sources of drought-related gene and gene regions to be used in the improvement of modern crop varieties. These include the natural progenitors of cultivated crops, and for wheat improvement, *Ae. tauschii*, which is more drought tolerant than *Triticum* and wild emmer wheat (*T. dicoccoides*), which harbors drought tolerance characteristics, lost during cultivation of modern lines, is of great importance [6].

Although development of higher-yielding crops under water-limited environments is the most viable solution to stabilizing and increasing wheat production under current climatic conditions, it is challenged by the nature of drought response as a trait and the complex genomic constitution of wheat [16]. However, recently, the utilization of drought tolerant wild species and the rapid advances in molecular biological, functional genomics, and transgenics technologies have facilitated drought-related studies, resulting in significant progress in the identification of related genes and gene regions and dissection of some of its molecular aspects. This paper summarizes the current state of drought-related research in *Triticum* species, focusing on the identification and functional characterization of drought-related molecules, analysis of their interactions in the complex network of drought response, and applications of these data to improve wheat cultivars utilizing molecular based-technologies.

2. *T. dicoccoides* and Drought Tolerance

Wild emmer wheat (*T. turgidum* ssp. *dicoccoides* (körn.) Thell) is the tetraploid ($2n = 4x = 28$; genome BBAA) progenitor of both domesticated tetraploid durum wheat (*T. turgidum* ssp. *durum* (Desf.) MacKey) and hexaploid ($2n = 6x = 42$; BBAADD) bread wheat (*T. aestivum* L.). It is thought to have originated and diversified in the Near East Fertile Crescent region through adaptation to a spectrum of ecological conditions. It is genetically compatible with durum wheat (*T. turgidum* ssp. *durum*) and can be crossed with bread wheat (*T. aestivum* L.) [17]. Wild emmer germplasm harbors a rich allelic pool, exhibiting a high level of genetic diversity, showing correlation with environmental factors, reported by population-wide analysis of allozyme and DNA marker variations [18–24].

Wild emmer wheat is important for its high drought tolerance, and some of *T. dicoccoides* genotypes are fully fertile in arid desert environments. Wild emmer wheat accessions were shown to thrive better under water-limited conditions in terms of their productivity and stability, compared to durum wheat. The wild emmer gene pool was shown to offer a rich allelic repertoire of agronomically important traits including drought tolerance [23, 25–28]. Hence, *T. dicoccoides* is an important source of drought-related genes and highly suitable as a donor for improving drought tolerance in cultivated wheat species.

Wild emmer wheat, being a potential reservoir of drought-related research, has been the source of several identified candidate drought-related genes with the development of "omics" approaches in the recent decades. In recent years, transcript profiling of leaf and root tissues from two *T. dicoccoides* genotypes, originating from Turkey,

TR39477 (tolerant variety), TTD-22 (sensitive variety), was performed by our group, in two separate studies, utilizing different methodologies. In one report, subtractive cDNA libraries were constructed from slow dehydration stressed plants, and over 13,000 ESTs were sequenced. In another study, Affymetrix GeneChip Wheat Genome Array was used to profile expression in response to shock drought stress [1, 29]. Wild emmer wheat was shown to be capable of engaging in known drought responsive mechanisms, harboring elements present in modern wheat varieties and also in other crop species. Additionally several genes or expression patterns, unique to tolerant wild emmer wheat, indicative of its distinctive ability to tolerate water deficiency, were also revealed. Transcript and metabolite profiling studies were also undertaken for two *T. dicoccoides* genotypes, originating from Israel, Y12-3 (tolerant variety) and A24-39 (sensitive variety), under drought stress and nonstress conditions. Leaf transcript profiling indicated differential multilevel regulation among cultivars and conditions [30]. Integration of root transcript and metabolite profiling data emphasized drought adaptation through regulation of energy related processes involving carbon metabolism and cell homeostasis (Table 1) [14]. Recently, in wild emmer wheat, our group also profiled drought induced expression of microRNA (miRNAs), small regulatory molecules known to be involved in several cellular processes including stress responses. In this study, leaf and root tissues of resistant wild emmer wheat varieties, TR39477 and TR38828, were screened via a microarray platform, and 13 differentially expressed miRNAs were found to be differentially expressed in response to drought (Table 1) [15].

Following the identification of *T. dicoccoides* drought-related gene candidates, as discussed previously, a number of these potential drought resistant genes were cloned and further characterized. In one of the recent reports, TdicTMPIT1 (integral transmembrane protein inducible by Tumor Necrosis Factor-α, TNF-α) was cloned from wild emmer root tissue and shown to be a membrane protein, associated with the drought stress response, exhibiting increased levels of expression, specifically in wild emmer wheat upon osmotic stress [31]. In a different study, TdicDRF1 (DRE binding factor 1), conserved between crop species, was cloned for the first time from wild emmer wheat. Its DNA binding domain, AP2/ERF (APETALA2/ethylene-responsive element binding factor), was shown to bind to drought responsive element (DRE), using an electrophoretic mobility shift assay (EMSA). It was revealed to exhibit cultivar and tissue specific regulation of its expression, through mechanisms involving alternative splicing [32]. Moreover, the relations between autophagy and drought response were analyzed in another line of research by the cloning of TdATG8 (autophagy related protein 8) and its further functional investigation with yeast complementation assay and virus induced gene silencing (VIGS) of plants. In this study, autophagy was shown to be induced in drought-stressed plants in an organ-specific mode, and silencing of ATG8 was shown to decrease drought tolerance of plants, revealing it as a positive regulator of drought stress [33] (Tables 2 and 3).

TABLE 1: Transcript, protein, metabolite profiling studies conducted in the last three years.

Species	Cultivars	Tissue	Drought stress application	Method	Reference
T. aestivum	Drought tolerance: Plainsman V: tolerant; Kobomugi: sensitive	Root	Moderate drought stress applied on tillering stage	cDNA microarray	[7]
T. aestivum	Drought tolerance: information can not be accessed	Grain	Short water shortage in early grain development	cDNA microarray	[8]
T. aestivum	Efficiency of stem reserve mobilization in peduncles: N49: tolerant; N14: sensitive	Stem	Progressive drought stress after anthesis	2D gel and MS	[9]
T. aestivum	Cultivar Vinjett	Grain	Drought applied at terminal spiklet or at anthesis	2D gel and MS	[10]
T. aestivum	Yield under drought: Excalibur: tolerant; RAC875: tolerant; Kukri: sensitive	Leaf	Cyclic drought applied after first flag leaf formation mimicking field conditions	SCX column HPLC and MS	[11]
T. durum	Able to acquire drought tolerance: Ofanto: tolerant	Leaf	Drought applied at booting stage (controls SWC: irrigated when it decreases %50 of field capacity; drought SWC: irrigated when it decreases %12.5 of field capacity)	cDNA-AFLP	[12]
T. durum	Drought tolerance: Om Rabia3: tolerant; Mahmoudi: sensitive	Embryo	Drought applied at final development stage of seed maturity	2D gel and HPRP column and MS	[13]
T. dicoccoides	Yield under drought conditions: Y12-3: tolerant; A24-39: sensitive	Leaf	Terminal drought applied at inflorescence emergence stage	Transcript profiling	[14]
T. dicoccoides	Yield under drought conditions: Y12-3: tolerant; A24-39: sensitive	Leaf	Drought applied after germination at five/six leaf stage	Transcript and metabolite profiling	[14]
T. dicoccoides	Drought tolerance: TR39477: tolerant; TR38828: tolerant	Leaf/root	Shock drought stress	miRNA profiling	[15]

T: *Triticum*; SWC: soil water content; 2D: 2-dimensional; SCX: strong cation exchange; HPLC: high performance liquid chromatography; MS: mass spectrometry; cDNA: complementary DNA; AFLP: amplified fragment length polymorphisms; HPRP: human prion protein.

3. Phenotyping for Drought Tolerance in Wheat with Physiological Traits

For screening out transgenic wheat lines with desirable drought tolerance, the physiological traits and processes which can be genetically manipulated to improve wheat adaptation to drought have to be taken into account. The genetic basis of drought tolerance in wheat is still elusive. At present the physiological traits (PTs) linked to heat tolerance appear to be a superlative accessible tool since they exhibit the favorable allele combination for drought tolerance. Such alleles interact with the environment and genetic background which includes variation in gene expression and hence are still poorly understood through the QTL approach [50]. Hybridization of heat tolerance PTs may not always have a predictable outcome related to net crop yield particularly in varying environmental conditions, but breeding such varieties with complementary PTs could augment the cumulative gene effect [51]. Thus the physiological phenotyping along with gene discovery can be valuable to pin down desired alleles and understand their genetic mechanism [50]. Cossani and Reynolds have proposed a model based on this concept of genetically characterized PT for improved drought tolerance of wheat [52]. The model focuses on 3 major genetic parameters of yield when water and nutrients are not limiting factors. The genetic parameters are discussed in the following.

3.1. Light Interception (LI) Traits

3.1.1. Canopy Architecture. Since increase in temperature is linked with a decrease in green area duration and leaf area index, light interception or LI traits can be manipulated by studying the variation in the rapid ground cover (RGC) and leaf senescence of wheat. RGC shows genotypic variability in relatively heritable and simple breeding targets such as embryo and grain size, specific leaf area, or seedling emergence rate [53]. Optimized distribution of light may improve radiation use efficiency (RUE) and LI traits since wheat displays a vast diversity in canopy structure. Furthermore leaves are more erect and smaller in size in many modern cultivars thereby facilitating RUE and allowing more light penetration to lower leaves.

3.1.2. Hindrance of Leaf Senescence. Leaf senescence during drought can be hindered by delayed expression of senescence related green thereby giving stay-green (SG) genotypes with normal photosynthesis [54]. Stay green is thus identified as an important adaptive PT for drought stress conditions,

TABLE 2: List of identified and characterized drought related genes in the last three years.

Gene	Function	Related mechanism/stress	Reference
TaPIMP1	Transcription factor: R2R3 type MYB TF	Drought	[34]
TaSRG	Transcription factor: *Triticum aestivum* salt response gene	Drought	[35]
TaMYB3R1	Transcription factor: MYB3R type MYB TF	drought	[36]
TaNAC (NAM/ATAF/CUC)	Transcription factor: plant specific NAC (NAM/ATAF/CUC) TF	Drought	[37]
TaMYB33	Transcription factor: R2R3 type MYB TF	Drought	[38]
TaWRKY2, TaWRKY19	Transcription factor: WRKY type TF	Drought	[39]
TdicDRF1	Transcription factor: DRE binding protein	Drought	[32]
TaABC1	Kinase: protein kinase ABC1 (activity of bc(1) complex)	Drought	[40]
TaSnRK2.4	Kinase: SNF1 type serine/threonine protein kinase	Drought	[41]
TaSnRK2.7	Kinase: SNF1 type serine/threonine protein kinase	drought	[42]
TdTMKP1	Phosphatase: MAP kinase phosphatase	Drought	[43]
TaCHP	CHP rich zinc finger protein with unknown function	ABA-dependent and -independent pathways	[44]
TaCP	Protein degradation: cysteine protease	Drought	[45]
TaEXPR23	Cell wall expansion: expansin	Water retention ability and osmotic potential	[46]
TaL5	Nucleocytoplasmic transport of 5S ribosomal RNA: ribosomal L5 gene	Drought	[47]
TdPIP1;1, TdPIP1;2	Protective protein: aquaporin	Drought	[48]
TdicATG8	Autophagy: autophagy related gene 8	Drought	[33]
TdicTMPIT1	Autophagy: integral transmembrane protein inducible by TNF-α	Drought	[31]
Era1, Sal1	Enhanced response to ABA, inositol polyphosphate 1-phosphatase	Drought	[49]

Ta: *Triticum aestivum*; Td: *Triticum durum*; Tdic: *Triticum dicoccoides*; DRE: drought related element; SNF: Sucrose nonfermenting; MAP: mitogen activated protein; ABA: abscisic acid; CHP: cysteine histidine proline; TNF-α: tumor necrosis factor α; PIMP: pathogen induced membrane protein; CP: cysteine protease; EXPR: expansin; PIP: plasma membrane intrinsic proteins.

but its role in improving grain yield in drought is still a matter of extensive research. However, some correlations were shown between SG and yield and identified QTLs in mapping populations [55]. Since chlorosis in plants is not expressed homogenously in plant organs aboveground, many approaches have been developed to estimate SG including spectral reflectance, but these also need to be more specific to functional SG.

3.2. Radiation Use Efficiency Traits

3.2.1. Photosynthesis and Photorespiration. According to Cossani and Reynolds, once the LI traits are optimized the focus on increased crop biomass will depend on RUE traits which include dark respiration, photorespiration, and other photosynthetic strategies. A central player of the photosynthetic pathway, Rubisco was observed to show lower affinity for CO_2 over O_2 in higher temperatures [52]. Thus, increasing the affinity of Rubisco is especially significant for adaptation to warm conditions. The importance of CO_2 fixation by Rubisco

for high temperature adaptation is also emphasised by the observation that C4 plants adapt to warm conditions by concentrating CO_2. Present transgenic attempts to convert C3 plants into C4 plants are still in progress and require more knowledge of the maintenance of the C4 pathway. Studies of the Rubisco kinetic properties of *Limonium gibertii* may be used in transgenics in wheat even though wheat Rubisco has an excellent CO_2 affinity. One model shows 12% increase in net assimilation when substrate specificity factor of wheat Rubisco was replaced from *L. gibertii* [56]. Rubisco activase active sites become inactive progressively under drought, thus associating the activase with heat shock chaperone cpn60β could provide Rubisco protection [57]. This has great potential since thermotolerant types of Rubisco in tropical species and diverse optimum temperature of Rubisco have been found in nature [58]. By exploiting this fact a chimeric enzyme was created thus increasing the heat resistance in Arabidopsis by combining the Arabidopsis Rubisco recognition domain and tobacco activase [55].

4. Identification of Drought-Related Genes and QTLs

Prior to focusing on individual drought-related components, drought response, due to its complex nature, must be viewed as a whole system, for which large scale identification of probable dehydration stress-related genes or QTLs is necessary. Potential markers for stress tolerance can be identified either through "omics" studies or QTL mapping of yield related traits under drought prone environments. In the long run, these markers can aid in screening cultivars for drought tolerance/sensitivity and/or improvement of drought tolerance in wheat.

4.1. Drought-Related Gene Identification by "Omics". "Omics" techniques examine all or a representative subset of an organism's genes, transcripts, proteins, or metabolites. As well as accumulating genomic sequence knowledge, data from profiling studies is also crucial in understanding the drought response, which is largely mediated by differential accumulations of drought-related components.

In the recent decades, high-throughput profiling techniques have been utilized for the identification of potential drought tolerance markers from different wheat species (Table 1). Some of the large scale profiling studies undertaken in wild emmer wheat were mentioned in Section 2 [14, 15, 59]. "Omics" studies were also performed to monitor dehydration induced transcripts and proteins of bread and durum wheat cultivars with differing sensitivities to drought, both in stress and nonstress conditions. Methodologies used in transcript profiling studies range from cDNA microarrays to cDNA-AFLP (amplified fragment length polymorphism). For differential protein identification, the common procedures used include 2D (2-Dimensional) gels, various chromatography techniques, and mass spectrometry. In these recent high-throughput studies, molecular mechanisms behind various drought induced physiological or morphological events were targeted, using related tissues and appropriate mode/timing/severity of stress treatments for each profiling experiment. In two of these studies, underlying molecular mechanisms of early grain development upon shock dehydration response and root functional responses upon moderate drought at tillering were investigated in bread wheat by transcript profiling [8, 60]. Proteome profiling was established in several bread wheat tissues: grain upon drought at terminal spikelet or at anthesis; leaf under field like cyclic drought conditions after first flag leaf formation; stem upon progressive drought stress after anthesis were established. The latter research was conducted to understand the underlying molecular mechanisms of mobilization of stem carbohydrate reserves to grains, a process that contributes to yield under terminal drought conditions and its findings pointed out to the involvement of senescence and protection against oxidative stress in effectiveness of the mobilization process [9]. In recent years, transcript profiling in durum wheat flag leaf upon field like drought at booting was performed [12]. In a different line of research, proteomic profiles of *T. durum* mature embryos were established, which is especially important since embryos are good model systems for drought studies, sustaining germination in extreme conditions of desiccation [13]. An overview of recently established profiling studies is provided in Table 1.

4.2. QTL Mapping. Dissection of drought tolerance, a complex quantitative phenotype, affected by multiple loci requires the identification of related quantitative trait loci (QTLs). QTL cloning is a large effort in terms of the technology, resources, and time required, but determination of QTLs is proceeded by great advantages in applications of marker-assisted selection (MAS) and better yielding cultivar development. Identification of QTLs takes advantage of molecular maps, developed by the use of DNA markers. The establishment of these molecular maps has been enabled by the recent advances in functional genomics, which have supplied bacterial artificial chromosomes (BACs), gene sequence data, molecular marker technology, and bioinformatic tools for comparative genomics. Mapping and fine mapping for the identification of candidate regions for a trait prior to positional cloning requires suitable mapping populations: recombinant inbred lines (RILs) and near isogenic lines (NILs), several of which have been established for wheat varieties. However, up to now, only a limited number of studies has succeeded in the positional cloning of wheat QTLs and none in the context of drought [2, 4].

In recent years, several yield QTLs were identified in wheat through linkage analysis and association mapping. Since yield is the most crucial trait to breeders, most QTLs for drought tolerance in wheat have been determined through yield and yield related measurements under water-limited conditions [64–68]. However, these studies are challenged by the factors that yield and drought are both complex traits, involving multiple loci and showing genotype and environment interactions. Yield is difficult to be described accurately with respect to water use, and its accurate phenotyping is a challenge since QTLs established in one environment may not be confirmed in other. For this reason, large scale phenotyping trials, carried out in multiple fields, taking into consideration the environmental varieties are crucial. Until now, a number of studies have identified QTLs associated with specific components of drought response using *T. durum*, *T. aestivum*, and *T.durum* X *T. dicoccoides* mapping populations; however the genomic regions associated with individual QTLs are still very large and unsuitable for screening in breeding programmes. However, in recent years, several yield related QTLs were mapped using (*T. aestivum* L.) RAC875/Kukri doubled haploid populations grown under a variety of environmental conditions including nonirrigated environments. In one study, inbred population was assessed under heat, drought, and high yield potential conditions to identify genetic loci for grain yield, yield components, and key morphophysiological traits [69]. In another study, regions associated with QTLs for grain yield and physical grain quality were assessed under 16 field locations and year combinations in three distinct seasonal conditions [70]. In a third study, QTLs were identified for days to ear emergence and flag leaf glaucousness

TABLE 3: List of genes confirmed to function in drought by transgenic studies in last three years.

Type of transgenic study	Source of the gene	Gene	Function	Related mechanism/ stress	Reference
Overexpression in *A. thaliana*	From *T. Aestivum*	WRKY2, WRKY19	Transcription factor: WRKY type TF	Drought	[39]
Overexpression in *A. thaliana*	From *T. Aestivum*	MYB33	transcription factor: R2R3 type MYB TF	Drought	[38]
Overexpression in *N. tabacum*	From *T. Aestivum*	PIMP1	Transcription factor: R2R3 type MYB TF	Drought	[34]
Overexpression in *N. tabacum*	From *T. Aestivum*	NAC (NAM/ATAF /CUC)	Transcription factor: plant-specific NAC (NAM/ATAF/CUC) TF	Drought	[37]
Overexpression in *A. thaliana*	From *T. Aestivum*	ABC1	Kinase: protein kinase ABC1 (activity of bc(1) complex)	Drought	[40]
Overexpression in *A. thaliana*	From *T. Aestivum*	SnRK2.4	Kinase: SNF1-type serine/threonine protein kinase	Drought	[41]
Overexpression in *A. thaliana*	From *T. Aestivum*	SnRK2.7	Kinase: SNF1-type serine/threonine protein kinase	Drought	[42]
Overexpression in *A. thaliana*	From *T. Aestivum*	CP	Protein degradation: cysteine protease	Drought	[45]
Overexpression in *A. thaliana*	From *T. Aestivum*	CHP	CHP rich zinc finger protein with unknown function	ABA-dependent and -independent pathways	[44]
Overexpression in *N. tabacum*	From *T. Aestivum*	EXPR23	Cell wall expansion: expansin	Water retention ability and osmotic potential	[46]
Overexpression in *A. thaliana*	From *T. Aestivum*	TaSIP	Salt induced protein with unknown function	Drought and salinity	[61]
Overexpression in *N. tabacum*	From *T. Durum*	PIP1;1, PIP1;2	Protective protein: aquaporin	Drought	[48]
Overexpression in *T. aestivum*	From *H. Vulgare*	HVAI	Protective protein: LEA	Drought	[62]
Transgenic ubiquitin: TaCHP	—	CHP	CHP rich zinc finger protein with unknown function	ABA-dependent and -independent pathways	[44]
TaABA08′OF1 deletion line	—	ABA08	ABA catabolism: ABA 8′-hydroxylase	Drought	[63]
VIGS silencing in *T. dicoccoides*	—	ATG8	Autophagy: autophagy related gene 8	Drought	[33]
VIGS silencing in *T. aestivum*	—	*Era1, Sal1*	Enhanced response to ABA, inositol polyphosphate 1-phosphatase	Drought	[49]

ABA: abscisic acid; CHP: cysteine histidine proline; SNF: sucrose nonfermenting; PIMP: pathogen induced membrane protein; CP: cysteine protease; EXPR: expansin; PIP: plasma membrane intrinsic proteins; LEA: late embryogenesis abundant; HVA: *Hordeum vulgare* aleurone; TaSIP: *Triticum aestivum* salt induced protein; VIGS: virus induced gene silencing.

under southern Australian conditions [71]. Another multi-environmental analysis provided a basis for fine mapping and cloning the genes linked to a yield related QTL [72]. These recent studies are promising, and along with the recent advances in DNA sequencing technology and new approaches of coupling linkage analysis with "omics" studies, these data will find their way into practical wheat breeding programmes in relation to drought [2, 4]. Drought-related QTLs identified in these studies are listed in Supplementary Table 1 (see Supplementary Materials available online at http://dx.doi.org/10.1155/2013/548246).

5. Identification of Molecular Mechanisms Related to Drought

Probable drought-related genes and QTLs, identified in "omics" and "QTL mapping" studies, should be further characterized, prior to their use in the development of better yielding cultivars. Elucidation of these components includes analyzing their gene and protein structure and determining their roles and interactions in the complex network of stress response signaling. Their functional relevance to drought should be shown and eventually confirmed with transgenic studies. This section summarizes the recent research regarding the characterization of drought-related genes, in detail, dissection of drought-related molecular pathways, and functional genomics studies.

5.1. Characterization of Drought-Related Genes. Prior to its utilization for drought tolerance improvement, for each putative drought-related gene region or molecule identified in "QTL mapping" or "omics" study, the immediate step is the cloning and in detail characterization of the gene and its protein. This process exploits a variety of *in silico* and basic molecular biology methods and involves several aspects that differ on the nature of the research, including analysis of gene and protein structure, phylogeny-based studies and determination of gene chromosomal localization, protein-protein and protein-DNA interactions, and transcript and protein subcellular localizations. Further characterization involves transcript and protein monitoring in response to stress conditions and functional analysis of the protein. Utilizing these strategies, in recent years several drought-related proteins were elucidated, the majority being stress-related transcription factors (TFs) and signal transducers.

Drought is known to be regulated at the transcriptional level, and TFs have been the focus of attention for the improvement of better yielding cultivars since targeting a single TF can affect several downstream-regulatory aspects of drought tolerance. Classically, two transcriptional regulatory circuits induced by drought have been studied: ABA-dependent and DREB-(dehydration-responsive element binding protein-) mediated (ABA-independent) pathways. These pathways are schematically depicted in Figure 1. One of the major classes of TFs involved in ABA-dependent stress responses is MYB TFs, and in the recent years, there has been a focus on the elucidation of bread wheat R2R3 and MYB3R type MYB TFs, known to be involved in ABA signaling of drought. In three different lines of research, drought responsive MYBs, TaPIMP1 (pathogen induced membrane protein), TaMYB33, and TaMYB3R1, were cloned and studied via the analysis of their domains, determination of their nuclear subcellular localizations, and assessment of transcriptional activation function to proteins [34, 36, 38]. Phylogenetic analysis of their protein sequences classified TaMYB3R1 as MYB3R type and the others as R2R3 type MYB TFs. The R2R3 type MYB TaPIMP1 was originally described as the first defense related MYB in wheat; however, detailed analyses indicated that TaPIMP1 is also induced by abiotic stresses, particularly drought. In addition, the induction of its expression by ABA and its inability to bind to the DRE-box element as indicated by EMSA suggest that TaPIMP1 acts in the ABA-dependent pathways of drought response [73]. Similarly, TaMYB33, another drought responsive R2R3 type MYB, was shown to be induced by ABA treatment, and the overexpression in *Arabidopsis* plants could not detect a significant increase in DREB2, suggesting that TaMYB33 is also involved in ABA-dependent mechanisms [38]. Both TaPIMP1 and TaMYB33 appear to enhance drought tolerance through ROS detoxification and reinforcement of osmotic balance. An elevated level of proline or proline synthesis common to both

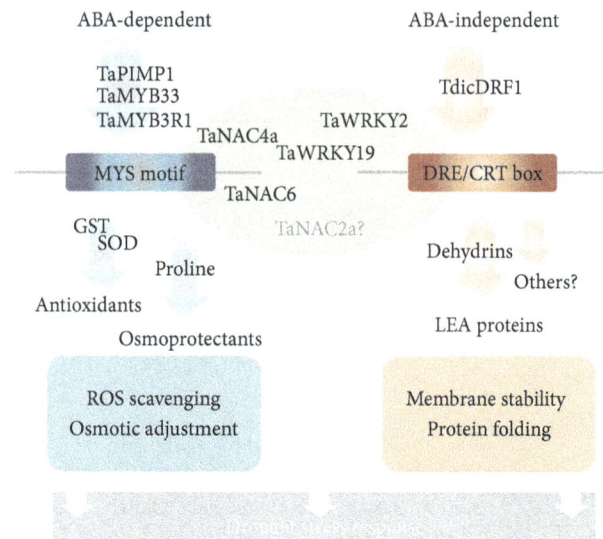

FIGURE 1: ABA-dependent and ABA-independent pathways of stress response. MYB and DREB TFs are given as examples to ABA-dependent and-independent routes. While ABA-dependent pathways appear to recruit antioxidant and osmoprotectant mechanisms, ABA-independent pathways generally involve protective proteins. NAC and WRKY TFs provide crosstalk between these pathways; where some members, such as TaNAC4 and TaNAC6, may predominantly act in an ABA-dependent fashion, some members may be closer to ABA-independent pathways. In several cases, such as TaWRKY19, both pathways are employed. It should be noted that both pathways are highly intermingled, and functions of several regulators, such as TaNAC2a, as well as entire pathways are yet to be elucidated.

TaPIMP1 and TaMYB33 mediated response is noteworthy [34, 38]. MYB3R type MYB TFs are less pronounced class of MYB proteins in stress response. TaMYB3R1, one of the few examples of MYB3R type MYBs in wheat, has been implicated in drought stress response and is also responsive to ABA, similar to TaPIMP1 and TaMYB33. However, the downstream events of TaMYB3R1-mediated drought stress response remains to be elucidated [36].

ABA-independent also called DREB-mediated pathways are largely governed by dehydration-responsive element-binding (DREB)/C-repeat-binding (CBF) proteins which recognize dehydration-responsive element (DRE)/C-repeat (CRT) motifs through a conserved AP2 domain. While DREB1 TFs are mainly responsive against cold stress, DREB2 TFs are more pronounced in drought stress response; although functional overlaps are possible where a certain DREB responds to multiple stresses [74]. It should be noted that DREB-mediated stress responses may also cooperate or overlap with ABA-dependent stress responses (Figure 1). A number of DREB homologs have deen identified in wheat and, although DREB2-mediated drought response is not fully elucidated yet, enhanced drought tolerance through DREB-mediated pathways is considered to involve LEA proteins [75]. As noted in Section 2, recently, a DREB2 homolog, TdicDRF1, was identified, cloned, and characterized for the first time in wild emmer wheat. Comparison of drought-stressed resistant and sensitive genotypes revealing differential expressions of TdicDRF1 suggested not only a conserved role on drought stress response but also a promising mechanism that can be utilized in improvement of wheat cultivars for drought tolerance [32].

In recent years, evidence has accumulated that there is crosstalk between classical ABA-dependent and ABA-independent pathways (Figure 1). The best known example of such occurrence is NAC TFs, which regulate drought stress response through both ABA-dependent and ABA-independent pathways. In a recent study, *T. aestivum* NAC (NAM/ATAF/CUC) TFs were identified *in silico*, phylogenetically classified and characterized, and their expression profiles were monitored in response to ABA and drought stress. In response to these treatments, TaNAC4a and TaNAC6 exhibited similar expression trends, suggesting an ABA-dependent regulation of drought, while in the case of TaNTL5 and TaNAC2a, the changes in the expression were not parallel [37]. In another study, WRKY type transcription factors (TaWRKY2 and TaWRKY19), which are known to be involved in plant abiotic stress response and ABA signaling were identified computationally, localized to the nucleus and shown to bind specifically to cis-element, W box. This report revealed that WRKY19 as a component of both ABA and DREB pathways, showing WRKY19 expression level, was responsive to ABA application, and in transgenic WRKY19 deficient plants, the expression levels of DREB pathway components were altered [39].

In addition to these known players of ABA-dependent and DREB-mediated pathways, other novel TFs are also discovered, and one such TF is the recently discovered *T. aestivum* salt response gene TF (TaSRG) which was shown to be induced in response to drought and ABA [35].

Other major targets of recent research have been enzymes that aid in reversible phosphorylation of signaling molecules in drought-related network of protein interactions. Although the major classes of stress-related kinases and phosphatases taking part in these cascades are known, namely, mitogen activated protein kinases (MAPKs), SNF-1-like kinases (SnRKs), calcium-dependent protein kinases (CDPKs), and MAP kinase phosphatases (MKPs), information regarding to these components is far from complete. For this purpose, known components should be further investigated since their exact positions and interactions in the complex signaling network are currently unknown. This has been applied recently, in two separate studies, in which SNF-1-like kinases, namely, TaSnRK2.4, TaSnRK2.7 from bread wheat, were characterized further in detail [41, 42]. In a different research, a MAP kinase phosphatase, TdTMKP1 was cloned from durum wheat, and its specific interaction with two MAPKs, TdTMPK3, and TdTMPK6, was verified, in accordance of its role as a negative regulator of MAPKs [43]. In an independent line of research, a priorly poorly characterized kinase TaABC1 (*T. aestivum* L. protein kinase) was investigated and shown to be involved in drought [40].

Although transcription factors and signal transducers have been the major focus of research in terms of characterization, in recent years other putative drought-related molecules were also isolated from wheat varieties, investigated, and supporting evidence for their roles in drought stress was obtained [31–33, 44–48]. Information regarding these studies is listed in Table 2. Most of the drought-related genes identified were confirmed either by using overexpressor plants or wheat deletion lines or silencing the gene of interest via virus induced gene silencing revealing its function [33, 39, 63] (Table 3).

5.2. Studies of Drought-Related Pathways. Plants in environments prone to drought stress have developed several tolerance strategies, resistance and avoidance mechanisms, which enable them to survive and reproduce under conditions of water scarcity.

The fundamental plant drought responses include growth limitation, changes in gene expression, altered hormonal levels, induced and suppressed signaling pathways, accumulation of compatible solutes and osmoprotectant proteins, suppression of metabolism, increased lipid peroxidation with higher levels of ROS, and counter-acting increased levels of antioxidant activity. Drought is regulated both at the transcriptional level and posttranscriptional level, the latter including the action of miRNAs and posttranslational modifications for proteosomal degradation [76] (Supplementary Table 2).

5.2.1. Compatible Solutes. Compatible solutes are nontoxic molecules that accumulate in the cytoplasm upon drought stress. Common compatible solutes are sugars, sugar alcohols, glycine betaine, amino acids, and proline. They are known to be involved in osmotic adjustment, function as ROS scavengers, protect proteins and cell structures, and exhibit adaptive value in metabolic pathways. In a recent study, compatible solutes in *T. aestivum* leaves were screened in response to water deficit at the reproduction stage. Major contributors to osmotic adjustment were revealed to be K^+ in the early stages of stress and molecules including glycinebetaine, proline, and glucose, in the late stress [77]. Recently, compatible solutes were also assessed in *T. aestivum* cultivars under different irrigation regimes. Drought application decreased the levels of inorganic solutes but increased the levels of organic solutes [78].

5.2.2. Protective Proteins. Protective proteins known to be involved in drought stress response include LEA proteins, aquaporins, heat shock proteins, and ion channels. Recently, in *T. aestivum* cultivars, changes in the transcript and protein levels of dehydrins and LEA proteins, in response to progressive drought applied at early vegetation and during its recovery, were monitored [79]. In a different study, two aquaporins were overexpressed providing direct evidence of their drought-related functions [48]. Additionally in another research, SNP and InDEL (insertion/deletion) repertoire of a Na^+/K^+ transporter, HKT-1, was assessed and observed to be compromised of mostly missense mutations, predominantly present in the tolerant wheat variety [80].

5.2.3. Signaling. Drought signaling includes signal perception and transduction. Known drought-related signal transducers and some recent reports on their characterization and functional assessment are summarized in Sections 5.1–5.3 [40–42]. Some of these studies were performed on calcium dependent protein kinases (CDPK), which sense and respond to Ca^{2+} an important secondary messenger of signal transduction cascades. In a recent study, two bread wheat CDPKs, CPK7 and CPK12, were revealed to be molecularly evolved through gene duplication followed by functional diversification. In this study, they were shown to contain different putative cis-element combinations in their promoters, and two CDPKs were shown to respond differently to drought, PEG, salt (NaCl), cold, hydrogen peroxide (H_2O_2), and ABA applications [81]. Other suggested signaling molecules of drought stress network are SA (salicylic acid) and NO (nitric oxide). In recent studies, NO was shown to be present in higher levels in drought tolerant wheat variety and may have a possible role in the drought induced limitation in root growth [82]. Recently, SA was shown to aid drought tolerance through increasing accumulation of solutes [78].

5.2.4. Photosynthesis and Respiration. Upon dehydration conditions, basal metabolic activities, including photosynthesis and respiration, are known to be altered in plants. Recently, *T. aestivum* and *T. durum* genotypes with differing sensitivities to osmotic stress were evaluated in terms of their gas exchange in response dehydration. Photosynthesis was analyzed in depth, in relation to ROS levels and physiological parameters, for elucidation of the mechanisms by which the tolerant cultivar sustains a high performance of water-use efficiency, maintaining respiration rate and photosynthesis even under stress conditions [83]. PS II (Photosystem II) is a key protein pigment complex of photosynthesis, which aids in light harvesting. Upon dehydration, PS II repair cycle is impaired affecting oxygen evolving process of PS II reaction center, leading to photooxidative damage. PS II has a biphasic primary phytochemistry kinetics, which is referred to as PS II heterogeneity. In recent studies, photosynthetic efficiency of PSII complex was measured in different wheat varieties under different environmental conditions and the extent and nature of this heterogeneity was assessed in detail, in relation to osmotic stress [84].

5.2.5. Growth. Stress response of plants in relation to growth differentiates based both on the tissue and the severity, timing, and mode of stress applied. The degree of osmotic stress induced limitation of plant organ growth also differs from cultivar to cultivar and does not have a correlation with drought tolerance [82, 85, 86]. Recently, dehydration induced retardation of root and leaf growth was studied in detail, in relation to several components of the drought response pathway. In some of these studies, cell wall-bound peroxidases, ROS and NO, were shown to be unfavorable for root expansion and suggested to have possible roles in retarding root cell wall extension [85]. Another report analyzed the underlying mechanisms of *T. aestivum* root elongation in detail, showing that even during plasmolysis

upon stress, although root elongation is retarded, new root hair cell formation is sustained [87]. In a different study, monitoring of *T. durum* leaves for ABA, water status, and leaf elongation rates upon drought stress and recovery revealed a retardation in leaf elongation even after stress recovery, suggesting that a rapid accumulation of ABA during stress may have caused the loss of cell wall extensibility [86].

Additionally, recently, expansin, known to be involved in cell wall loosening, was overexpressed in tobacco, confirming its role in plant water retention ability and osmotic potential [83].

5.2.6. Transcriptional Regulation and Posttranscriptional/ Translational Modifications. The expression of several drought-related gene products is regulated at the transcriptional level. Known drought-related TFs and some recent reports on their characterization and functional assessment are summarized in Sections 5.1–5.3 [32, 34–39]. Additional studies undertaken in the recent years include preliminary research, drought induced expression profiling of MYB transcripts, revealing TaMYBsdu1 as a potential drought-related TF and characterization of SNP and INDEL repertoire of *T. durum* DREB1 and WRKY1 [80, 88].

As well as on the genomic level, drought stress is also regulated at the posttranscriptional and posttranslational levels. The major players of posttranscriptional regulation are miRNAs, which have been identified in a variety of crops, including *Triticeae* species, using both computational and experimental approaches, and their expression profiling was performed in wild emmer wheat and crop model species including *Brachypodium* [15, 76, 89–92]. Posttranslational modifications include protein degradation, mostly via ubiquitination. In a recent study, the leaves of resistant *T. aestivum* cultivar were shown to exhibit a relatively small increase in cysteine protease function upon dehydration, limiting protein loss of this cultivar. In this study, a cysteine protease present only under drought conditions was detected [93]. In a different report, the functional role of a cysteine protease protein was confirmed by its overexpression in *Arabidopsis* [45].

5.2.7. ROS and Antioxidants. Upon drought stress, ROS generation occurs mainly in the chloroplast and mitochondria, which results in oxidative damage and lipid peroxidation. Nonenzymatic and enzymatic antioxidants are produced by the plant to detoxify ROS. In a recent report, drought induced ROS generation and antioxidant activity was screened comparatively in the root whole cells and mitochondria of drought acclimated and nonacclimated *T. aestivum* seedlings. A special role of mitochondrial scavenging mechanisms was highlighted by the finding that in specifically nonacclimated seedlings, mitochondrial antioxidants were found to be predominantly active. In the same study, a quick increase in antioxidant mechanisms was observed with stress recovery, thus it was proposed that accumulation of high levels of H_2O_2 upon stress can inhibit antioxidant mechanisms [94]. Another study was undertaken in two *T. aestivum* cultivars, differing in their drought tolerance, focusing especially on the level and activity of ascorbate glutathione cycle related enzymes in stem and leaf tissues. It was shown that ascorbate processing and oxidation were differentially changed in the two cultivars [7]. Recently several other studies of ROS were performed to determine its role in the drought response network in relation to photosynthesis, NO, root growth, ABA, and BABA (β-aminobutyric acid) [82, 95, 96].

5.2.8. Abscisic Acid.
ABA is a plant hormone that is known to be involved in plant developmental processes and was also shown to be an inducer of stress-related pathways. Recently the role of ABA catabolism in drought was supported by direct evidence from a wheat deletion line [63]. ABA is a major hot topic of drought research, and recently several studies are performed to determine its role in drought in relation to several drought-related molecules: protective proteins, NO, ROS, leaf and root growth, osmotic adjustment, BABA, ROS, and antioxidants [79, 82, 86, 95]. Additionally, in a recent study alginate oligosaccharides (AOS) prepared from degradation of alginate were shown to play a role in enhancing drought resistance in *T. aestivum* growth period by upregulating the drought tolerance related genes involved in ABA signal pathway, such as LEA1, SnRK2, and pyrroline-5-carboxylate synthetase gene (P5CS) [97].

5.3. Transgenic Studies for Identification of Gene Function.
Transgenic plants provide the most straightforward way to demonstrate the functional relevance of the potential drought-related gene. Using functional genomics methods, modification of a single gene can be achieved in an identical genetic background. Analyzing the functional role of a protein of interest is achieved by the creation of overexpressor plants or loss-of-function mutants. These studies are most often carried out in model species, *Arabidopsis thaliana* or *Nicotiana benthamiana*. The advantages of these species are basically their rapid reproduction time and the ease of genetic transformation. However, other transgenic model systems, which hold the advantage of being phylogenetically more similar to monocot crop species, are also being developed, most importantly *Brachypodium distachyon*. The use of even a phylogenetically very close model plant in function verification of a gene of interest does not exclude the possibility that its role can differ in the crop of interest. Therefore transgenic studies applied directly on wheat are being developed but currently still more time and labor consuming. Other systems, which hold the advantage of straightforward analysis of gene function in the target crops are deletion lines and virus induced gene silencing (VIGS), which aids in functional characterization through silencing of targeted transcripts.

5.3.1. Overexpression Studies.
Direct evidence for the functional role of several drought response candidates was established via their overexpression in *A. thaliana* or tobacco by *Agrobacterium* mediated transformation. To gain insight into the mechanisms the molecule of concern exerts a role in drought, and these studies are often coupled by the profiling of stress-related genes and characterization of the overexpressor plants in terms of their altered morphological and physiological properties. In recent years, drought-related molecular function of several transcription factors, signal transducers, and some other proteins were confirmed via their overexpression in *Arabidopsis* or *N. tabacum*. Information regarding these studies is summarized in Table 3.

Recently, in four independent studies, overexpression of transcription factors (TaWRKY2, TaWRKY19, TaMYB33, TaPIMP1, and TaNAC) was shown to confer elevated drought tolerance in target model organisms [34, 37–39]. With the use of these overexpressor plants, WRKY proteins were shown to be involved in the DREB pathway [39], and MYB proteins were revealed to function in ROS detoxification [34, 38]. MYB TaPIMP1 overexpressors were also shown to exhibit increased levels of ABA synthesis and its restricted signaling [38]. Additionally, recently introgression of three kinases (TaABC1, TaSnRK2.7, and TaSnRK2.4) into *Arabidopsis* in separate studies was shown to improve drought tolerance evident by the water content and related measurements of overexpressor plants. All three kinases were observed to improve photosynthetic efficiency. TaABC1 was shown to be involved in DREB and ABA pathways [40]. TaSnRK2.7 was revealed to be involved in carbohydrate metabolism and mechanisms involving root growth [42]. TaSnRK2.4 overexpressors displayed differences in development and showed strengthened cell membrane stability [41]. In addition to transcription factors and kinases, overexpression studies of other proteins were also performed in model organisms, in relation to drought and drought-related capabilities. Cysteine protease, TaCP transgenics were observed to have higher survival rates under drought [45]. CHP rich zinc finger protein TaCHP was revealed to be involved in ABA-dependent and -independent signaling pathways [44]. Expansin, TaEXPR23, was shown to increase water retention ability and decrease osmotic potential [46]. Durum wheat aquaporins (TdPIP1; 1 and TdPIP2; 1) were revealed to regulate root and leaf growth [48]. Besides, in a recent study, *T. aestivum* salt induced protein (TaSIP) was shown to have a role in drought and salt tolerance via overexpression of the gene in *A. thaliana* resulted superior physiological properties [61] (Table 3).

5.3.2. Wheat Deletion Lines and Virus Induced Gene Silencing.
In recent years, a number of other transgenic studies were also performed for function determination. In such a study, *T. aestivum* deletion lines which lack ABA 8'-hydroxylase gene, involved in ABA catabolism, were used to study the role of ABA metabolism in the reproductive stage drought tolerance of cultivars. This study revealed a parallel between sensitivity to osmotic stress and higher spike ABA levels [63]. In a different study, ubiquitin:TaCHP transgenic wheat lines were used in studies in relation to the role of CHP rich zinc finger protein in drought stress [44]. Additionally, VIGS via barley stripe mosaic virus (BSMV) derived vectors was undertaken to silence ATG8 in wild emmer wheat, revealing it as a positive regulator of drought stress [33] (Table 3). VIGS was also used in another research in order to investigate the roles of *Era1* (enhanced response to abscisic acid), *Sal1*

(inositol polyphosphate 1-phosphatase), and *Cyp707a* (ABA 8'-hydroxylase) in response to limiting water conditions in wheat. When subjected to limiting water conditions, VIGS-treated plants for *Era1* and *Sal1* resulted in increased relative water content (RWC), improved water use efficiency (WUE). Compared to other tested genes *Era1* was found to be the most promising as a potential target for drought tolerance breeding in wheat [49].

6. Improvement of Modern Wheat

Recent advances in molecular biological, functional, and comparative tools open up new opportunities for the molecular improvement of modern wheat. Recently developed techniques enable faster identification and characterization of drought-related gene(s) and gene region(s). Natural variants of modern species harbor a large repertoire of potential drought-related genes and hold a tremendous potential for wheat improvement. Introduction of drought-related components of wheat can be performed either with breeding through marker-assisted selection or transgenic methods. Recent increase in sequence availability due to recently developed high-throughput sequencing strategies has provided several high quality genetic markers for breeding. Transgenic strategies with enhanced transformation and selection methods are currently being developed.

6.1. Marker-Assisted Selection. Molecular breeding approaches based on specific traits utilize molecular markers for the screening of drought tolerance in cultivars. Loci that are targeted in marker-assisted selection (MAS) are most often derived from QTL mapping studies of quantitative traits [98]. MAS is most often performed based on physiomorphological characteristics related to yield under drought conditions. Markers that are utilized in such a context include SSR (simple sequence repeat) markers, *Xgwm136,* and NW3106, which are linked to genes that effect tillering capacity and coleoptile length, respectively [99]. Other selection markers are linked to Rht (reduced height) genes, which are known to be associated with harvest index. Additionally, transcription factor-derived markers, especially DREB proteins hold a great potential as PCR-based selection markers that can be useful in MAS [100]. However, the isolation of transcription factors is a challenge since they belong to large gene families containing members with high sequence similarities. Identification and successful isolation of a single drought-related loci is compelling also in general due to the complex genomic structure of wheat. However, in the near future, completion of wheat genome sequencing will pace identification of specific loci and the development of markers to be used in selection during breeding processes [98, 101].

6.2. Use of Transgenics. An alternative to ongoing breeding programmes is transgenic methods, which enable the transfer of only the desired loci from a source organism to elite wheat cultivars, avoiding possible decrease in yield due to the cotransfer of unwanted adjacent gene segments. Until now, transcription factors have been the most appealing

targets for transgenic wheat improvement, due to their role in multiple stress-related pathways. In two different lines of research, overexpression of cotton and *A. thaliana* DREB was performed in wheat, resulting in transgenic lines with improved drought tolerance [102–104]. In another study, a barley LEA protein, *HVA1*, was also overexpressed in wheat, and overexpressors were observed to have better drought tolerance [105]. Transgenic wheat obtained with *Arabidopsis* DREB and *HVA1* protein overexpression was also shown to produce higher yield in the field under drought conditions, but further studies are required to confirm their performance under different environments [105].

It is not unreasonable to predict in the following decades: GM (genetically modified) wheat will be transferred to the fields as a common commercial crop. However, to pace this process, new transgenics methodologies should be developed since the current methods are laborious and time consuming. In a recent study, drought enhancement of bread wheat was established with the overexpression of barley *HVA1*, using a novel technique, which combines doubled haploid technology and *Agrobacterium* mediated genetic transformation [62] (Table 3).

6.3. Use of Proteomics. Despite the impressive technological breakthroughs in the genomics of drought resistant cultivars the overall scenario is not so promising, and new dimensions have to be explored for the exact elucidation of the wheat drought response process. Hence new studies are focusing to study wheat tolerance at the proteomic level to target different proteins and understand their role in stress. One particular study during grain development used comparative proteomic analysis and used 2 varieties of wheat resistant (CIMMYT wheat variety Kauz) and sensitive (Janz to drought). They applied linear and nonlinear 2-DE and MALDI-TOF mass spectrometry and elucidated that non-linear 2-DE showed a high resolution and identifies 153 spots of proteins that were differentially expressed, 122 of which were detected by MALDI-TOF. The characterized proteins were primarily metabolism proteins (26% carbohydrate metabolism), proteins involved in defense and detoxification (23%), and the rest of 17% were storage proteins. The study successfully showed the differential expression of various proteins in drought resistant and tolerant varieties. Kauz wheat variety showed high expression of LEA and alpha-amylase inhibitors and catalase isozyme 1, WD40 repeat protein, whereas these proteins were either unchanged or downregulated in Janz variety. Vice versa ascorbate peroxidase G beta-like protein and ADP glucose pyrophosphorylase remained unchanged in Kauz but were all downregulated in Janz. Proteins such as triticin precursor and sucrose synthase showed a considerably higher expression in Kauz water deficit variety compared to Janz water deficit plants. Thus the differential expression shows that biochemical and protein level expression could be a simpler approach to understanding and manipulating drought stress in plants [106].

A parallel approach to understanding the protein expression and posttranslational modification in wheat was carried out by Budak et al. in which 2 wild varieties of emmer wheat

Triticum turgidum ssp. *dicoccoides* TR39477 and TTD22 were used along with one modern wheat cultivar *Triticum turgidum* ssp. *durum* cv. Kızıltan. The complete leaf proteome profiles of all three genotypes were compared by 2-DE gel electrophoresis and nanoscale liquid chromatographic electrospray ionization tandem mass spectrometry. Instead of using only drought tolerant and drought resistant varieties another third intermediate variety (modern) was also used. Although many proteins were common in all 3 cultivars both modern and durum but 75 differentially expressed proteins were detected [107]. Consequently comparative proteomics may provide a clearer picture and alternate way to evaluate and characterize drought resistant genes and proteins in wheat varieties.

7. Conclusion and Future Perspectives

Drought stress is one of the major limitations to crop production. To develop improved cultivars with enhanced tolerance to drought stress, identification of osmotic stress-related molecules and determination of their roles and locations in several physiological, biochemical, and gene regulatory networks is necessary. Several QTLs for key morphopysiological characteristics and yield were identified under water-limited conditions through creation of linkage maps using parentals with different drought coping abilities. In recent decades, application of high-throughput screening, "omics" strategies on *Triticum* species with differential drought tolerance coping abilities, has revealed several stress-related candidate gene(s) or gene block(s). Furthermore, using a variety of bioinformatics, molecular biology, and functional genomics tools, drought-related candidates were characterized, and their roles in drought tolerance were studied. Major drought-related molecules were revealed to be signal transduction pathway components and transcription factors. Several osmoprotectants, compatible solutes, ROS, and antioxidants were shown to accumulate in response to dehydration. Drought stress was found to alter various ongoing metabolic processes, such as growth, photosynthesis, and respiration.

Analysis of drought response has been complicated in the absence of wheat genomic sequence data. However, with the recent advances in sequencing technologies, genome sequence of bread wheat is almost complete by the efforts of ITMI (The International *Triticeae* Mapping Initiative) and IWGSC (International Wheat Genome Sequencing Consortium). Availability of whole wheat genome sequence will contribute to the ongoing studies of exploring the extensive reservoir of alleles in drought tolerant wild germplasm, and this also enables better marker development, genome analysis and large scale profiling experiments. "Omics" strategies have especially contributed to drought research since osmotic stress response is not only genomic based but also regulated at the posttranscriptional and posttranslational levels. Advances in transformation/selection strategies have paced molecular transformation of wheat, which has an advantage to conventional and marker-assisted breeding for targeted introduction of only the desired loci.

It is reasonable to predict that in the following years higher yielding wheat under drought conditions will be developed through breeding or molecular transformation of novel genes obtained from screening of wheat germplasms and will be commercially grown to balance the production with the consumption of the increasing human population. Research exploiting recent advances in genomics technologies has made it possible to dissect and resynthesize molecular regulation of drought and manipulate crop genomes for drought tolerance. The future efforts will be to integrate and translate these resources into practical higher yielding field products.

References

[1] N. Z. Ergen and H. Budak, "Sequencing over 13 000 expressed sequence tags from six subtractive cDNA libraries of wild and modern wheats following slow drought stress," *Plant, Cell & Environment*, vol. 32, no. 3, pp. 220–236, 2009.

[2] D. Fleury, S. Jefferies, H. Kuchel, and P. Langridge, "Genetic and genomic tools to improve drought tolerance in wheat," *Journal of Experimental Botany*, vol. 61, no. 12, pp. 3211–3222, 2010.

[3] M. Tester and P. Langridge, "Breeding technologies to increase crop production in a changing world," *Science*, vol. 327, no. 5967, pp. 818–822, 2010.

[4] D. Z. Habash, Z. Kehel, and M. Nachit, "Genomic approaches for designing durum wheat ready for climate change with a focus on drought," *Journal of Experimental Botany*, vol. 60, no. 10, pp. 2805–2815, 2009.

[5] M. P. Reynolds, "Drought adaptation in wheat," in *Drought Tolerance in Cereals*, J. M. Ribaut, Ed., chapter 11, pp. 402–436, Haworth's Food Products Press, New York, NY, USA, 2006.

[6] M. Ashraf, M. Ozturk, and H. R. Athar, *Salinity and Water Stress: Improving Crop Efficiency*, Berlin, Germany, Springer, 2009.

[7] M. Sečenji, E. Hideg, A. Bebes, and J. Györgyey, "Transcriptional differences in gene families of the ascorbate-glutathione cycle in wheat during mild water deficit," *Plant Cell Reports*, vol. 29, no. 1, pp. 37–50, 2010.

[8] A. Szucs, K. Jäger, M. E. Jurca et al., "Histological and microarray analysis of the direct effect of water shortage alone or combined with heat on early grain development in wheat (*Triticum aestivum*)," *Physiologia Plantarum*, vol. 140, no. 2, pp. 174–188, 2010.

[9] M. M. Bazargani, E. Sarhadi, A. A. Bushehri et al., "A proteomics view on the role of drought-induced senescence and oxidative stress defense in enhanced stem reserves remobilization in wheat," *Journal of Proteomics*, vol. 74, no. 10, pp. 1959–1973, 2011.

[10] F. Yang, A. D. Jørgensen, H. Li et al., "Implications of high-temperature events and water deficits on protein profiles in wheat (*Triticum aestivum* L. cv. Vinjett) grain," *Proteomics*, vol. 11, no. 9, pp. 1684–1695, 2011.

[11] K. L. Ford, A. Cassin, and A. Bacic, "Quantitative proteomic analysis of wheat cultivars with differing drought stress tolerance," *Frontiers in Plant Science*, vol. 2, article 44, 2011.

[12] P. Rampino, G. Mita, P. Fasano et al., "Novel durum wheat genes up-regulated in response to a combination of heat and drought stress," *Plant Physiology and Biochemistry*, vol. 56, pp. 72–78, 2012.

[13] S. Irar, F. Brini, A. Goday, K. Masmoudi, and M. Pagès, "Proteomic analysis of wheat embryos with 2-DE and liquid-phase chromatography (ProteomeLab PF-2D)—a wider perspective

of the proteome," *Journal of Proteomics*, vol. 73, no. 9, pp. 1707–1721, 2010.

[14] T. Krugman, Z. Peleg, L. Quansah et al., "Alteration in expression of hormone-related genes in wild emmer wheat roots associated with drought adaptation mechanisms," *Functional & Integrative Genomics*, vol. 11, no. 4, pp. 565–583, 2011.

[15] M. Kantar, S. J. Lucas, and H. Budak, "miRNA expression patterns of *Triticum dicoccoides* in response to shock drought stress," *Planta*, vol. 233, no. 3, pp. 471–484, 2011.

[16] S. Farooq, "Triticeae: the ultimate source of abiotic stress tolerance improvement in wheat," in *Salinity and Water Stress*, M. Ashraf, Ed., chapter 7, pp. 65–71, Springer, Berlin, Germany, 2009.

[17] M. Feldman and E. R. Sears, "The wild genetic resources of wheat," *Scientific American*, vol. 244, pp. 102–112, 1981.

[18] P. Dong, Y. M. Wei, G. Y. Chen et al., "EST-SSR diversity correlated with ecological and genetic factors of wild emmer wheat in Israel," *Hereditas*, vol. 146, no. 1, pp. 1–10, 2009.

[19] T. Fahima, M. S. Röder, K. Wendehake, V. M. Kirzhner, and E. Nevo, "Microsatellite polymorphism in natural populations of wild emmer wheat, *Triticum dicoccoides*, in Israel," *Theoretical and Applied Genetics*, vol. 104, no. 1, pp. 17–29, 2002.

[20] T. Fahima, G. L. Sun, A. Beharav, T. Krugman, A. Beiles, and E. Nevo, "RAPD polymorphism of wild emmer wheat populations, *Triticum dicoccoides*, in Israel," *Theoretical and Applied Genetics*, vol. 98, no. 3-4, pp. 434–447, 1999.

[21] E. Nevo and A. Beiles, "Genetic diversity of wild emmer wheat in Israel and Turkey—structure, evolution, and application in breeding," *Theoretical and Applied Genetics*, vol. 77, no. 3, pp. 421–455, 1989.

[22] E. Nevo, E. Golenberg, A. Beiles, A. H. D. Brown, and D. Zohary, "Genetic diversity and environmental associations of wild wheat, *Triticum dicoccoides*, in Israel," *Theoretical and Applied Genetics*, vol. 62, no. 3, pp. 241–254, 1982.

[23] Z. Peleg, Y. Saranga, T. Krugman, S. Abbo, E. Nevo, and T. Fahima, "Allelic diversity associated with aridity gradient in wild emmer wheat populations," *Plant, Cell & Environment*, vol. 31, no. 1, pp. 39–49, 2008.

[24] J. R. Wang, Y. M. Wei, X. Y. Long et al., "Molecular evolution of dimeric α-amylase inhibitor genes in wild emmer wheat and its ecological association," *BMC Evolutionary Biology*, vol. 8, no. 1, article 91, 2008.

[25] Z. Peleg, T. Fahima, S. Abbo et al., "Genetic diversity for drought resistance in wild emmer wheat and its ecogeographical associations," *Plant, Cell & Environment*, vol. 28, no. 2, pp. 176–191, 2005.

[26] J. Peng, D. Sun, and E. Nevo, "Domestication evolution, genetics and genomics in wheat," *Molecular Breeding*, vol. 28, no. 3, pp. 281–301, 2011.

[27] J. Peng, D. Sun, and E. Nevo, "Wild emmer wheat, *Triticum dicoccoides*, occupies a pivotal position in wheat domestication process," *Australian Journal of Crop Science*, vol. 5, no. 9, pp. 1127–1143, 2011.

[28] J. H. Peng, D. F. Sun, Y. L. Peng, and E. Nevo, "Gene discovery in *Triticum dicoccoides*, the direct progenitor of cultivated wheats," *Cereal Research Communications*, vol. 41, no. 1, pp. 1–22, 2013.

[29] N. Z. Ergen, J. Thimmapuram, H. J. Bohnert, and H. Budak, "Transcriptome pathways unique to dehydration tolerant relatives of modern wheat," *Functional and Integrative Genomics*, vol. 9, no. 3, pp. 377–396, 2009.

[30] T. Krugman, V. Chagué, Z. Peleg et al., "Multilevel regulation and signalling processes associated with adaptation to terminal drought in wild emmer wheat," *Functional and Integrative Genomics*, vol. 10, no. 2, pp. 167–186, 2010.

[31] S. Lucas, E. Dogan, and H. Budak, "TMPIT1 from wild emmer wheat: first characterisation of a stress-inducible integral membrane protein," *Gene*, vol. 483, no. 1-2, pp. 22–28, 2011.

[32] S. Lucas, E. Durmaz, B. A. Akpnar, and H. Budak, "The drought response displayed by a DRE-binding protein from *Triticum dicoccoides*," *Plant Physiology and Biochemistry*, vol. 49, no. 3, pp. 346–351, 2011.

[33] D. Kuzuoglu-Ozturk, O. Cebeci Yalcinkaya, B. A. Akpinar et al., "Autophagy-related gene, TdAtg8, in wild emmer wheat plays a role in drought and osmotic stress response," *Planta*, vol. 236, no. 4, pp. 1081–1092, 2012.

[34] H. Liu, X. Zhou, N. Dong, X. Liu, H. Zhang, and Z. Zhang, "Expression of a wheat MYB gene in transgenic tobacco enhances resistance to *Ralstonia solanacearum*, and to drought and salt stresses," *Functional & Integrative Genomics*, vol. 11, no. 3, pp. 431–443, 2011.

[35] X. He, X. Hou, Y. Shen, and Z. Huang, "TaSRG, a wheat transcription factor, significantly affects salt tolerance in transgenic rice and *Arabidopsis*," *FEBS Letters*, vol. 585, no. 8, pp. 1231–1237, 2011.

[36] H. Cai, S. Tian, C. Liu, and H. Dong, "Identification of a MYB3R gene involved in drought, salt and cold stress in wheat (*Triticum aestivum* L.)," *Gene*, vol. 485, no. 2, pp. 146–152, 2011.

[37] Y. Tang, M. Liu, S. Gao et al., "Molecular characterization of novel TaNAC genes in wheat and overexpression of TaNAC2a confers drought tolerance in tobacco," *Plant Physiology*, vol. 144, no. 3, pp. 210–224, 2012.

[38] Y. Qin, M. Wang, Y. Tian, W. He, L. Han, and G. Xia, "Over-expression of TaMYB33 encoding a novel wheat MYB transcription factor increases salt and drought tolerance in *Arabidopsis*," *Molecular Biology Reports*, vol. 39, no. 6, pp. 7183–7192, 2012.

[39] C. F. Niu, W. Wei, Q. Y. Zhou et al., "Wheat WRKY genes TaWRKY2 and TaWRKY19 regulate abiotic stress tolerance in transgenic *Arabidopsis* plants," *Plant, Cell & Environment*, vol. 35, no. 6, pp. 1156–1170, 2012.

[40] C. Wang, R. Jing, X. Mao, X. Chang, and A. Li, "TaABC1, a member of the activity of bc1 complex protein kinase family from common wheat, confers enhanced tolerance to abiotic stresses in *Arabidopsis*," *Journal of Experimental Botany*, vol. 62, no. 3, pp. 1299–1311, 2011.

[41] X. Mao, H. Zhang, S. Tian, X. Chang, and R. Jing, "TaSnRK2.4, an SNF1-type serine/threonine protein kinase of wheat (*Triticum aestivum* L.), confers enhanced multistress tolerance in *Arabidopsis*," *Journal of Experimental Botany*, vol. 61, no. 3, pp. 683–696, 2010.

[42] H. Zhang, X. Mao, R. Jing, X. Chang, and H. Xie, "Characterization of a common wheat (*Triticum aestivum* L.) TaSnRK2.7 gene involved in abiotic stress responses," *Journal of Experimental Botany*, vol. 62, no. 3, pp. 975–988, 2011.

[43] I. Zaïdi, C. Ebel, M. Touzri et al., "TMKP1 is a novel wheat stress responsive MAP kinase phosphatase localized in the nucleus," *Plant Molecular Biology*, vol. 73, no. 3, pp. 325–338, 2010.

[44] C. Li, J. Lv, X. Zhao et al., "TaCHP: a wheat zinc finger protein gene down-regulated by abscisic acid and salinity stress plays a positive role in stress tolerance," *Plant Physiology*, vol. 154, no. 1, pp. 211–221, 2010.

[45] Q. W. Zang, C. X. Wang, X. Y. Li et al., "Isolation and characterization of a gene encoding a polyethylene glycol-induced cysteine protease in common wheat," *Journal of Biosciences*, vol. 35, no. 3, pp. 379–388, 2010.

[46] Y. Y. Han, A. X. Li, F. Li, M. R. Zhao, and W. Wang, "Characterization of a wheat (*Triticum aestivum* L.) expansin gene, TaEXPB23, involved in the abiotic stress response and phytohormone regulation," *Plant Physiology and Biochemistry*, vol. 54, pp. 49–58, 2012.

[47] G. Z. Kang, H. F. Peng, Q. X. Han, Y. H. Wang, and T. C. Guo, "Identification and expression pattern of ribosomal L5 gene in common wheat (*Triticum aestivum* L.)," *Gene*, vol. 493, no. 1, pp. 62–68, 2012.

[48] M. Ayadi, D. Cavez, N. Miled, F. Chaumont, and K. Masmoudi, "Identification and characterization of two plasma membrane aquaporins in durum wheat (*Triticum turgidum* L. subsp. durum) and their role in abiotic stress tolerance," *Plant Physiology and Biochemistry*, vol. 49, no. 9, pp. 1029–1039, 2011.

[49] H. Manmathan, D. Shaner, J. Snelling, N. Tisserat, and N. Lapitan, "Virus-induced gene silencing of Arabidopsis thaliana gene homologues in wheat identifies genes conferring improved drought tolerance," *Journal of Experimental Botany*, vol. 64, no. 5, pp. 1381–1392, 2013.

[50] M. Reynolds and R. Tuberosa, "Translational research impacting on crop productivity in drought-prone environments," *Current Opinion in Plant Biology*, vol. 11, no. 2, pp. 171–179, 2008.

[51] M. Reynolds and G. Rebetzke, "Application of plant physiology in wheat breeding," in *The World Wheat Book: A History of Wheat Breeding*, A. P. Bonjean, W. J. Angus, and M. van Ginkel, Eds., vol. 2, Paris, France, 2011.

[52] C. M. Cossani and M. P. Reynolds, "Physiological traits for improving heat tolerance in wheat," *Plant Physiology*, vol. 160, no. 4, pp. 1710–1718, 2012.

[53] R. A. Richards and Z. Lukacs, "Seedling vigour in wheat—sources of variation for genetic and agronomic improvement," *Australian Journal of Agricultural Research*, vol. 53, no. 1, pp. 41–50, 2002.

[54] P. O. Lim, H. J. Kim, and H. G. Nam, "Leaf senescence," *Annual Review of Plant Biology*, vol. 58, pp. 115–136, 2007.

[55] U. Kumar, A. K. Joshi, M. Kumari, R. Paliwal, S. Kumar, and M. S. Röder, "Identification of QTLs for stay green trait in wheat (*Triticum aestivum* L.) in the "Chirya 3" × "Sonalika" population," *Euphytica*, vol. 174, no. 3, pp. 437–445, 2010.

[56] M. A. J. Parry, M. Reynolds, M. E. Salvucci et al., "Raising yield potential of wheat. II. Increasing photosynthetic capacity and efficiency," *Journal of Experimental Botany*, vol. 62, no. 2, pp. 453–467, 2011.

[57] M. E. Salvucci and S. J. Crafts-Brandner, "Inhibition of photosynthesis by heat stress: the activation state of Rubisco as a limiting factor in photosynthesis," *Physiologia Plantarum*, vol. 120, no. 2, pp. 179–186, 2004.

[58] I. Kurek, K. C. Thom, S. M. Bertain et al., "Enhanced thermostability of *Arabidopsis* rubisco activase improves photosynthesis and growth rates under moderate heat stress," *The Plant Cell*, vol. 19, no. 10, pp. 3230–3241, 2007.

[59] T. Krugman, V. Chagué, Z. Peleg et al., "Multilevel regulation and signalling processes associated with adaptation to terminal drought in wild emmer wheat," *Functional and Integrative Genomics*, vol. 10, no. 2, pp. 167–186, 2010.

[60] M. Sečenji, Á. Lendvai, P. Miskolczi et al., "Differences in root functions during long-term drought adaptation: comparison of

[61] H.-Y. Du, Y.-Z. Shen, and Z.-J. Huang, "Function of the wheat TaSIP gene in enhancing drought and salt tolerance in transgenic *Arabidopsis* and rice," *Plant Molecular Biology*, vol. 81, no. 4-5, pp. 417–429, 2013.

[62] H. Chauhan and P. Khurana, "Use of doubled haploid technology for development of stable drought tolerant bread wheat (*Triticum aestivum* L.) transgenics," *Plant Biotechnology Journal*, vol. 9, no. 3, pp. 408–417, 2011.

[63] X. Ji, B. Dong, B. Shiran et al., "Control of abscisic acid catabolism and abscisic acid homeostasis is important for reproductive stage stress tolerance in cereals," *Plant Physiology*, vol. 156, no. 2, pp. 647–662, 2011.

[64] A. A. Diab, R. V. Kantety, N. Z. Ozturk, D. Benscher, M. M. Nachit, and M. E. Sorrells, "Drought—inducible genes and differentially expressed sequence tags associated with components of drought tolerance in durum wheat," *Scientific Research and Essays*, vol. 3, no. 1, pp. 9–27, 2008.

[65] M. Maccaferri, M. C. Sanguineti, S. Corneti et al., "Quantitative trait loci for grain yield and adaptation of durum wheat (*Triticum durum* Desf.) across a wide range of water availability," *Genetics*, vol. 178, no. 1, pp. 489–511, 2008.

[66] K. L. Mathews, M. Malosetti, S. Chapman et al., "Multi-environment QTL mixed models for drought stress adaptation in wheat," *Theoretical and Applied Genetics*, vol. 117, no. 7, pp. 1077–1091, 2008.

[67] C. L. McIntyre, K. L. Mathews, A. Rattey et al., "Molecular detection of genomic regions associated with grain yield and yield-related components in an elite bread wheat cross evaluated under irrigated and rainfed conditions," *Theoretical and Applied Genetics*, vol. 120, no. 3, pp. 527–541, 2010.

[68] Z. Peleg, T. Fahima, T. Krugman et al., "Genomic dissection of drought resistance in durum wheat × wild emmer wheat recombinant inbreed line population," *Plant, Cell & Environment*, vol. 32, no. 7, pp. 758–779, 2009.

[69] D. Bennett, M. Reynolds, D. Mullan et al., "Detection of two major grain yield QTL in bread wheat (*Triticum aestivum* L.) under heat, drought and high yield potential environments," *Theoretical and Applied Genetics*, vol. 125, no. 7, pp. 1473–1485, 2012.

[70] D. Bennett, A. Izanloo, M. Reynolds, H. Kuchel, P. Langridge, and T. Schnurbusch, "Genetic dissection of grain yield and physical grain quality in bread wheat (*Triticum aestivum* L.) under water-limited environments," *Theoretical and Applied Genetics*, vol. 125, no. 2, pp. 255–271, 2012.

[71] D. Bennett, A. Izanloo, J. Edwards et al., "Identification of novel quantitative trait loci for days to ear emergence and flag leaf glaucousness in a bread wheat (*Triticum aestivum* L.) population adapted to southern Australian conditions," *Theoretical and Applied Genetics*, vol. 124, no. 4, pp. 697–711, 2012.

[72] J. Bonneau, J. Taylor, B. Parent et al., "Multi-environment analysis and improved mapping of a yield-related QTL on chromosome 3B of wheat," *Theoretical and Applied Genetics*, vol. 126, no. 3, pp. 747–761, 2013.

[73] Z. Zhang, X. Liu, X. Wang et al., "An R2R3 MYB transcription factor in wheat, TaPIMP1, mediates host resistance to Bipolaris sorokiniana and drought stresses through regulation of defense- and stress-related genes," *New Phytologist*, vol. 196, no. 4, pp. 1155–1170, 2012.

active gene sets of two wheat genotypes," *Plant Biology*, vol. 12, no. 6, pp. 871–882, 2010.

[74] P. K. Agarwal, P. Agarwal, M. K. Reddy, and S. K. Sopory, "Role of DREB transcription factors in abiotic and biotic stress tolerance in plants," *Plant Cell Reports*, vol. 25, no. 12, pp. 1263–1274, 2006.

[75] C. Egawa, F. Kobayashi, M. Ishibashi, T. Nakamura, C. Nakamura, and S. Takumi, "Differential regulation of transcript accumulation and alternative splicing of a DREB2 homolog under abiotic stress conditions in common wheat," *Genes & Genetic Systems*, vol. 81, no. 2, pp. 77–91, 2006.

[76] M. Kantar, S. J. Lucas, and H. Budak, "Drought stress: molecular genetics and genomics approaches," *Advances in Botanical Research*, vol. 57, pp. 445–493, 2011.

[77] S. A. Nio, G. R. Cawthray, L. J. Wade, and T. D. Colmer, "Pattern of solutes accumulated during leaf osmotic adjustment as related to duration of water deficit for wheat at the reproductive stage," *Plant Physiology and Biochemistry*, vol. 49, no. 10, pp. 1126–1137, 2011.

[78] N. Loutfy, M. A. El-Tayeb, A. M. Hassanen, M. F. Moustafa, Y. Sakuma, and M. Inouhe, "Changes in the water status and osmotic solute contents in response to drought and salicylic acid treatments in four different cultivars of wheat (*Triticum aestivum*)," *Journal of Plant Research*, vol. 125, no. 1, pp. 173–184, 2012.

[79] I. I. Vaseva, B. S. Grigorova, L. P. Simova-Stoilova, K. N. Demirevska, and U. Feller, "Abscisic acid and late embryogenesis abundant protein profile changes in winter wheat under progressive drought stress," *Plant Biology*, vol. 12, no. 5, pp. 698–707, 2010.

[80] L. Mondini, M. Nachit, E. Porceddu, and M. A. Pagnotta, "Identification of SNP mutations in DREB1, HKT1, and WRKY1 genes involved in drought and salt stress tolerance in durum wheat (*Triticum turgidum* L. var durum)," *OMICS*, vol. 16, no. 4, pp. 178–187, 2012.

[81] S. Geng, Y. Zhao, L. Tang et al., "Molecular evolution of two duplicated CDPK genes CPK7 and CPK12 in grass species: a case study in wheat (*Triticum aestivum* L.)," *Gene*, vol. 475, no. 2, pp. 94–103, 2011.

[82] I. Tari, A. Guóth, J. Benyó, J. Kovács, P. Poór, and B. Wodala, "The roles of ABA, reactive oxygen species and Nitric Oxide in root growth during osmotic stress in wheat: comparison of a tolerant and a sensitive variety," *Acta Biologica Hungarica*, vol. 61, no. 1, pp. 189–196, 2010.

[83] V. Vassileva, C. Signarbieux, I. Anders, and U. Feller, "Genotypic variation in drought stress response and subsequent recovery of wheat (*Triticum aestivum* L.)," *Journal of Plant Research*, vol. 124, no. 1, pp. 147–154, 2011.

[84] R. Singh-Tomar, S. Mathur, S. I. Allakhverdiev, and A. Jajoo, "Changes in PS II heterogeneity in response to osmotic and ionic stress in wheat leaves (*Triticum aestivum*)," *Journal of Bioenergetics and Biomembranes*, vol. 44, no. 4, pp. 411–419, 2012.

[85] J. Csiszar, A. Gallé, E. Horváth et al., "Different peroxidase activities and expression of abiotic stress-related peroxidases in apical root segments of wheat genotypes with different drought stress tolerance under osmotic stress," *Plant Physiology and Biochemistry*, vol. 52, pp. 119–129, 2012.

[86] M. Mahdid, A. Kameli, C. Ehlert, and T. Simonneau, "Rapid changes in leaf elongation, ABA and water status during the recovery phase following application of water stress in two durum wheat varieties differing in drought tolerance," *Plant Physiology and Biochemistry*, vol. 49, no. 10, pp. 1077–1083, 2011.

[87] M. Volgger, I. Lang, M. Ovečka, and I. Lichtscheidl, "Plasmolysis and cell wall deposition in wheat root hairs under osmotic stress," *Protoplasma*, vol. 243, no. 1–4, pp. 51–62, 2010.

[88] M. Rahaie, G. P. Xue, M. R. Naghavi, H. Alizadeh, and P. M. Schenk, "A MYB gene from wheat (*Triticum aestivum* L.) is up-regulated during salt and drought stresses and differentially regulated between salt-tolerant and sensitive genotypes," *Plant Cell Reports*, vol. 29, no. 8, pp. 835–844, 2010.

[89] T. Unver and H. Budak, "Conserved micrornas and their targets in model grass species brachypodium distachyon," *Planta*, vol. 230, no. 4, pp. 659–669, 2009.

[90] H. Budak and A. Akpinar, "Dehydration stress-responsive miRNA in Brachypodium distachyon: evident by genome-wide screening of microRNAs expression," *OMICS*, vol. 15, no. 11, pp. 791–799, 2011.

[91] S. J. Lucas and H. Budak, "Sorting the wheat from the chaff: identifying miRNAs in genomic survey sequences of *Triticum aestivum* chromosome 1AL," *PLoS One*, vol. 7, no. 7, Article ID e40859, 2012.

[92] M. Kantar, B. A. Akpınar, M. Valárik et al., "Subgenomic analysis of microRNAs in polyploid wheat," *Functional & Integrative Genomics*, vol. 12, no. 3, pp. 465–479, 2012.

[93] L. Simova-Stoilova, I. Vaseva, B. Grigorova, K. Demirevska, and U. Feller, "Proteolytic activity and cysteine protease expression in wheat leaves under severe soil drought and recovery," *Plant Physiology and Biochemistry*, vol. 48, no. 2-3, pp. 200–206, 2010.

[94] D. S. Selote and R. Khanna-Chopra, "Antioxidant response of wheat roots to drought acclimation," *Protoplasma*, vol. 245, no. 1-4, pp. 153–163, 2010.

[95] Y.-L. Du, Z.-Y. Wang, J.-W. Fan, N. C. Turner, T. Wang, and F.-M. Li, "Beta-Aminobutyric acid increases abscisic acid accumulation and desiccation tolerance and decreases water use but fails to improve grain yield in two spring wheat cultivars under soil drying," *Journal of Experimental Botany*, vol. 63, no. 13, pp. 4849–4860, 2012.

[96] I. M. Huseynova, "Photosynthetic characteristics and enzymatic antioxidant capacity of leaves from wheat cultivars exposed to drought," *Biochimica et Biophysica Acta*, vol. 1817, no. 8, pp. 1516–1523, 2012.

[97] H. Liu, Y. H. Zhang, H. Yin, W. X. Wang, X. M. Zhao, and Y. G. Du, "Alginate oligosaccharides enhanced *Triticum aestivum* L. tolerance to drought stress," *Plant Physiology and Biochemistry*, vol. 62, pp. 33–40, 2013.

[98] J. R. Witcombe, P. A. Hollington, C. J. Howarth, S. Reader, and K. A. Steele, "Breeding for abiotic stresses for sustainable agriculture," *Philosophical Transactions of the Royal Society B*, vol. 363, no. 1492, pp. 703–716, 2008.

[99] S. Gulnaz, M. Sajjad, I. Khaliq, A. S. Khan, and S. H. Khan, "Relationship among coleoptile length, plant height and tillering capacity for developing improved wheat varieties," *International Journal of Agriculture and Biology*, vol. 13, no. 1, pp. 130–133, 2011.

[100] B. Wei, R. Jing, C. Wang et al., "Dreb1 genes in wheat (*Triticum aestivum* L.): development of functional markers and gene mapping based on SNPs," *Molecular Breeding*, vol. 23, no. 1, pp. 13–22, 2009.

[101] E. Nevo and G. Chen, "Drought and salt tolerances in wild relatives for wheat and barley improvement," *Plant, Cell & Environment*, vol. 33, no. 4, pp. 670–685, 2010.

[102] P. Guo, M. Baum, S. Grando et al., "Differentially expressed genes between drought-tolerant and drought-sensitive barley

genotypes in response to drought stress during the reproductive stage," *Journal of Experimental Botany*, vol. 60, no. 12, pp. 3531–3544, 2009.

[103] A. Pellegrineschi, M. Reynolds, M. Pacheco et al., "Stress-induced expression in wheat of the *Arabidopsis* thaliana DREB1A gene delays water stress symptoms under greenhouse conditions," *Genome*, vol. 47, no. 3, pp. 493–500, 2004.

[104] D. Hoisington and R. Ortiz, "Research and field monitoring on transgenic crops by the Centro Internacional de Mejoramiento de Maíz y Trigo (CIMMYT)," *Euphytica*, vol. 164, no. 3, pp. 893–902, 2008.

[105] A. Bahieldin, H. T. Mahfouz, H. F. Eissa et al., "Field evaluation of transgenic wheat plants stably expressing the HVA1 gene for drought tolerance," *Physiologia Plantarum*, vol. 123, no. 4, pp. 421–427, 2005.

[106] S.-S. Jiang, X. N. Liang, X. Li et al., "Wheat drought-responsive grain proteome analysis by linear and nonlinear 2-DE and MALDI-TOF mass spectrometry," *International Journal of Molecular Sciences*, vol. 13, no. 12, pp. 16065–16083, 2012.

[107] H. Budak, B. A. Akpinar, T. Unver, and M. Turktas, "Proteome changes in wild and modern wheat leaves upon drought stress by two-dimensional electrophoresis and nanoLC-ESI-MS/MS," *Plant Molecular Biology*, 2013.

6

Genetic Variation and Population Structure in Jamunapari Goats Using Microsatellites, Mitochondrial DNA, and Milk Protein Genes

P. K. Rout,[1] K. Thangraj,[2] A. Mandal,[1] and R. Roy[1]

[1] Central Institute for Research on Goats, Makhdoom, Farah, Mathura 281122, India
[2] Centre for Cellular and Molecular Biology, Uppal Road, Hyderabad 500007, India

Correspondence should be addressed to P. K. Rout, rout_ctc@hotmail.com

Academic Editor: Martien Groenen

Jamunapari, a dairy goat breed of India, has been gradually declining in numbers in its home tract over the years. We have analysed genetic variation and population history in Jamunapari goats based on 17 microsatellite loci, 2 milk protein loci, mitochondrial hypervariable region I (HVRI) sequencing, and three Y-chromosomal gene sequencing. We used the mitochondrial DNA (mtDNA) mismatch distribution, microsatellite data, and bottleneck tests to infer the population history and demography. The mean number of alleles per locus was 9.0 indicating that the allelic variation was high in all the loci and the mean heterozygosity was 0.769 at nuclear loci. Although the population size is smaller than 8,000 individuals, the amount of variability both in terms of allelic richness and gene diversity was high in all the microsatellite loci except ILST 005. The gene diversity and effective number of alleles at milk protein loci were higher than the 10 other Indian goat breeds that they were compared to. Mismatch analysis was carried out and the analysis revealed that the population curve was unimodal indicating the expansion of population. The genetic diversity of Y-chromosome genes was low in the present study. The observed mean M ratio in the population was above the critical significance value (Mc) and close to one indicating that it has maintained a slowly changing population size. The mode-shift test did not detect any distortion of allele frequency and the heterozygosity excess method showed that there was no significant departure from mutation-drift equilibrium detected in the population. However, the effects of genetic bottlenecks were observed in some loci due to decreased heterozygosity and lower level of M ratio. There were two observed genetic subdivisions in the population supporting the observations of farmers in different areas. This base line information on genetic diversity, bottleneck analysis, and mismatch analysis was obtained to assist the conservation decision and management of the breed.

1. Introduction

Genetic diversity, the primary component of adaptive evolution, is essential for the long-term survival probability of a population [1–4]. Genetic diversity within domesticated species depends on several factors such as changing agricultural practices, breed replacement, and cross breeding. Genetic diversity has been analysed by using protein polymorphism, mitochondrial diversity, and microsatellite marker in both domestic and wild species [4–11]. Jamunapari goat, the majestic milk-producing goat breed of India, has suffered a reduction in numbers in its home tract [12, 13] and is considered as an endangered breed [14]. The Indian Jamunapari goat is one of the ancestors of the American Nubian and has been used in India and adjacent countries as an improver breed. The breed possesses several unique characteristics such as higher kidding rate despite its large body size. The breed inhabits isolated ravines in the Chakarnagar area of Etawah (Uttar Pradesh, India) (Figure 1), and geographical isolation has contributed towards the evolution of this unique breed. The breed is gradually declining in number due to land reclamation, decrease in grazing area, breed replacement, and the population size is less than 8,000 [12]; therefore, there is an urgent need to define strategies for conservation of this breed in its natural habitat.

Genetic Variation and Population Structure in Jamunapari Goats Using Microsatellites, Mitochondrial DNA, and Milk Protein Genes

55

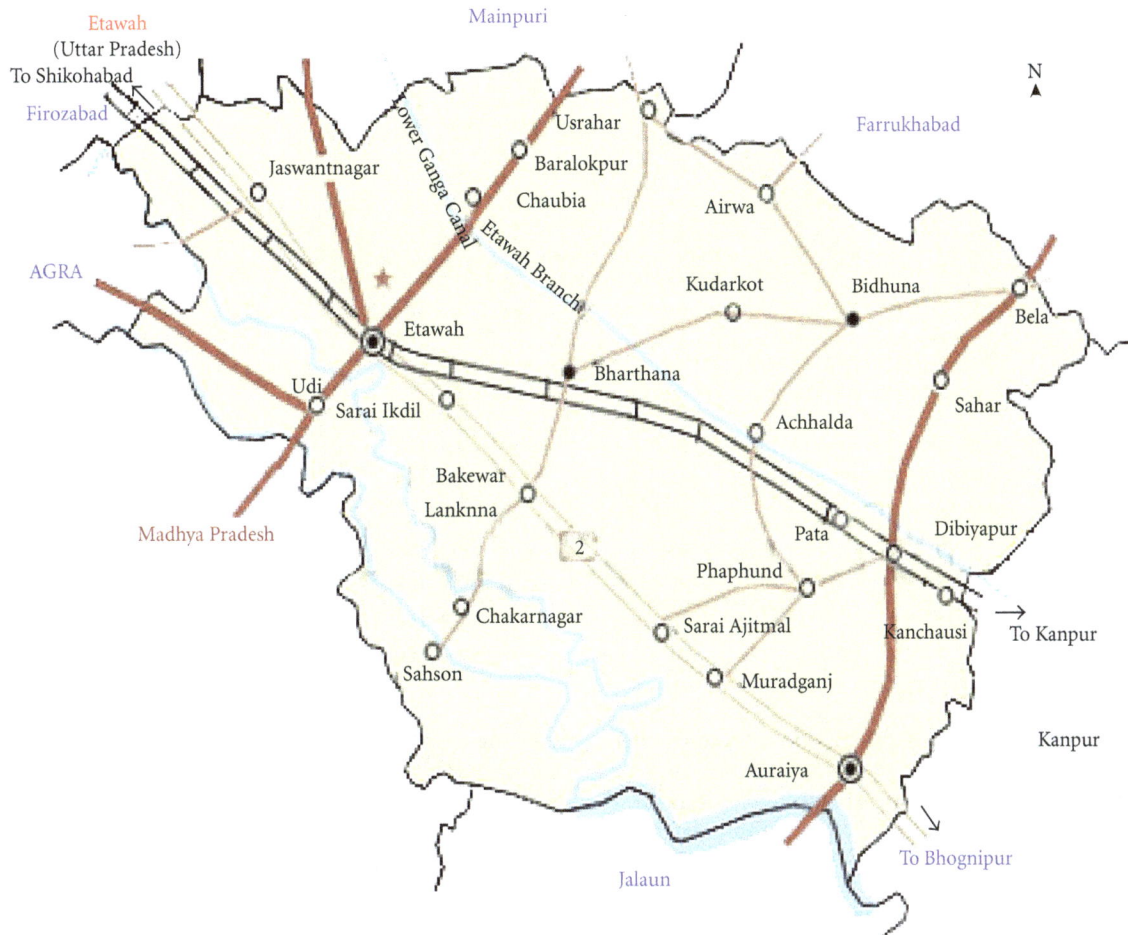

FIGURE 1: Maps of Chakarnagar (Etawah, UP) showing the home tract of Jamunapari goats.

In this study, we sample the Jamunapari goat population to analyse the genetic variation due to locus-specific events (selective sweep) as well as genome wide events (bottlenecks). Microsatellite markers are highly polymorphic and have been extensively used for breed diversity analysis. Mitochondrial DNA (mt DNA) and Y-chromosome region are usually sensitive to genetic drift and can be useful for detecting effects of bottlenecks in the population. Nonneutral markers are also being used to analyse population diversity, and milk protein gene has been used as the region is directly involved for the survival of the individual and under strong selective pressure. By integrating data from multiple markers, we provide the possible factors affecting the genetic consequence of population reduction in this breed.

2. Materials and Methods

Fifty blood samples were collected in 10 villages in which the breed has a major concentration. Samples were collected from the individuals exhibiting typical breed characteristics such as white colour, Roman nose, and pendulous ear (farmers are not selecting for these traits) and at least two samples were collected from each village. An effort was made to collect samples from unrelated individuals based on informa-tion provided by farmers. The breeding buck is available with one or two farmers in every village, and some farmers also maintain breeding bucks during breeding season, disposing of them after the breeding season. Blood samples were collected from each animal using EDTA vacutainer and stored at −20°C till further use.

Microsatellite analysis was carried out to test for signatures of recent population bottlenecks in Jamunapari goats. This analysis was carried out on 49 DNA samples with 17 microsatellite markers (Table 1) as reported by Rout et al. [11]. For these 17 loci, genetic variation was quantified using measures of the total number of alleles, number of polymorphic loci, observed and expected heterozygosity per locus, and allelic richness using GENEPOP (Version 3.4; [15]), FSTAT2.93 [16], and AGA$_{rst}$ [17]. Heterozygosity was measured as the mean observed heterozygosity (Ho) and the mean expected heterozygosity (H_E) based on Hardy-Weinberg assumptions. We tested genotypic linkage disequilibrium between all pairs of loci in each population with GENEPOP (Version 3.4; [15]) based on Markov chain method with 10,000 iterations and 100 batches. We also used FSTAT software to assess 95% confidence intervals of Weir and Cockerham's f, which measures deviation from the Hardy-Weinberg equilibrium (HWE) for populations and

TABLE 1: Microsatellite markers and chromosomal location, total number of alleles and genetic diversity in the Jamunapari goats.

Markers	Chromosome number	Observed number of alleles	Allele size range (bp)	Gene diversity	Allelic richness	F_{IS}
BM4621	6	15	106–140	0.862	15.00	0.652
NRAMP	2	10	224–248	0.807	10.00	0.554
OarAE101	6	8	92–108	0.809	8.00	0.555
IDVGA7	25	15	210–240	0.890	15.00	0.573
ILSTS005	10	3	178–188	0.497	3.00	0.235
BM6526	27	9	154–178	0.801	9.00	0.500
ETH225	14	6	140–152	0.703	6.00	0.089
OarHH56	23	9	152–168	0.818	9.00	0.560
INRABERN192	7	10	178–208	0.823	10.00	0.417
OarFCB48	17	8	150–164	0.831	8.00	0.351
OarHH62	20	5	108–118	0.719	5.00	0.499
TGLA40		10	174–198	0.782	10.00	0.540
BM143	6	8	96–118	0.741	8.00	0.514
SRCRSP 5	21	8	160–178	0.794	8.00	0.748
SRCRSP6	19	10	138–158	0.680	10.00	0.530
SRCRSP9		11	120–144	0.877	11.00	0.247
SRCRSP10	8	9	260–276	0.836	9.00	0.785

corresponds to Wright's within-population inbreeding coefficient F_{IS}.

Milk protein genes, which are expected to be nonneutral markers, were also used to analyse the population variability. Two milk protein genes, namely, β-LG gene and CSN1S1 (αs_1-casein) were analysed using PCR-RFLP to observe genetic variability in 35 individuals. The αs_1-casein (CSN1S1) gene produced an amplified fragment of 223 bp which was digested with the XmnI restriction enzyme. The β-LG gene produced an amplified product of 426 bp, and RFLP analysis was carried out with the SacII restriction enzyme. The PCR-RFLP analysis was carried out as described by Kumar et al. [18, 19], and the data were analysed separately for mean number of alleles, expected heterozygosity and Hardy-Weinberg equilibrium (HWE) using POPGENE software [20].

mtDNA HVRI sequencing was carried out as described by Joshi et al. [10]. Four hundred and fifty-seven base pairs from the mtDNA HVRI regions of 50 individuals were aligned using CLUSTAL X. We used mismatch distribution [21] to analyse the population expansion as implemented in ARLEQUIN 3.1 [22]. Fu's F value was calculated from mtDNA haplotypes to test for deviations from neutral equilibrium condition [23]. The qualitative and quantitative aspect of the population's genetic history may be uncovered by the analysis of frequency distributions of pairwise sequence mismatches. Mismatch analysis (the distribution of all pair-wise nucleotide differences between sequences) was carried out to test the deviation of the observed data from neutral predictions expected in constant-sized populations.

Genetic divergence was analysed by selecting three primers from ovine male-specific region (AMLEY, SRY, and ZFY gene) [24]. PCR was carried out in a 50 μL reaction volume containing 100 ng of DNA, 20 pM of each primer, 200 μM of dNTP, 2 mM Mgcl$_2$, and %U of Taq DNA polymerase (New India Biolab, MA, USA). The samples were subjected to seq-

uencing after purifying the PCR product by gene elute PCR clean up kit. Individual PCR amplified products were subjected to sequencing in 12 samples. PCR products were sequenced on both the strands directly using 50 ng (2.0 μL) of PCR product and 4 pM (1.0 μL) of primer, 4 μL of Big Dye Terminator ready reaction kit (Perkin Elmer, Foster City, USA), and 3.0 μL of double distilled water to adjust the volume to 10.0 μL. Cycle sequencing was carried out in a Gene Amp 9600 thermal cycler (Perkin Elmer) employing the PCR conditions. Extended products were purified by alcohol precipitation followed by washing with 70% alcohol. Purified samples were dissolved in 10 μL of 50% Hi-Di formamide and analysed in an ABI 3700 automated DNA Analyzer (Perkin Elmer, USA). Nucleotide diversity, expected heterozygosity, Tajima's D, and Fu's Fs values were estimated in ARLEQUIN 3.1 [22].

Genetic bottleneck was detected using microsatellite data by three approaches, heterozygote excess, mode-shift, and M ratio test. We first used the M ratio (the mean ratio of the number of alleles to total range in allele size) [25] as implemented in AGA$_{rst}$ [17], because of its consistent performance in identifying populations with known bottlenecks. M ratio calculates the changes that occur after a bottleneck in the distribution of allele sizes relative to the number of alleles in a population. It has been established that an M ratio less than 0.71 signifies a bottleneck [25].

The BOTTLENECK programme [26] was used as an alternative measure of genetic bottlenecks to test for excess gene diversity relative to that expected under mutation-drift equilibrium. The heterozygosity excess method exploits the fact that allele diversity is reduced faster than heterozygosity during a bottleneck, because rare alleles are lost rapidly and have little effect on heterozygosity, thus producing a transient excess in heterozygosity relative to that expected in a population of constant size with the same number of alleles

Genetic Variation and Population Structure in Jamunapari Goats Using Microsatellites, Mitochondrial DNA, and
Milk Protein Genes

57

[26, 27]. To determine the population "genetic reduction signatures" characteristic of recent reductions in effective population size (Ne), the Wilcoxon's heterozygosity excess test [26] and the allele frequency distribution mode shift analysis [28] were performed using BOTTLENECK [26]. The heterozygosity excess method was used to analyse the population, and the data for the heterozygosity excess test were examined under the two-phased model (TPM; 95% stepwise mutation model with 5% multistep mutations and a variance among multiple steps of 12), which is considered best for microsatellite data [26, 29]. We also analysed the allele frequency distribution for gaps. A qualitative descriptor of allele frequency distribution (the mode-shift indicator), which is reported to discriminate between bottlenecked and stable population [28], was obtained using the programme BOTTLENECK.

We used an individual-based clustering approach (STRUCTURE 2.1, [30]) to determine the most likely number of genetic clusters (k) in the Jamunapari populations. STRUCTURE software sorts individual genotypes into clusters that maximize the fit of the data to theoretical expectation. Based on preliminary analyses, we evaluated the likelihood of $k = 2$ and $k = 3$, with 5 runs performed for each k, and a burn-in length of 500,000 and 100,000 MCMC replicates for each run. We assumed an admixture model and correlated allele frequencies among populations [30].

3. Results and Discussion

The markers with their chromosome number, number of alleles identified, and allele size range have been described in Table 1. Among the polymorphic markers, BM4621 and IDVGA7 showed highest number of alleles (15) at each locus. The number of alleles per microsatellite marker was above 6 for all markers except for ILSTS005 and Oar HH62. The total number of alleles was 153 over the 17 loci. The allelic richness ranged from 3.00 to 15.00 across the microsatellite markers (Table 1) and the mean number of alleles per locus was 9.0. Allelic richness was identical to allele frequency implying that there was no bias based on sample size. The average gene diversity ranged from 0.489 to 0.866 over the loci. The mean expected and observed heterozygosity was 0.769 and 0.386 (Table 1). All the loci showed higher gene diversity than ILSTS005 in the analysed samples. The high mean number of alleles per locus and expected heterozygosities indicated that the overall gene diversity was high in the population. Heterozygosity and allele number are aligning with high diversity score in the population. Takezaki and Nei (1996) suggested that microsatellite loci can be included diversity analysis having heterozygosity from 0.3 to 0.8 in the population. Two loci departed significantly from the Hardy-Weinberg equilibrium (HWE). In the analysed samples, 18 microsatellite locus pairs demonstrated linkage disequilibrium (LD) with P value <0.05. The LD was significant in 13.21% of the locus pair combinations in the population. The overall excess of homozygosity for the population as a whole varied from 0.089 to 0.785 over the loci and the average was 0.500. The high levels of allelic diversity are coupled with very high F_{is} indicating that the population is experiencing high levels

TABLE 2: Bottleneck detection in the Jamunapari goats.

Marker	Heq*	SD	*(He-Heq)/SD	He excess	M ratio*
BM4621	0.894	.021	−1.825	+	0.833
NRAMP	0.834	.036	−0.886	+	0.692
OarAE101	0.787	.050	0.337	−	0.889
IDVGA7	0.894	.021	−0.436	+	0.938
ILSTS005	0.441	.151	0.364	−	0.364
BM6526	0.808	.048	−0.244	+	0.692
ETH225	0.714	.074	−0.159	+	0.857
OarHH56	0.813	.043	−0.007	+	0.857
INRABERN192	0.833	.038	−0.359	+	0.625
OarFCB48	0.784	.053	0.845	−	1.00
OarHH62	0.653	.093	0.672	−	0.833
TGLA40	0.831	.041	−1.275	+	0.769
BM143	0.786	.053	−0.934	+	0.667
SRCRSP5	0.786	.052	0.051	−	0.800
SRCRSP6	0.830	.037	−4.182	+	0.909
SRCRSP9	0.849	.034	0.763	−	0.846
SRCRSP10	0.813	.041	0.393	−	1.000

* Heq is the heterozygosity expected at equilibrium obtained through coalescent simulation under the "two-phase mutation model". (He-Heq)/SD: the standardized difference for each locus, M ratio: the number of allele/(range in allele size + 1).

of nonrandom mating in the breeding tract but simultaneously maintaining allelic diversity over the entire range of the breed. Gour et al. [31] also observed high inbreeding in Jamunapari goats; further, high estimates of inbreeding have been reported for Asian goat populations by Barker et al. [8].

Genetic variation at $CSN1S1$ and β-LG was 0.395 and 0.107. The effective number of alleles was 1.653 and 1.20 at $CSN1S1$ and β-LG loci, respectively. The β-LG locus showed significant departure from HW equilibrium. The gene diversity and effective number of alleles at milk protein loci were higher than for 10 other Indian goat breeds, supporting the fact that the breed maintains higher genetic variability [18, 19].

The population was examined for allele frequency distribution for gaps, and M ratios are presented in Table 2. The M ratios ranged from 0.364 to 1.00 with an average of 0.815, which was significantly higher than the critical value. The M ratio was less than 0.71, diagnostic value of genetic bottlenecks, in the case of ILSTS005, INRABERN192, and BM143. The observed M ratio for all other markers in the population was very high and close to one indicating that it is a very slowly changing population (at least not showing the sign of bottleneck).

Bottleneck detection in Jamunapari goat was presented in Table 2. The mode shift test did not detect any distortion of allele frequency and showed a normal "L" shaped distribution which is a typical property of a population in equilibrium (Figure 2). The heterozygosity excess method was carried out to analyse historical bottlenecks. Out of 17 loci, 7 loci showed heterozygosity excess (Wilcoxon signed rank test, $P = 0.8487$, one tail for heterozygosity excess), and there

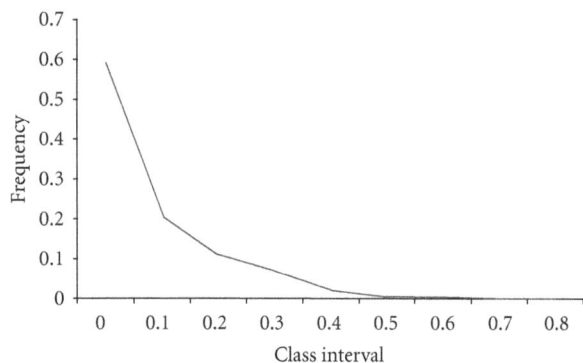

FIGURE 2: L-shaped mode shift graph showing the absence of bottleneck in Jamunapari goats.

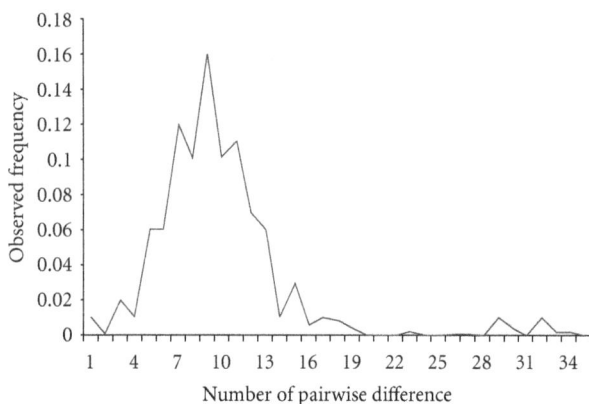

FIGURE 3: Mismatch distribution in Jamunapari goats analysed for mt-DNA control region.

and the Sum of Square Difference (SSD) value was 0.0048. The SSD value showed large population expansion in Jamunapari goats. Figure 3 depicts the mismatch distribution of Jamunapari goat. The shape of the distribution of number of observed differences between pairs of DNA sequences showed almost unimodal curve for the Jamunapari goat. The curve showed very small second mode towards the end indicating that some minor population expansion at some stage might have occurred in various geographical areas. Mismatch distribution has been extensively used to estimate the demographic parameters of past population expansion or contraction as it leaves a recognizable signature in the pattern of molecular diversity [21, 32, 33]. The unimodal distribution of pairwise differences in the breed (Figure 3) and Fu's Fs value indicated a sudden demographic population expansion and origin of the breed from a limited number of founder populations. Migration had an effect on the shape of curve, and intermediate migration rate would have led to a multimodal curve [34]. However, the factors like inbreeding and admixture with other population could affect the shape of mismatch distribution.

The amplified product for amelogenin gene (AMELY), SRY gene, and ZFY gene was 733 bp, 632 bp, and 584 bp, respectively. No diversity was observed in SRY and AMELY genes. Nucleotide diversity and expected heterozygosity were 0.135 ± 0.089 and 0.1384, respectively, for ZFY gene. Similarly Tajima's D value and Fu's Fs value were zero for SRY and AMELY gene. Tajima's D value and Fu's Fs value were 0.6797 and 2.539, respectively and non-significant for ZFY gene. Transition and transversion ratio was 47 : 90 for ZFY gene. ZFY gene showed low gene diversity and positive Tajima's D value and Fu's Fs value. Y-linked nucleotide diversity was found low in human, wolf, cattle, reindeer and Lynx indicating that reduced levels of Y-chromosome polymorphism may be a generalized feature of mammalian genome [35]. Y chromosome variability is expected to be lowest as compared to autosomes and X-chromosome. The major factors explain that the low levels of Y chromosome variability are selection, mating system, or migration patterns, or other mechanisms lowering male effective population size. The low levels of Y chromosome variability that we found in goat could be attributed to a strong sex bias in breeding.

The proportion of membership of individuals in the Jamunapari population placed 12.6% of population into one cluster group in both STRUCTURE analyses ($k = 2$ or 3). The cluster defined by STRUCTURE showed clear membership of individuals in two clusters indicating a genetic subdivision within the breed. There was a difference between expected and observed heterozygosity supporting the genetic subdivision within the breed. It has been also observed during survey that there are two strains of Jamunapari locally known as "Kathey" and "Kandhan". Kathey-type goats have long thick broad and less folded ears with thick and flat neck. Kandhan-type goats have long, soft, and folded ears, and the neck is thin and cylindrical. Kandhan is restricted to river banks of Chambal and Kathey is distributed throughout the breeding tract. (CIRG Report).

The average M ratio was large and above the critical significance value (Mc) in the population suggesting that it has

was no significant departure from mutation-drift equilibrium detected in the population. Under the two-phase mutation model equilibrium, each individual has a roughly equal chance of having heterozygote deficiency or excess. The analysis indicated that the population has not suffered any bottleneck recently and was a constant size population. The present study agrees well with the observation of Gour et al. [31].

The sequence analysis of the 457 bp mitochondrial HVRI region identified 34 mtDNA haplotypes, and the overall haplotype diversity was 0.984. The mtDNA variation detected relatively high number of haplotypes from a total 50 individuals, and haplotype diversity was quite high. Fu's Fs value is based on the probability of recovering a number of haplotypes greater than or equal to the observed number in a sample drawn from a stationery population with the same mean number of pairwise differences as the observed sample. Fu's Fs value was -15.53. The significant negative Fs values indicated the large and sudden expansion in the population at geographical locations. The population showed a significant negative Fs value indicating excess of rare mutations, a pattern commonly attributed to a normal growing population.

Mismatch distribution analysis revealed the genetic signature of population expansion of Jamunapari goat. The tau value with 95% confidence interval was 8.20 (5.22–10.68),

Genetic Variation and Population Structure in Jamunapari Goats Using Microsatellites, Mitochondrial DNA, and Milk Protein Genes

59

not suffered severe or long-lasting genetic bottlenecks [25, 36]. Heterozygosity excess test and the mode shift indicators demonstrated that there was no sign of recent reductions in effective population size (Ne) in the population. Moreover, the milk protein loci also exhibited higher gene diversity at both the loci as compared to other Indian goat breeds [18, 19]. Moreover, the mtDNA analysis supported population expansion model as evidenced from Fu's Fs value, unimodal distribution of pairwise differences (mismatch curve). Genetic variation is often reduced due to demographic reduction in population and can also show other signs of population bottlenecks. However, this breed showed retention of diversity in the face of population reduction. A similar type of trend had been observed in salmon population [37] and turtle [38]. Genetic responses to bottlenecks depend on life history and the life-span of the species, the severity of the demographic decline, level of present gene flow, and nature of demographic rebound [25, 39]. Several factors might be affecting the population leading to maintenance of dramatic genetic variation despite severe demographic declines. The variation may be due to exchange of animals between different areas and selection criteria of farmers to use the individuals as parents for the next generation. The dispersal of local populations between farmers and selling to outside agencies may have served to maintain genetic variability in the population. The breeding buck is available with one or two farmers in every village and all the farmers mate their goats with the available bucks showing the nonrandom mating in the breeding tract but maintaining allelic diversity over the entire range of the breed. Again the existence of genetic subdivision supports the existing gene diversity in the population. Historically there was no evidence of a major earthquake or climate change in the area. Theoretically, the impact of even severe bottlenecks can be small if the bottleneck is followed by the rapid flush of growth in which most genetic variability is maintained [39]. Again, the population growth is evident but also animals are supplied to outside the home tract as a medium of remunerative income to farmers. Additionally, the overlapping generations in domestic goats buffer them from long-term losses of genetic variability in comparison to species with discrete generations. More importantly, the high reproductive rate of this large goat breed slows down the losses due to genetic drift from the base population. Small populations are generally considered to be susceptible to a number of genetic problems like low level of variability, inbreeding depression, and the ability to overcome disease agents; however, the population did not exhibit any such effect over the years in the adopted villages [12, 13]. The population has not shown any noticeable physiological sign of inbreeding depression as there was no reduction in fecundity (kidding rate); and the mortality in the breeding tract over the year was also low (<7.5%, [12]). It was observed that the rate of decline in genetic diversity was relatively slow in the population in its natural breeding tract; however, the low genetic diversity in Y chromosomal genes, as well as the effects of genetic bottlenecks in some loci due to decreased heterozygosity and lower level of M ratio, supports population reduction in breeding tract. Conservation will be much more difficult when the population becomes genetically impoverished and is effective and easy to implement when the populations are genetically stable. Therefore, it is necessary to initiate necessary steps to conserve the breed for future use.

Acknowledgment

The authors are grateful to the Director of Central Institute for Research on Goats for providing facilities to carry out the work.

References

[1] J. C. Avise, "Mitochondrial DNA polymorphism and a connection between genetics and demography of relevance to conservation," *Conservation Biology*, vol. 9, no. 3, pp. 686–690, 1995.

[2] D. W. Coltman, W. D. Bowen, and J. M. Wright, "Birth weight and neonatal survival of harbour seal pups are positively correlated with genetic variation measured by microsatellites," *Proceedings of the Royal Society B*, vol. 265, no. 1398, pp. 803–809, 1998.

[3] D. H. Reed and R. Frankham, "Correlation between fitness and genetic diversity," *Conservation Biology*, vol. 17, no. 1, pp. 230–237, 2003.

[4] E. H. Harley, I. Baumgarten, J. Cunningham, and C. O'Ryan, "Genetic variation and population structure in remnant populations of black rhinoceros, Diceros bicornis, in Africa," *Molecular Ecology*, vol. 14, no. 10, pp. 2981–2990, 2005.

[5] I. Tapio, S. Värv, J. Bennewitz et al., "Prioritization for conservation of northern European cattle breeds based on analysis of microsatellite data," *Conservation Biology*, vol. 20, no. 6, pp. 1768–1779, 2006.

[6] J. Kantanen, I. Olsaker, L.-E. Holm et al., "Genetic diversity and population structure of 20 North European cattle breeds," *Journal of Heredity*, vol. 91, no. 6, pp. 446–457, 2000.

[7] T. Pastor, J. C. Garza, P. Allen, W. Amos, and A. Aguilar, "Low genetic variability in the highly endangered mediterranean monk seal," *Journal of Heredity*, vol. 95, no. 4, pp. 291–300, 2004.

[8] J. S. F. Barker, S. G. Tan, S. S. Moore, T. K. Mukherjee, J. L. Matheson, and O. S. Selvaraj, "Genetic variation within and relationships among populations of Asian goats (Capra hircus)," *Journal of Animal Breeding and Genetics*, vol. 118, no. 4, pp. 213–233, 2001.

[9] M. H. Li, S. H. Zhao, C. Bian et al., "Genetic relationships among twelve Chinese indigenous goat populations based on microsatellite analysis," *Genetics Selection Evolution*, vol. 34, no. 6, pp. 729–744, 2002.

[10] M. B. Joshi, P. K. Rout, A. Mandal, C. Thangaraj, C. Tyler-Smith, and L. Singh, "Phylogeography and origin of Indian domestic goats," *Molecular Biology and Evolution*, vol. 21, no. 3, pp. 454–462, 2004.

[11] P. K. Rout, M. B. Joshi, A. Mandal, D. Laloe, L. Singh, and K. Thangaraj, "Microsatellite-based phylogeny of Indian domestic goats," *BMC Genetics*, vol. 9, article 11, 2008.

[12] P. K. Rout, V. K. Saxena, B. U. Khan et al., "Characterization of Jamunapari goats in their hometract," *Animal Genetic Resource Information*, vol. 27, pp. 43–53, 2000.

[13] P. K. Rout, M. K. Singh, R. Roy, N. Sharma, and G. F. W. Haen-lein, "Jamunapari-a diary goat breed in India," *Dairy Goat Journal*, vol. 82, no. 3, pp. 37–39, 2004.

[14] "World watch list for domestic animal diversity," in *World Watch List*, B. D. Scherf, Ed., FAO, Rome, Italy, 2000.

[15] M. Raymond and F. Rousset, "GENEPOP (version 1.2): population genetics software for exact tests and ecumenicism," *Journal of Heredity*, vol. 86, no. 3, pp. 248–249, 1995.

[16] J. Goudet, "FSTAT (version 1.2): a computer program to calculate F-statistics," *Journal of Heredity*, vol. 86, no. 6, pp. 485–486, 1995.

[17] E. H. Harley, $AGAR_{ST}$, *Version 2.8, A Program for Calculating Allele Frequencies, GST and R_{ST} from Microsatellite Data. Wild Life Genetics Unit 2002*, University of Care Town, Johannesburg, South Africa, 2002.

[18] A. Kumar, P. K. Rout, A. Mandal, and R. Roy, "Identification of the CSN1S1 allele in Indian goats by the PCR-RFLP method," *Animal*, vol. 1, no. 8, pp. 1099–1104, 2007.

[19] A. Kumar, P. K. Rout, and R. Roy, "Polymorphism of β-lacto globulin gene in Indian goats and its effect on milk yield," *Journal of Applied Genetics*, vol. 47, no. 1, pp. 49–53, 2006.

[20] F. Yeh, R.-C. Yang, and T. Boyle, *POPGENE Version 1.31*, University of Alberta, Edmonton, Canada, 1999.

[21] A. R. Rogers and H. Harpending, "Population growth makes waves in the distribution of pairwise genetic differences," *Molecular Biology and Evolution*, vol. 9, no. 3, pp. 552–569, 1992.

[22] L. Excoffier, P. Smouse, and J. Quattro, "Analysis of molecular variance inferred from metric distances among DNA haplotypes: application to human mitochondrial DNA restriction data," *Genetics*, vol. 131, no. 2, pp. 479–491, 1992.

[23] X.-Y. Fu, "Statistical tests of neutrality of mutations against population growth, hitchhiking and background selection," *Genetics*, vol. 147, no. 2, pp. 915–925, 1997.

[24] J. R. S. Meadows, R. J. Hawken, and J. W. Kijas, "Nucleotide diversity on the ovine Y chromosome," *Animal Genetics*, vol. 35, no. 5, pp. 379–385, 2004.

[25] J. C. Garza and E. Williamson, "Detection of reduction in population size using data from microsatellite loci," *Molecular Ecology*, vol. 10, no. 2, pp. 305–318, 2001.

[26] S. Piry, G. Luikart, and J. M. Cornuet, "BOTTLENECK: a computer program for detecting recent reductions in the effective population size using allele frequency data," *Journal of Heredity*, vol. 90, no. 4, pp. 502–503, 1999.

[27] J. M. Cornuet and G. Luikart, "Description and power analysis of two tests for detecting recent population bottlenecks from allele frequency data," *Genetics*, vol. 144, no. 4, pp. 2001–2014, 1996.

[28] G. L. Luikart, F. W. Allendorf, J. M. Cornuet, and W. B. Sherwin, "Distortion of allele frequency distributions provides a test for recent population bottlenecks," *Journal of Heredity*, vol. 89, no. 3, pp. 238–247, 1998.

[29] A. Di Rienzo, A. C. Peterson, J. C. Garza, A. M. Valdes, M. Slatkin, and N. B. Freimer, "Mutational processes of simple-sequence repeat loci in human populations," *Proceedings of the National Academy of Sciences of the United States of America*, vol. 91, no. 8, pp. 3166–3170, 1994.

[30] J. K. Pritchard, M. Stephens, and P. Donnelly, "Inference of population structure using multilocus genotype data," *Genetics*, vol. 155, no. 2, pp. 945–959, 2000.

[31] D. S. Gour, G. Malik, S. P. S. Ahlawat et al., "Analysis of genetic structure of Jamunapari goats by microsatellite markers," *Small Ruminant Research*, vol. 66, no. 1–3, pp. 140–149, 2006.

[32] L. Excoffier and S. Schneider, "Why hunter-gatherer populations do not show signs of Pleistocene demographic expansions," *Proceedings of the National Academy of Sciences of the United States of America*, vol. 96, no. 19, pp. 10597–10602, 1999.

[33] L. M. Vigilant, M. Stoneking, H. Harpending, K. Hawkes, and A. C. Wilson, "African populations and the evolution of human mitochondrial DNA," *Science*, vol. 253, no. 5027, pp. 1503–1507, 1991.

[34] P. Marjoram and P. Donnelly, "Pairwise comparisons of mitochondrial DNA sequences in subdivided populations and implications for early human evolution," *Genetics*, vol. 136, no. 2, pp. 673–683, 1994.

[35] L. Hellborg and H. Ellegren, "Low levels of nucleotide diversity in mammalian Y chromosomes," *Molecular Biology and Evolution*, vol. 21, no. 1, pp. 158–163, 2004.

[36] K. C. Doerner, W. Braden, J. Cork et al., "Population genetics of resurgence: white-tailed deer in Kentucky," *Journal of Wildlife Management*, vol. 69, no. 1, pp. 345–355, 2005.

[37] H. Neville, D. Isaak, R. Thurow, J. Dunham, and B. Rieman, "Microsatellite variation reveals weak genetic structure and retention of genetic variability in threatened Chinook salmon (Oncorhynchus tshawytscha) within a Snake River watershed," *Conservation Genetics*, vol. 8, no. 1, pp. 133–147, 2007.

[38] C. H. Kuo and F. J. Janzen, "Genetic effects of a persistent bottleneck on a natural population of ornate box turtles (Terrapene ornata)," *Conservation Genetics*, vol. 5, no. 4, pp. 425–437, 2004.

[39] M. Nei, T. Maruyama, and R. Chakraborty, "The bottleneck effect and genetic variability in populations," *Evolution*, vol. 29, no. 1, pp. 1–10, 1975.

7

Genetic Diversity and Variability in Endangered Pantesco and Two Other Sicilian Donkey Breeds Assessed by Microsatellite Markers

Salvatore Bordonaro, Anna Maria Guastella, Andrea Criscione,
Antonio Zuccaro, and Donata Marletta

DISPA, Sezione di Scienze delle Produzioni Animali, Università degli studi di Catania, Via Valdisavoia 5, 95123 Catania, Italy

Correspondence should be addressed to Donata Marletta, d.marletta@unict.it

Academic Editors: M. Ota and G. A. Rocha

The genetic variability of Pantesco and other two Sicilian autochthonous donkey breeds (Ragusano and Grigio Siciliano) was assessed using a set of 14 microsatellites. The main goals were to describe the current differentiation among the breeds and to provide genetic information useful to safeguard the Pantesco breed as well as to manage Ragusano and Grigio Siciliano. In the whole sample, that included 108 donkeys representative of the three populations, a total of 85 alleles were detected. The mean number of alleles was lower in Pantesco (3.7), than in Grigio Siciliano and Ragusano (4.4 and 5.9, resp.). The three breeds showed a quite low level of gene diversity (He) ranging from 0.471 in Pantesco to 0.589 in Grigio. The overall genetic differentiation index (Fst) was quite high; more than 10% of the diversity was found among breeds. Reynolds' (D_R) genetic distances, correspondence, and population structure analysis reproduced the same picture, revealing that, (a) Pantesco breed is the most differentiated in the context of the Sicilian indigenous breeds, (b) within Ragusano breed, two well-defined subgroups were observed. This information is worth of further investigation in order to provide suitable data for conservation strategies.

1. Introduction

The donkey (*Equus asinus*) was domesticated about 6000 years ago starting from one or two subspecies of African wild asses (*E. africanus*) [1, 2]. For many centuries donkey has been used as beast of burden in many cultures. Today donkeys and mules are still essential for transportation of heavy load, people, and possessions in poor, arid, and rough regions of the world [3], but in the developed country, this pack animal is no longer required. As a consequence, not only individual breeds are endangered, but also the whole species is heading for extinction [4].

In Italy, six donkey breeds are already extinct; in contrast during the last few years, due to the exploitation of donkey's products (milk and meat) many local breeds and populations are growing in census. Moreover some small populations are undergoing conservation programmes.

Pantesco breed represents an emblematic case: this ancient breed was imported by the Arabs to the isle of Pantelleria [5] in the Sicilian channel. For long time it was employed in agriculture on the rugged paths of the island. Pantesco has a short haired and fine brown coat, with white belly, muzzle, and eye rings. The withers height is about 125–130 cm; the type is dolichomorphic. This donkey, able to move in the fast and sure "Tölt-gait", is well adapted to harsh environment [6].

The employment of Pantesco stallions in Sicilian stud farms is documented since 1926. This breed was also used in the breeding of mules, employed in the army, and exported to USA and Greece.

After the Second World War, because of the mechanization in agriculture, the exportation, and the establishment of the stud book of Ragusano, Pantesco became severely threatened with extinction.

In the last twenty years the Sicilian Forest Administration (Azienda Foreste Demaniali di Trapani) carried out a morphologic and genetic selection on more than 200 potential Pantesco crossbreed reared in Egadi Island and Trapani Province. Nine donkeys (three males and six females) were identified and used as founders to reconstitute the breed. Embryo transfer (ET) was also employed as a tool in the conservation project [7].

Today about 47 recorded Pantesco donkeys are bred in the San Matteo Farm of Erice (Trapani). Within the frame of breed conservation, genetic characterization is important with regard to breed integrity and represents an essential prerequisite for handling genetic resources [8]. Microsatellite markers proved to be a reliable and frequently used tool to quantify genetic variation within and among breeds and useful for the conservation management of animal populations [9].

A preliminary characterization of the Sicilian donkeys, mainly focused on the genetic analysis of Pantesco, was performed using a set of 11 microsatellites [10] and genealogical data [11].

The aim of this study was to measure the genetic diversity and variability in the three Sicilian indigenous donkey breeds (Pantesco, Ragusano, and Grigio Siciliano), using molecular information supplied by a set of 14 microsatellite markers, in order to provide suitable data for breeding schemes and conservation strategies.

2. Material and Methods

2.1. Sampling and DNA Extraction. Blood (10 mL) was sampled in K3-EDTA tubes from 108 Sicilian donkeys (39 Pantesco, 53 Ragusano, 16 Grigio Siciliano) reared all over Sicily. Sampling was achieved among minimally related donkeys by using pedigree information when available and avoiding first- and second-order relatives. In the case of Pantesco breed, the sample consisted of nearly the entire Stud Book-registered population (47 heads) reared in the "San Matteo Farm" in Erice (Trapani). Forty six horses belonging to Sicilian Oriental Purebred (*E. caballus*) were added to the data set and used as out-group in phylogenetic analysis.

DNA was extracted using "Illustra blood genomic Prep Mini Spin" kit (GE Healthcare, Little Chalfont, UK) and then checked for quality and concentration by NanoDrop ND 1000 spectrophotometer (Thermo Fisher Scientific, Wilmington, USA).

2.2. Microsatellite Amplification and Analysis. The whole sample (108 donkeys and 46 horses) was genotyped through a set of 14 microsatellite markers (AHT4, AHT5, ASB23, HMS2, HMS3, HMS5, HMS6, HMS7, HTG4, HTG6, HTG7, HTG10, HTG15, VHL20) amplified in three PCR multiplex reactions, using a PE GeneAmp PCR 9600 system thermocycler (Applied Biosystems, Foster City, CA, USA). Fluorescent-labelled PCR products were diluted, mixed with an internal size standard, and analysed by the automatic AB3130 DNA Sequencer equipped with GeneScan and Genotyper software (Applied Biosystems, Foster City, CA, USA). Microsatellite markers were chosen, on the base of their degree of information obtained on a smaller sample, among those reported in the literature leading with donkey [12] and horse [13] biodiversity.

2.3. Statistical Analysis. Individual multilocus genotypes were processed by means of GENALEX v.6.4 platform [14] to perform file conversions and calculate the main parameters of genetic variability. For each *locus* and breed and on the whole sample, the allele frequencies, private alleles (A_p), and observed (H_o) and unbiased expected (H_e) heterozygosities were calculated.

The polymorphism information content (PIC) for each *locus* and breed was calculated [15].

Hardy-Weinberg equilibrium was tested by the software Genepop v.4.0 [16] which was used to perform the score test per *locus* and breed and global tests across *loci* and across sample; tests were implemented using the Markov chain algorithm (10000 dememorizations, 5000 batches, and 5000 iterations per batch).

The presence of null alleles was tested with MICRO-CHECKER v.2.2.3 [17], using the methods by Chakraborty et al. [18] and Brookfield [19].

FSTAT v.2.9.3 software [20] was used to estimate the *F*-statistics [21] and their significance as well as the rarefacted number of alleles (Ar) based on the minimum sample size.

The significance levels obtained from multiple tests, carried out for HW-Equilibrium and *F*-statistics, were corrected by the sequential Bonferroni method [22] to reduce the occurrence of type I error.

In order to measure the short-term divergence of the donkey breeds, the Reynolds' (D_R) pairwise genetic distances [23] were calculated by PHYLIP ver.3.69 package [24]. Moreover, the Neighbour-Joining algorithm was implemented on D_R and the strength of the nodes was based on 1000 bootstrap resamplings of the allelic frequencies.

The model-based approach proposed in the software STRUCTURE 2.3 [25] was used to assess the genomic clustering of the sample. As suggested by the authors for populations with possible mixed ancestry, the admixture model associated to the option of correlated allele frequencies [26] was implemented to infer the populations' structure using no prior information. Running length was set to 500000 burn-ins followed by 500000 iterations. The range of possible clusters (K) tested was from 1 to 10 and 10 different runs were carried out for each K. The number of clusters fitting best our data was established by plotting the mean $\ln \Pr(X \mid K)$ over the multiple independent runs for each K, as suggested by the authors.

The correspondence analysis in which the Chi-square distances measure the proximity of the taxa was performed by GENETIX v.4.05 software [27] and breeds and individuals were spatially plot in accordance with allele frequencies.

3. Results

The 14 microsatellite markers resulted to be polymorphic in the whole sample and in each breed but for HMS5 in Pantesco (Table 1). A total of 85 alleles were observed in the three Sicilian donkey breeds, with the number of alleles ranging from 2 to 11 (6.07 on average). The observed heterozygosity (H_o) varied between 0.154 (HMS5) and 0.736

Genetic Diversity and Variability in Endangered Pantesco and Two Other Sicilian Donkey Breeds Assessed by
Microsatellite Markers

63

TABLE 1: Number of alleles, observed (H_o) and expected heterozygosity (H_e), and PIC values per *locus* inferred on the whole sample of three Sicilian donkey breeds.

Locus	N° of alleles	H_o	H_e	PIC
AHT4	8	0.533	0.747	0.703
AHT5	10	0.581	0.822	0.792
ASB23	5	0.704	0.740	0.692
HMS2	8	0.271	0.582	0.533
HMS3	8	0.463	0.611	0.558
HMS5	2	0.154	0.159	0.146
HMS6	4	0.250	0.483	0.393
HMS7	6	0.222	0.310	0.291
HTG10	7	0.736	0.794	0.760
HTG15	3	0.514	0.553	0.482
HTG4	3	0.178	0.194	0.176
HTG6	5	0.556	0.674	0.608
HTG7	11	0.648	0.817	0.796
VHL20	5	0.561	0.652	0.591
Average	6.07	0.455	0.581	0.537
SE	±0.722	±0.054	±0.059	±0.06

TABLE 2: F-statistics (Fit, Fst, and Fis) per *locus* and overall the three Sicilian donkey breeds.

Locus	Fit	Fst	Fis
AHT4	0.322**	0.123**	0.227**
AHT5	0.342**	0.153**	0.223**
ASB23	0.082	0.091**	−0.01
HMS2	0.547**	0.067**	0.515**
HMS3	0.315**	0.244**	0.093
HMS5	0.056	0.065*	−0.009
HMS6	0.496**	0.064	0.461**
HMS7	0.292**	0.032	0.268*
HTG10	0.093	0.053**	0.042
HTG15	0.096	0.071**	0.027
HTG4	0.081	−0.009	0.089
HTG6	0.234**	0.18**	0.066
HTG7	0.229**	0.071**	0.171**
VHL20	0.182*	0.124**	0.066
All	0.251†††	0.108†††	0.161†††

Adjusted nominal levels (Bonferroni): *$P < 0.05/14$; **$P < 0.01/14$; ***$P < 0.001/14$; †††$P < 0.001$.

TABLE 3: Number of individuals, mean number of alleles (MNAs), allelic richness (Ar), observed (H_o) and expected (H_e) heterozygosities, and Fis inferred per breed in three Sicilian donkey sample.

Breed	N	MNA	Ar	H_o	H_e	Fis
Ragusano	53	5.857	3.8	0.496	0.579	0.144***
SE		±0.69		±0.058	±0.057	
Pantesco	39	3.714	3.0	0.385	0.471	0.185***
SE		±0.47		±0.073	±0.066	
Grigio Siciliano	16	4.429	3.7	0.496	0.589	0.162***
SE		±0.48		±0.057	±0.057	

Adjusted nominal levels (Bonferroni): ***$P < 0.001/3$.

The highly significant ($P < 0.001$) Fis value (0.161) revealed a rather high inbreeding degree within breeds. In particular, six loci gave a relevant significant contribution to the total inbreeding index (Fit = 0.251), with a high heterozygote deficit within breeds.

Mean number of alleles (MNA), allelic richness, Fis, and heterozygosities per breed are reported in Table 3. Ragusano breed showed the highest number of alleles (82), Pantesco the lowest (52). Pantesco highlighted the lowest genetic variability for all the parameters inferred per breed.

A total of 18 breed-specific alleles were observed: 15 in Ragusano, 2 in Pantesco, and 1 in Grigio Siciliano, always at a frequency lower than 10% (data not shown).

Fis values, calculated per breed, indicated a moderate deficit of heterozygosity in all the three genetic types, probably due to a departure from random mating.

In the whole sample a significant deviation from HW-equilibrium was observed ($P < 0.001$). At breed level an excess of homozygotes was detected at 3 *loci* in Pantesco, 3 in Ragusano, and 1 in Grigio Siciliano. Only one out of 14 microsatellite markers (HMS2) was not consistently in Hardy-Weinberg equilibrium in all the three breeds, so that it was excluded from the clustering analysis.

The topology of the Neighbour Joining tree (Figure 1), built on D_R genetic distances, clearly highlighted the genetic differentiation of Pantesco breed (average $D_R = 0.104$) in comparison with Ragusano and Grigio Siciliano which were the closest breeds ($D_R = 0.032$) and significantly clustered (98.3% node support).

The clustering analysis using the Bayesian model approach was conducted on 13 microsatellite markers and under the hypothesis that donkey breeds had an ancestral common origin. The mean estimated ln probability of data ($\ln \Pr(X \mid K)$) for each inferred cluster was plotted (data not shown) and suggested $K = 5$ as the number of ancestral clusters that captures most of the structure in the sample. Breeds' genome fractions versus the five inferred clusters are reported in Figure 2. For $K = 5$, Pantesco's genome is mainly distributed into two clusters (1 and 5) with a total membership close to 90%, Ragusano's resulted to be equally shared into the clusters 2, 3, and 4 (total 93.5%), while Grigio Siciliano breed mostly belonged to the clusters 2 and 4 (73.84%).

The results of the admixture analysis, reported for $K = 5$, highlighted the clear differentiation of Pantesco breed

(HTG10), while the expected heterozygosity (H_e) ranged between 0.159 (HMS5) and 0.822 (AHT5).

The polymorphism information content (PIC) per *locus* showed only two markers with values under the 20% and an average of 0.537 (Table 1).

The significant overall *loci* F_{ST} index revealed that 10.8% of the total genetic variation observed in the sample is explained by population differences, whereas the remaining is due to the differences within subpopulations. The *locus* which contributed most to the differentiation of the samples was HMS3, while HTG4 resulted to be a nondiscriminating marker (Table 2).

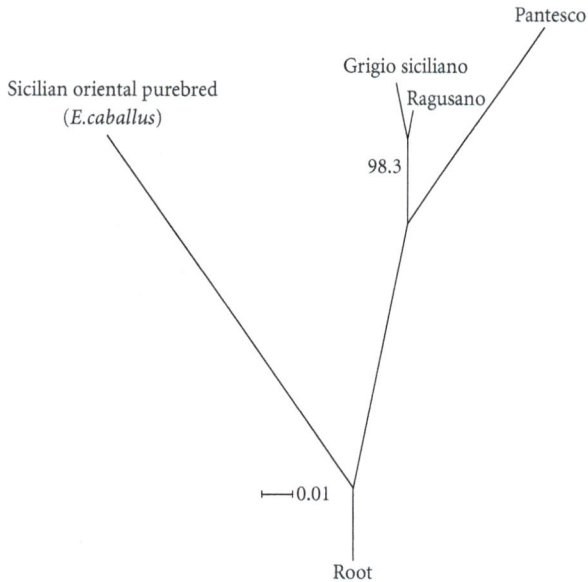

FIGURE 1: Neighbour-Joining rooted tree built on D_R genetic distances. This is a consensus tree out of the trees resulted from 1000 bootstrap resamplings of the allele frequencies at 13 *loci*'s.

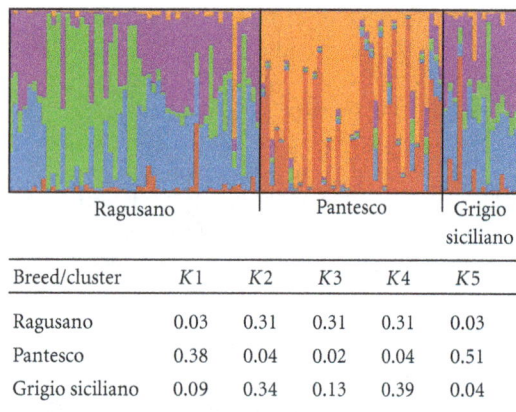

Breed/cluster	K1	K2	K3	K4	K5
Ragusano	0.03	0.31	0.31	0.31	0.03
Pantesco	0.38	0.04	0.02	0.04	0.51
Grigio siciliano	0.09	0.34	0.13	0.39	0.04

FIGURE 2: Breeds' genome distribution into the $K = 5$ inferred clusters. Analysis, conducted on the allele frequencies of 13 *loci*.

from the other two Sicilian breeds and at the same time a substructuring within the Ragusano in comparison to the rest of the sample. These outcomes were apparent along the analysis range from $K = 3$ to $K = 10$, while in correspondence with $K = 2$ only the originality of Pantesco was visible.

Correspondence analysis provided an alternative spatial representation of breeds and individuals scattered in the metric space (Figure 3). The first axis, which contributed mostly (86.23%) to the total inertia, led the Pantesco donkeys to form a well-defined group, while Ragusano and Grigio Siciliano showed their close relationship.

4. Discussion

Genetic characterization studies dealing on donkey species are scant and focused mainly on Mediterranean and Asian breeds. In terms of mean number of alleles, the genetic variability observed in the three Sicilian donkey breeds was lower than that reported in five Spanish breeds [12] and three Croatian breeds [28], but higher than that observed in the Amiata donkey from Italy [29]. In our sample, expected heterozygosity was lower than that inferred in European breeds in the above-mentioned studies [12–29] and in eight Chinese breeds [30]. This outcome is reasonable if we consider the presence of Pantesco: this breed is undergoing genetic recovery, starting from a small nucleus of 9 founders but the actual total number of heads makes it as an endangered breed with a low genetic variability. Notwithstanding this, the overall Fst index showed a good rate of differentiation at population level: the value of 10.8% was higher than that reported for Catalana breed [31] and five Spanish breeds [23].

Molecular characterization of Sicilian breeds revealed a high degree of internal structuring, highlighted by the high and significant Fst indexes per *locus*. This evidence can be mainly imputable to Pantesco's structure which clearly differentiated from Ragusano's and Grigio Siciliano's. The marked differentiation of Pantesco seems to be a sign of the appropriate plan of genetic management carried out so far.

The observed *loci* polymorphism, despite the low PIC values, made the breeds' differentiation possible. The pairwise genetic distances and the Neighbour-Joining tree highlighted the close genetic relationship between Ragusano and Grigio Siciliano, which defined a distinct cluster from Pantesco. This result is supported by the historical records which reports that until 1950 stallions reared in Sicily had both bay and grey coat color (the coat colors of the current Ragusano and Grigio Siciliano, resp.) and the morphological differentiation was based only on body size with respect to the breeding area [32]. Only starting from 1953, when Ragusano's Stud-book was established, Sicilian donkeys characterized by the grey coat color were excluded from selection and they were consigned to a marginal role.

The admixture analysis led to the identification of an interesting and useful result regarding the genomic structure of the analyzed sample. With regard to the presented results ($K = 5$), anyhow consistent along the analyzed K range, Ragusano breed presented about 30% of its genome's fraction which belong to a exclusive cluster: this might represent the most selected nucleus of the breed; at the same time, the remaining genome's structure is in common with that of Grigio Siciliano, confirming the occurred gene flow between them and their common origin. Structure analysis clearly shows that Pantesco's genome is grouped in exclusive clusters for almost the 90%. This data strengthens the originality of this breed in the context of the Sicilian indigenous donkey breeds.

Sicilian donkey breeds and populations are already classified as endangered. The low genetic variability, observed in Ragusano, Grigio Siciliano, and particularly expected in Pantesco, makes further safeguard and management plans compelling. Exploitation management should be realized by increasing the number of official stallions and reduce as low as possible the inbreeding rate at mating, particularly in those

Genetic Diversity and Variability in Endangered Pantesco and Two Other Sicilian Donkey Breeds Assessed by Microsatellite Markers

65

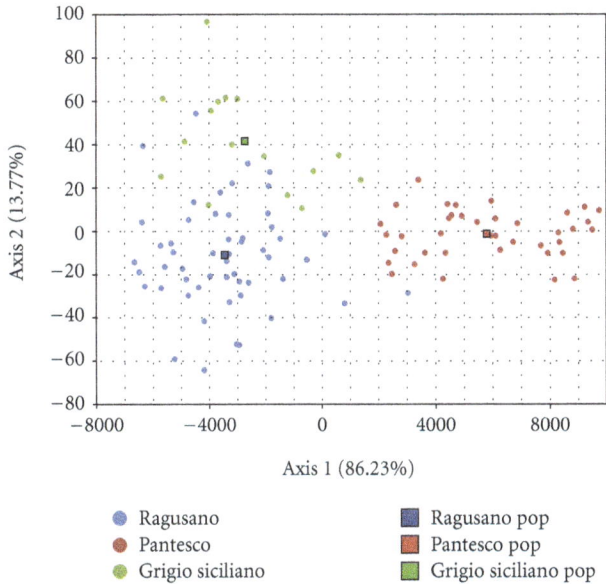

FIGURE 3: Plot of the correspondence analysis on allele frequencies of 13 *loci* of the three Sicilian donkey breeds.

farms specialized in milk production which account for more than 50 females.

In the case of Pantesco breed, the possibility of admitting in its selection schemes reproducing females from Ragusano breed appears advantageous in order to widen the current genetic pool.

Safeguard protocols need to be accomplished before the inbreeding rate brings a marked fitness reduction and leads to an increasing frequency of genetic diseases, reproductive disorders, and a general drop of vital and productive performance [33].

Acknowledgment

The authors declare that they have not any direct or indirect financial relation with the Applied Biosystems company (Foster City, CA, USA).

References

[1] A. Beja-Pereira, P. R. England, N. Ferrand et al., "African origins of the domestic donkey," *Science*, vol. 304, no. 5678, article 1781, 2004.

[2] S. Rossel, F. Marshall, J. Peters, T. Pilgram, M. D. Adams, and D. O'Connor, "Domestication of the donkey: timing, processes, and indicators," *Proceedings of the National Academy of Sciences of the United States of America*, vol. 105, no. 10, pp. 3715–3720, 2008.

[3] P. Starkey, in *The Origins and Development of African Livestock: Archaeology, Genetics, Linguistics and Ethnography*, K. C. McDonald and R. M. Blench, Eds., pp. 478–502, University College London Press, London, UK, 2000.

[4] http://dad.fao.org.

[5] E. Mascheroni, *Nuova Enciclopedia Agraria Italiana*, vol. 1 of *Equini*, Unione Tipografico-Editrice Torinese, 1929.

[6] R. Baroncini, in *L'asino, il Mulo e il Bardotto*, Edagricole, Ed., pp. 188–189, 1987.

[7] F. Camillo, D. Panzani, C. Scollo et al., "Embryo recovery rate and recipients' pregnancy rate after nonsurgical embryo transfer in donkeys," *Theriogenology*, vol. 73, no. 7, pp. 959–965, 2010.

[8] A. Zuccaro, S. Bordonaro, A. Criscione et al., "Genetic diversity and admixture analysis of Sanfratellano and three other Italian horse breeds assessed by microsatellite markers," *Animal*, vol. 2, no. 7, pp. 991–998, 2008.

[9] M. W. Bruford, D. G. Bradley, and G. Luikart, "DNA markers reveal the complexity of livestock domestication," *Nature Reviews Genetics*, vol. 4, no. 11, pp. 900–910, 2003.

[10] A. M. Guastella, A. Zuccaro, S. Bordonaro, A. Criscione, D. Marletta, and G. D'Urso, "Genetic diversity and relationship among the three autochthonous Sicilian donkey populations assessed by microsatellite markers," *Italian Journal of Animal Science*, vol. 6, supplement 1, article 143, 2007.

[11] S. Bordonaro, A. M. Guastella, A. Zuccaro, A. Criscione, F. Tidona, and D. Marletta, "Genetic diversity and inbreeding in the pantesco donkey," *Italian Journal of Animal Science*, vol. 8, supplement 2, p. 186, 2009.

[12] J. Aranguren-Méndez, J. Jordana, and M. Gomez, "Genetic diversity in Spanish donkey breeds using microsatellite DNA markers," *Genetics Selection Evolution*, vol. 33, no. 4, pp. 433–442, 2001.

[13] M. L. Glowatzki-Mullis, J. Muntwyler, W. Pfister et al., "Genetic diversity among horse populations with a special focus on the Franches-Montagnes breed," *Animal Genetics*, vol. 37, no. 1, pp. 33–39, 2006.

[14] R. Peakall and P. E. Smouse, "GENALEX 6: genetic analysis in excel. Population genetic software for teaching and research," *Molecular Ecology Notes*, vol. 6, no. 1, pp. 288–295, 2006.

[15] D. Botstein, R. L. White, M. Skolnick, and R. W. Davis, "Construction of a genetic linkage map in man using restriction fragment length polymorphisms," *The American Journal of Human Genetics*, vol. 32, no. 3, pp. 314–331, 1980.

[16] F. Rousset, "GENEPOP'007: a complete re-implementation of the GENEPOP software for Windows and Linux," *Molecular Ecology Resources*, vol. 8, no. 1, pp. 103–106, 2008.

[17] C. Van Oosterhout, W. F. Hutchinson, D. P. M. Wills, and P. Shipley, "MICRO-CHECKER: software for identifying and correcting genotyping errors in microsatellite data," *Molecular Ecology Notes*, vol. 4, no. 3, pp. 535–538, 2004.

[18] R. Chakraborty, M. de Andrade, S. P. Daiger, and B. Budowle, "Apparent heterozygote deficiencies observed in DNA typing data and their implications in forensic applications," *Annals of Human Genetics*, vol. 56, no. 1, pp. 45–57, 1992.

[19] J. F. Y. Brookfield, "A simple new method for estimating null allele frequency from heterozygote deficiency," *Molecular Ecology*, vol. 5, no. 3, pp. 453–455, 1996.

[20] J. Goudet, "FSTAT, a program to estimate and test gene diversities and fixation indices (version 2.9.3)," 2001, http://www2.unil.ch/popgen/softwares/fstat.htm.

[21] B. S. Weir and C. C. Cockerham, "Estimating F-statistics for the analysis of population structure," *Evolution*, vol. 38, no. 6, pp. 1358–1370, 1984.

[22] W. R. Rice, "Analyzing tables of statistical tests," *Evolution*, vol. 43, no. 1, pp. 223–225, 1989.

[23] J. Reynolds, B. S. Weir, and C. C. Cockerham, "Estimation of the coancestry coefficient: basis for a short-term genetic distance," *Genetics*, vol. 105, no. 3, pp. 767–779, 1983.

[24] J. Felsenstein, PHYLIP (Phylogeny Inference Package) version 3.6. Distributed by the author. Department of Genome Sciences, University of Washington, Seattle, 2005, http://evolution.genetics.washington.edu/phylip/getme.html.

[25] J. K. Pritchard, M. Stephens, and P. Donnelly, "Inference of population structure using multilocus genotype data," *Genetics*, vol. 155, no. 2, pp. 945–959, 2000.

[26] D. Falush, M. Stephens, and J. K. Pritchard, "Inference of population structure using multilocus genotype data: linked loci and correlated allele frequencies," *Genetics*, vol. 164, no. 4, pp. 1567–1587, 2003.

[27] K. Belkhir, P. Borsa, L. Chikhi, N. Raufaste, and F. Bonhomme, (1996–2004) GENETIX 4.05, logiciel sous Windows pour la génétique des populations. Laboratoire Génome, Populations, Interactions, CNRS UMR 5000, Université de Montpellier II, Montpellier, France.

[28] A. Ivankovic, T. Kavar, P. Caput, B. Mioc, V. Pavic, and P. Dovc, "Genetic diversity of three donkey populations in the Croatian coastal region," *Animal Genetics*, vol. 33, no. 3, pp. 169–177, 2002.

[29] R. Ciampolini, F. Cecchi, E. Mazzanti, E. Ciani, M. Tancredi, and B. De Sanctis, "The genetic variability analysis of the Amiata donkey breed by molecular data," *Italian Journal of Animal Science*, vol. 6, no. 1, pp. 78–80, 2007.

[30] W. Zhu, M. Zhang, M. Ge et al., "Microsatellite analysis of genetic diversity and phylogenetic relationship of light donkey breeds in china," *Scientia Agricultura Sinica*, vol. 02, 2006, Abstract in English—Article in Chinese http://en.cnki.com.cn/Article_en/CJFDTotal-ZNYK200602025.htm.

[31] J. Jordana, P. Folch, and J. A. Aranguren, "Microsatellite analysis of genetic diversity in the Catalonian donkey breed," *Journal of Animal Breeding and Genetics*, vol. 118, no. 1, pp. 57–63, 2001.

[32] T. Bonadonna, "Etnologia zootecnica," vol. VI–UTET, p. 673, 1955.

[33] S. A. Meszaros, R. G. Banks, and J. H. J. van der Werf, "Optimizing breeding structure in sheep flocks when inbreeding depresses genetic gain through effects on reproduction," in *Proceedings of the 6th World Congress on Genetics Applied to Livestock Production*, vol. 25, p. 415, Armidale, Australia, 1998.

8

Evolutionary Relationship between Two Firefly Species, *Curtos costipennis* and *C. okinawanus* (Coleoptera, Lampyridae), in the Ryukyu Islands of Japan Revealed by the Mitochondrial and Nuclear DNA Sequences

Masahiko Muraji,[1] Norio Arakaki,[2] and Shigeo Tanizaki[3]

[1] Division of Insect Sciences, National Institute of Agrobiological Sciences, Ibaraki 305-8634, Japan
[2] Plant Disease and Insect Pest Management Section, Okinawa Prefectural Agricultural Research Center, Okinawa 901-0336, Japan
[3] Arakawa 2357-11, Ishigaki, Okinawa 907-0024, Japan

Correspondence should be addressed to Masahiko Muraji, mmuraji@affrc.go.jp

Academic Editors: J. L. Elson and G.-C. Fang

The phylogenetic relationship, biogeography, and evolutionary history of closely related two firefly species, *Curtos costipennis* and *C. okinawanus*, distributed in the Ryukyu Islands of Japan were examined based on nucleotide sequences of mitochondrial (2.2 kb long) and nuclear (1.1-1.2 kb long) DNAs. In these analyses, individuals were divided among three genetically distinct local groups, *C. costipennis* in the Amami region, *C. okinawanus* in the Okinawa region, and *C. costipennis* in the Sakishima region. Their mtDNA sequences suggested that ancestral *C. costipennis* population was first separated between the Central and Southern Ryukyu areas, and the northern half was then subdivided between *C. costipennis* in the Amami and *C. okinawanus* in the Okinawa. The application of the molecular evolutionary clocks of coleopteran insects indicated that their vicariance occurred 1.0–1.4 million years ago, suggesting the influence of submergence and subdivision of a paleopeninsula extending between the Ryukyu Islands and continental China through Taiwan in the early Pleistocene.

1. Introduction

The Ryukyu Islands form a chain of more than 200 islands extending for about 1,200 km between the Japanese mainland and Taiwan. The faunae of this area are divided among three portions, the Northern, Central, and Southern Ryukyu areas, and this pattern is considered to have been strongly influenced by changes of the land configuration of this area during the Neogene and Quaternary Periods that occurred due to the interaction between tectonic changes of the Ryukyu ridge and changes of the sea level [1–3]. Paleolands or land bridges connecting this area with Taiwan and continental China, super islands connecting neighboring islands, and geologically long-standing channels that emerged during these periods are considered to be major factors, as they acted as corridors and barriers to biological dispersal [3, 4].

The genus *Curtos* Motschulsky is a group of fireflies including 16 species mainly distributed in southeastern Asia [5, 6]. In Japan, only two species, *Curtos costipennis* (Gorham) and *C. okinawanus* Matsumura, are known to exist in the Ryukyu Islands. The former species, also extant on the Chinese continent and in Taiwan, is distributed widely in the Northern, Central, and Southern Ryukyu Islands excluding the Okinawa region in the southern half of the Central Ryukyu Islands, while the range of the latter species is restricted to the Okinawa region, interrupting the range of the former species. Based on morphological and behavioral observations of the two species, Ohba and Goto [7] suggested that *C. okinawanus* might have derived from *C. costipennis*. However, it is not clear where and how these species have evolved or how *C. okinawanus* came to occupy the Okinawa region in the middle of the range of *C. costipennis*. Otherwise, *C. costipennis* might have entered the Ryukyu Islands from

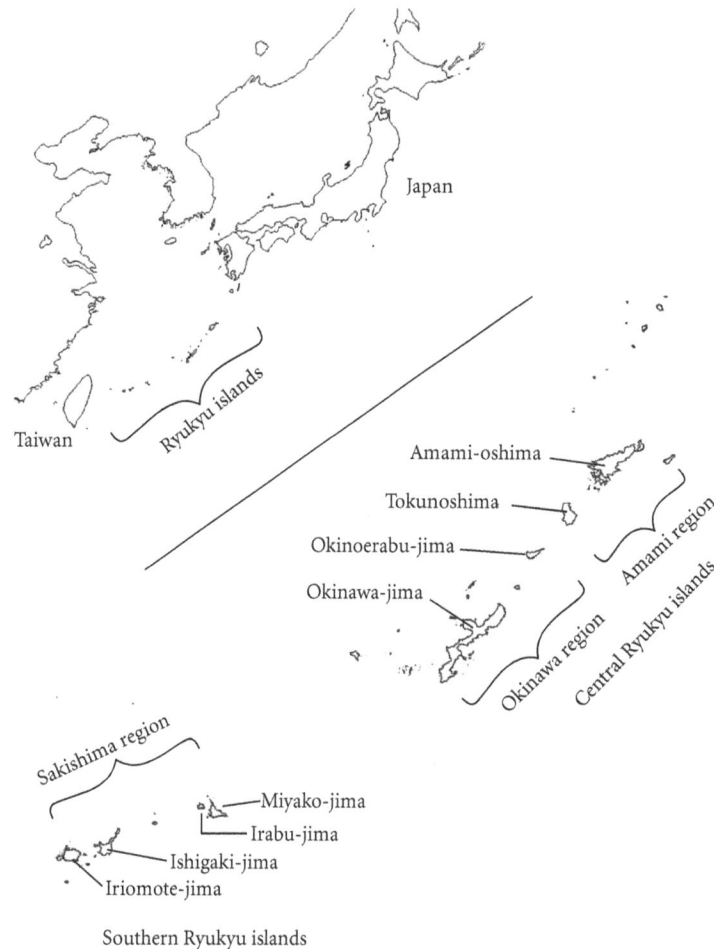

FIGURE 1: Map showing islands where *Curtos* fireflies were collected.

Taiwan or the Chinese continent, before and after the occurrence of *C. okinawanus* in the Okinawa region, while it is not clear whether or not many paleolands emerged during the evolutionary history of the two species. To answer these questions, we performed molecular phylogenetic analyses of the two species.

To do this, we collected specimens of the two species from 51 localities in the Ryukyu Islands, sequenced three sections of the mitochondrial and nuclear DNA of individual insects, and then constructed phylogenetic trees based on the sequences. Our objective was to examine (1) the phylogenetic relationship among local populations of the two species, (2) the geographic distribution patterns of genetically separated groups, and (3) the evolutionary history of the two firefly species. The results strongly suggested that *C. costipennis* is paraphyletic and that *C. okinawanus* have evolved from populations of *C. costipennis* that have been isolated in the Central Ryukyu Islands.

2. Materials and Methods

Adults of *C. costipennis* and *C. okinawanus* were collected from 51 localities (Table 1) in the Ryukyu Islands (Figure 1). Although several isles in the Tokara Islands of the Northern Ryukyu Islands are also known to be habitats of the former species [5], we could not obtain samples from these islets. For an outgroup taxon in phylogenetic analyses, the firefly *Luciola kuroiwae* Matsumura collected in Okinawa-jima Island was used. Specimens were stored in 99.9% ethanol. Template DNA was extracted from the insect body excluding the head and pronotum using a Wizard Genomic DNA Purification Kit (Promega Co., Madison, WI, USA) and dissolved in $100 \, \mu L$ sterilized distilled water.

For template DNAs extracted from individual insects, three DNA fragments, the mtDNA fragment containing the tRNA^{leu} gene and portions of the cytochrome oxidase subunit I and II genes (CO), the fragment containing the 16S ribosomal RNA gene (rDNA), and the fragment containing internal transcribed spacer 2 (ITS2) of the nuclear rDNA, were amplified using the primer sets FFMT2210F/FFMT3578R, AAMT12948F/FFMT13911R, and ITS2F/ITS2R, respectively (Table 2). The primers FFMT2210F, FFMT3578R, and FFMT13911R were designed based on sequences obtained in the preliminary experiments using several firefly species. Numerals in the name of primers for COI-COII and 16S rDNA indicate nucleotide position in the total mtDNA sequence of Drosophila yakuba [8] corresponding to the 5′ end of the primer. The primers ITS2F

Evolutionary Relationship between Two Firefly Species, Curtos costipennis and C. okinawanus (Coleoptera, Lampyridae), in the Ryukyu Islands of Japan Revealed by the Mitochondrial and Nuclear DNA Sequences

69

TABLE 1: Materials used in this study.

Species	Locality	Date	Code
Curtos costipennis			
	Amami region		
	Amami-oshima Island		
	Uragami, naze, Amami city, Kagoshima Pref.	2010.VI.20	Amami A
	Chinase, Naze, Amami city, Kagoshima Pref.	2010.VI.20	Amami B
	Yamatohama, Yamato town, Kagoshima Pref.	2010.VI.21	Amami C
	Arangachi, Uken Vil., Kagoshima Pref.	2010.VI.22	Amami D
	Koniya, Setouchi town, Kagoshima Pref.	2010.VI.23	Amami E
	Wase, Sumiyo, Amami city, Kagoshima Pref.	2010.VI.24	Amami F
	Tokunoshima Island		
	Todoroki, Tokunoshima town, Kagoshima Pref.	2010.VI.25	Tokuno A
	Mt. Gusuku-yama, Tokunishima town, Kagoshima Pref.	2010.VI.25	Tokuno B
	Agon, Isen town, Kagoshima Pref.	2010.VI.26	Tokuno C
	Akirigami, Setaki, Amagi town, Kagoshima Pref.	2010.VI.26	Tokuno D
	Sakishima region		
	Miyako-jima Island		
	Higashinakasonezoe, Miyakojima city, Okinawa Pref.	2009.V.27	Miyako A
	Nobaru, Ueno, Miyakojima city, Okinawa Pref.	2009.VI.25	Miyako B
	Yonaha, Shimoji, Miyakojima city, Okinawa Pref.	2010.IX.30	Miyako C
	Yoshino, Gusukube, Miyakojima city, Okinawa Pref.	2010.X.16	Miyako D
	Minafuku, Gusukube, Miyakojima city, Okinawa Pref.	2010.X.16	Miyako E
	Irabu-jima Island		
	Kuninaka, Irabu, Miyakojima city, Okinawa Pref.	2010.X.9	Irabu A
	Shimojishima, Irabu, Miyakojima city, Okinawa Pref.	2010.X.9	Irabu B
	Ishigaki-jima Island		
	Mt. Maesedake, Ishigaki city, Okinawa Pref.	2009.V.20	Ishigaki A
	Kuura, Ishigaki city, Okinawa Pref.	2009.V.25	Ishigaki B
	Ibaruma, Ishigaki city, Okinawa Pref.	2009.V.27	Ishigaki C
	Inoda, Ishigaki city, Okinawa Pref.	2009.V.29	Ishigaki D
	Fukai, Ishigaki city, Okinawa Pref.	2009.VI.1	Ishigaki E
	Omoto, Ishigaki city, Okinawa Pref.	2009.VI.8	Ishigaki F
	Ozato, Ishigaki city, Okinawa Pref.	2009.VI.10	Ishigaki G
	Yasura, Ishigaki city, Okinawa Pref.	2009.VI.18	Ishigaki H
	Iriomote-jima Island		
	Ootomi, Taketomi town, Okinawa Pref.	2009.V.22	Iriomote A
	Uehara A, Taketomi town, Okinawa Pref.	2009.V.23	Iriomote B
	Funauki, Taketomi town, Okinawa Pref.	2009.VI.19	Iriomote C
	Utara, Taketomi town, Okinawa Pref.	2009.VI.20	Iriomote D
	Aira, Taketomi town, Okinawa Pref.	2009.VI.21	Iriomote E
	Kanokawa, Taketomi town, Okinawa Pref.	2010.IV.18	Iriomote F
	Uehara B, Taketomi town, Okinawa Pref.	2010.V.22	Iriomote G
Curtos okinawanus			
	Okinawa region		
	Okinoerabu-jima Island		
	Point A, China town, Kagoshima Pref.	2009.VII.3	Erabu A
	Point B, China town, Kagoshima Pref.	2009.VII.3	Erabu B
	Point C, Wadomari town, Kagoshima Pref.	2009.VII.4	Erabu C
	Point D, Wadomari town, Kagoshima Pref.	2009.VII.4	Erabu D
	Point E, Wadomari town, Kagoshima Pref.	2009.VII.4	Erabu E

TABLE 1: Continued.

Species	Locality	Date	Code
	Okinawa-jima Island		
	Benoki, Kunigami vil., Okinawa Pref.	2009.VI.5	Okinawa A
	Arume, Higashi vil., Okinawa Pref.	2009.VI.9	Okinawa B
	Kawakami, Nago city, Okinawa Pref.	2009.VI.9	Okinawa C
	Gokayama, Nakijin vil., Okinawa Pref.	2009.VI.14	Okinawa D
	Kin, Kin town, Okinawa Pref.	2009.VI.22	Okinawa E
	Maebaru, Ginoza vil., Okinawa Pref.	2009.VI.23	Okinawa F
	Gushiken, Motobu town, Okinawa Pref.	2010.V.19	Okinawa G
	Hija, Yomitan vil., Okinawa Pref.	2009.VI.4	Okinawa H
	Makabe, Itoman city, Okinawa Pref.	2009.VI.8	Okinawa I
	Ishikawa-Agariyama, Uruma city, Okinawa Pref.	2009.VI.15	Okinawa J
	Kochinda, Yaese town, Okinawa Pref.	2009.VI.30	Okinawa K
	Nakadomari, Onna vil., Okinawa Pref.	2010.V.11	Okinawa L
	Zakimi, Yomitan vil., Okinawa Pref.	2010.V.18	Okinawa M
	Noborikawa, Okinawa city, Okinawa Pref.	2010.V.20	Okinawa N
Luciola kuroiwae (outgroup)			
	Okinawa-jima Island		
	Arime, Higashi vil., Okinawa Pref.	2009.VI.9	L. *kuroiwae* A
	Kawakami, Nago city, Okinawa Pref.	2009.VI.9	L. *kuroiwae* B
	Yoza, Itoman city, Okinawa Pref.	2009.VI.13	L. *kuroiwae* C

TABLE 2: PCR primers used in this study shown in 5′–3′ direction.

Gene	Name[1]	Sequence
COI-COII	FFMT2210F	TAC CAG GAT TTG GTA TAA TTT CTC AT
	FFMT3578R	GGA TAG TTC ATG AGT GGA TTA CAT C
	FFMT3140R	ATT GTT CTA TTA AAG GTG AAA TTC T
16S rDNA	AAMT12948F	ATC CAA CAT CGA GGT CGC AAA CT[2]
	FFMT13911R	GTA GTT TTG TAC CTT GTG TAT CAG GGT
ITS2	ITS2F	TGT GAA CTG CAG GAC ACA TG
	ITS2R	CCT GTT CGC TCG CAG CTA CT
	FFITS2F	GGT GAG CTC GTC CCC GCA TCG
	FFITS2R	GTG TAA TAT CAT TTG ATA TCG

[1] Numerals in the name of primers for COI-COII and 16S rDNA indicate nucleotide position in the total mtDNA sequence of *Drosophila yakuba* [8] corresponding to the 5′ end of the primer.
[2] AAMT12948F was designed in our previous study [9].

and ITS2R were designed based on the aligned homologous sequences of several insects downloaded from the DDBJ nucleotide sequence database. Amplified nuclear ITS2 was cloned using a TOPO TA cloning Kit (Invitrogen, Carlsbad, CA, USA) according to the manufacturer's instructions. One or two colonies were picked from individual insect and directly used for PCR and nucleotide sequencing. In addition to the primers used for PCR, FFMT3140R, FFITS2F, and FFITS2R designed based on sequences obtained in this study were used for sequencing. PCR, labeling, and sequencing were performed as described in Muraji et al. [9]. Sequences of representative individuals were submitted to the DDBJ/EMBL/GenBank nucleotide sequence databases (accession numbers: AB671246-AB671262, AB672623-AB672630).

Sequences were aligned as described in Muraji et al. [9] and used to compute basic statistical data and to generate phylogenetic trees based on the neighbor-joining method using MEGA ver. 4.1 software [10]. Maximum parsimony analyses were performed with PAUP* ver. 4.0b10 [11], using a heuristic search procedure with TBR swapping and 100 max. tree options.

3. Results

Nucleotide sequences of mtDNA (CO: 1,300–1,304 bp long, rDNA: 889–895 bp long) were determined for 57 individuals. As in the mtDNA of many other insects, these sequences were biased toward A and T (A + T%: 74.3 in CO and 84.4 in rDNA). The frequency of nucleotide substitutions

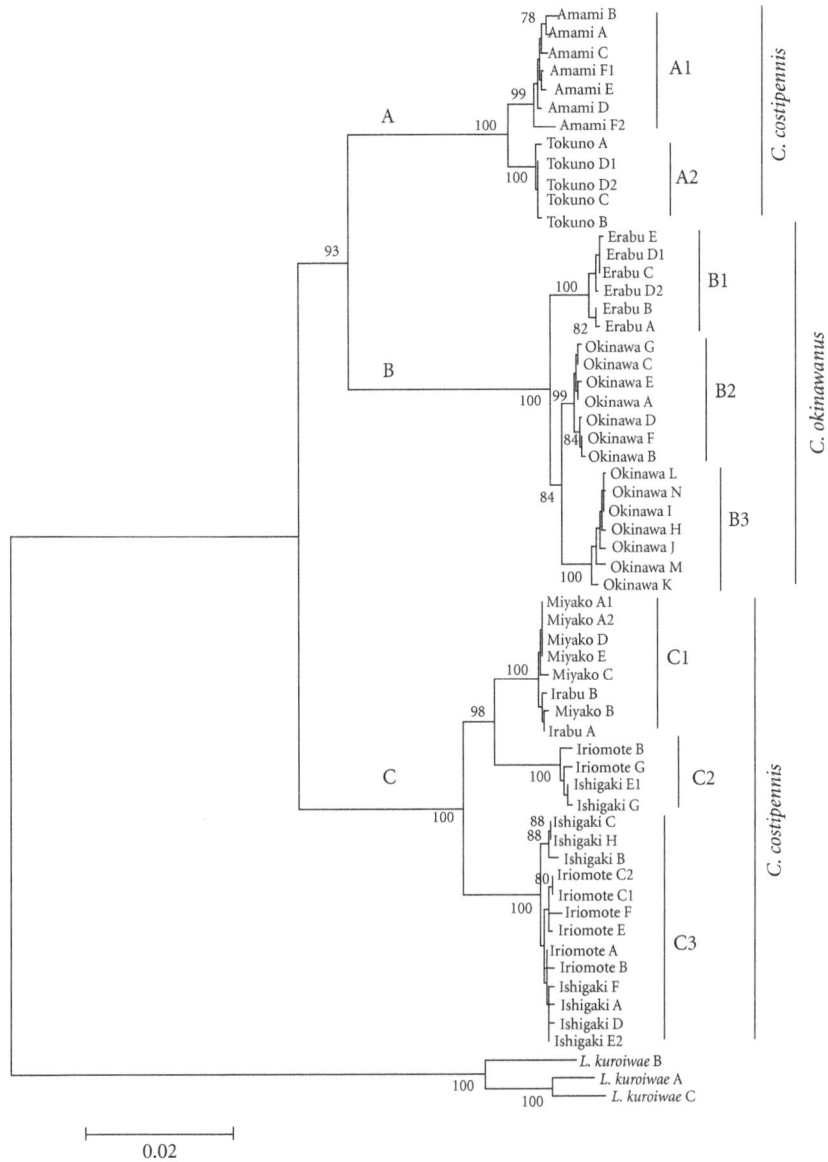

FIGURE 2: Neighbor-joining tree based on the Jukes-Cantor distances calculated using the 2,206 bp long combined data set including both mitochondrial CO and rDNA sequences. Sequences of different individuals collected from the same locality are indicated by a numeral in the taxon name. Bootstrap confidence levels calculated based on 1000 replications are shown near the nodes.

was higher in CO (polymorphic sites: 25.8%, parsimony informative sites: 23.7%) than in rDNA (polymorphic sites: 15.7%, parsimony informative sites: 14.4%).

In the phylogenetic analyses using the neighbor-joining method, the CO, rDNA, and combined (CO + rDNA) data sets produced the same topology in terms of the relationships among the haplotypic groups irrespective of the method used to calculate the genetic distances. The same topology was also recognized in all of the equally parsimonious trees generated using CO (length: 512; CI: 0.801; RI: 0.972; RC: 0.778), rDNA (length: 184; CI: 0.886; RI: 0.982; RC: 0.870), and combined data sets (length: 697; CI: 0.882; RI: 0.974; RC: 0.801). At the basal node of these trees, individuals were divided into two major lineages, group A + B and group C (Figure 2), and the former group was then subdivided into groups A and B.

Groups A, B, and C were specific to the Amami (islands of Amami-oshima and Tokunoshima), Okinawa (Okinoerabu-jima and Okinawa-jima), and Sakishima regions (Miyako-jima, Ishigaki-jima, and Iriomote-jima), respectively. The ranges of groups A and C coincided with that of C. costipennis and that of B coincided with that of C. okinawanus. At the terminal nodes of the phylogenetic trees, 2, 3, and 3 minor haplotypic groups were also detected in groups A, B, and C, respectively. All of these groups were strongly supported by the bootstrap analyses (93–100%).

The nucleotide sequence of ITS2 was determined for 50 clones obtained from 29 individuals. The length and sequence of the fragments were variable among and within the populations of the Amami (1,150–1,171 bp long), Okinawa (1,166–1,213 bp long), and Sakishima regions

FIGURE 3: Neighbor-joining tree based on the Jukes-Cantor distances calculated using the 856 bp long data set of the nuclear rDNA ITS2 sequence. Different sequences obtained from the same individuals were distinguished by lower-case letters, a and b, included in the taxon name. Bootstrap confidence levels calculated based on 1000 replications are shown near the nodes. This tree was unrooted.

(1,089–1160 bp long), and, due to insertion/ deletion mutations, the sequences could not be aligned in several sections. Thus, a 856 bp long data set generated excluding 14 (315–329 bp in total length), 15 (322–369 bp), and 13 (247–314 bp) sections from the Amami, Okinawa, and Sakishima populations, respectively, was used for phylogenetic analyses. In these analyses, the topology of the neighbor-joining trees was highly consistent among the different methods used to calculate the genetic distances (Figure 3). The topology also agreed with that of the equally parsimonious 100 trees (length: 269; CI: 0.903; RI: 0.990; RC: 0.894) obtained by the maximum parsimony method. In these trees, three groups, corresponding to groups A, B, and C in Figure 2, diverged simultaneously at the basal node. Subgroups at the lower level were recognized only in the populations that originated

in the Sakishima region. All these groupings were supported strongly by the bootstrap analyses (99-100%).

4. Discussion

In this study, we found that Japanese *Curtos* fireflies were divided among three genetically separated local populations, corresponding to *C. costipennis* in the Amami region, *C. okinawanus* in the Okinawa region, and *C. costipennis* in the Sakishima region (Figures 2 and 3). Although the nucleotide sequences of *C. okinawanus* formed a monophyletic group (group B), those of *C. costipennis* were separated between two distinct groups, A and C, and the monophyly of this species was not recognized in analyses using both mtDNA

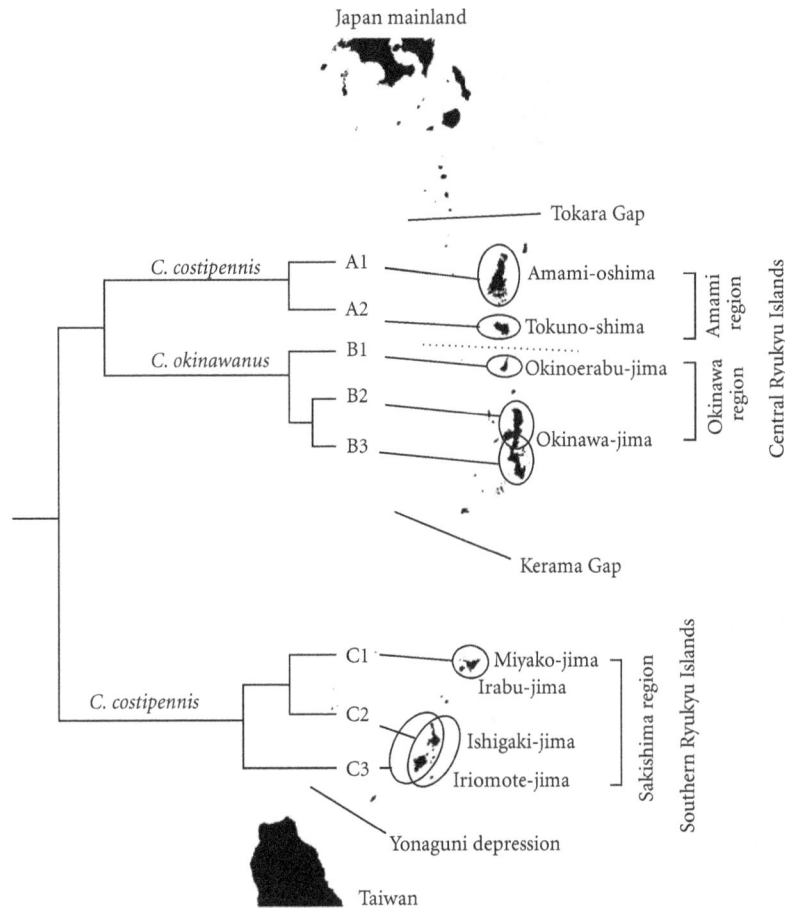

FIGURE 4: Phylogeography of *Curtos* fireflies in the Central and Southern Ryukyu Islands.

and nuclear DNA sequences (Figures 2 and 3). In addition, the mtDNA haplotypes of *C. okinawanus* (group B) were positioned between those of *C. costipennis* (groups A and C) (Figure 2). These results indicate that the currently recognized *C. costipennis* species is paraphyletic. Otherwise, *C. okinawanus* must be a subspecies or a geographic strain included within the single species *C. costipennis*.

Although the two species closely resemble each other in both morphology and behavior [7], they are apparently distinct in terms of elytral coloration (*C. costipennis*: dark yellow except for elytral apex, *C. okinawanus*: entirely brownish black), and this characteristic is recognized as diagnostically important [6]. However, the present study revealed that such a characteristic did not reflect evolutionary relationships correctly. On the other hand, several researchers recognized variations in body size, the shape of the male genital organ, and the number of punctures on the elytra among local populations of *C. costipennis* [5, 6]. Reevaluation of these characteristics is needed to confirm the taxonomic status of the two species.

Among the three major groups detected in this study, a closer relationship was recognized between *C. costipennis* and *C. okinawanus* distributed in neighboring regions (the Amami and Okinawa regions in the Central Ryukyu Islands)

than between the populations of *C. costipennis* separated between remote areas (the Amami and Sakishima regions) (Figure 4). This phenomenon seems to be consistent with the view of Ohba and Goto [7] that *C. okinawanus* might have derived from *C. costipennis* in the Central Ryukyu Islands. As discussed below, their evolutionary history can be well understood by considering the influence of the Pleistocene paleogeography of this area.

Among the combined mtDNA sequences obtained in this study, mean substitution rates of 6.8% (6.1–7.4%) and 5.7% (5.4–6.0%) were estimated between groups A + B and C, and between A and B, respectively. From these values, divergence times of 1.2–1.4 and 1.0-1.1 million years were estimated by applying an mtDNA evolutionary clock, 5–5.7% per million years, calibrated for several coleopteran insect groups [12–14]. For these periods, the hypothesized paleogeography of the Ryukyu Islands [1, 15] postulated a large paleopeninsula extending from the Chinese continent to the Northern Ryukyu Islands through Taiwan (1.2 and 1.7 million years ago). Recent studies of the paleomarine environment indicated that this peninsula began to break around 1.6 million years ago [16]. Subsequently, the peninsula submerged to form a chain of small islands, which has remained for more than one million years. Although the land configuration in the late Pleistocene is rather controversial, the seabed topography suggests the emergence of several large paleo-islands connecting neighboring islands [2, 4].

If such a hypothesis is accepted, the common ancestor of major groups is considered to have dispersed through the Ryukyu Islands along the large paleopeninsula, and their separation is considered to have occurred during the course of the gradual subsidence and subdivision of the paleopeninsula. The divergence times estimated above suggest that the channel between the Central and Southern Ryukyu areas (Kerama Gap in Figure 4) and that between the Amami and Okinawa regions opened sequentially around 1.3 and 1.0 million years ago, respectively. After a long period of isolation among small islands, the local populations must have been connected again by superislands that emerged in the late Pleistocene. Because Miyako-jima Island is known to have submerged beneath the sea in the mid-Pleistocene [17], the occurrence of fireflies on this island suggests a land connection in the late Pleistocene. The integrity of the nucleotide sequences within the respective regions (Figures 2 and 3) suggest that the super islands in this period were also separated among the Amami, Okinawa, and Sakishima regions. Minor haplotypes at the terminal nodes of the phylogenetic trees are considered to be due to the subdivision of the super islands caused by the sea level rise thereafter. The coexistence of two different minor haplotypes on the islands of Ishigaki-jima, Iriomote-jima, and Okinawa-jima (Figure 4) suggests that super islands emerged several times in this period.

The fauna of the Ryukyu Islands is known to be separated among the Northern, Central, and Southern Ryukyu areas by geologically long-standing channels at the Tokara and Kerama Gaps (Figure 4). These channels are considered to have long acted as barriers to biological dispersal [3], and their role was reflected in the relictual distribution and endemism of many terrestrial animals including mammals, birds, reptiles, amphibians, and insects [3, 18, 19] in the Central Ryukyu Islands. Although the faunal borderline is ambiguous between the Amami and Okinawa regions within the Central Ryukyu area, differences in species, subspecies, and mtDNA sequences have also been seen in many terrestrial animals among the regions [3, 19–22]. As in the fireflies examined in this study, *Neolucanus* beetles [22] showed sequential population vicariances in which the mtDNA haplotype was first separated between the areas north and south of the Kerama Gap, and the northern half was then further subdivided between the Amami and Okinawa regions, suggesting that the channel between the Amami and Okinawa regions (>500 m depth) opened after the opening of the Kerama Gap (>1,000 m depth) and had slightly less importance in determining faunal differences.

We believe that the speciation of *C. okinawanus* must be understood in this context. Considering the paleogeography of this area, the repetitive entry of fireflies into the Central Ryukyu Islands from neighboring areas is unlikely. Our conclusion is that the species derived in the Okinawa region from a population of *C. costipennis* isolated in the Central Ryukyu area. The fact that the two species can engage in interspecific copulation under laboratory conditions [23]

suggests that the speciation was based on the geographic segregation between the Amami and Okinawa regions.

In this study, we examined the evolutionary scenario of *Curtos* fireflies distributed in the Ryukyu Islands. However, because this study did not treat specimens that originated from the Northern Ryukyu Islands and Taiwan, we could not evaluate the function of two geologically long-standing channels at the Tokara Gap and the Yonaguni Depression (Figure 4). To complete the evolutional scenario of this species group, further studies using these populations as well as populations from the Chinese continent are needed.

Acknowledgments

The authors thank Takashi Fukaishi of Ishigaki city, Okinawa Prefecture, and Dr. Sadao Wakamura of Kyoto Gakuen University, for kind cooperation with the collection of the fireflies. This study was supported in part by a Grant-in-Aid for Scientific Research from the Ministry of Education, Culture, Sports, Science and Technology of Japan to M. Muraji (no. 22580061).

References

[1] M. Kimura, "Establishment and paleogeography of the Ryukyu arc," in *The Formation of the Ryukyu Arc and Migration of Biota*, M. Kimura, Ed., pp. 19–54, Okinawa Times, Okinawa, Japan, 2002.

[2] K. Kizaki and I. Oshiro, "The origin of the Ryukyu Islands," in *Natural History of Ryukyu*, K. Kizaki, Ed., pp. 8–37, Tsukiji-Shokan, Tokyo, Japan, 1980.

[3] H. Ota, "Geographic patterns of endemism and speciation in amphibians and reptiles of the Ryukyu Archipelago, Japan, with special reference to their paleogeographical implications," *Researches on Population Ecology*, vol. 40, no. 2, pp. 189–204, 1998.

[4] S. M. Lin, C. A. Chen, and K. Y. Lue, "Molecular phylogeny and biogeography of the grass lizards genus *Takydromus* (Reptilia: Lacertidae) of East Asia," *Molecular Phylogenetics and Evolution*, vol. 22, no. 2, pp. 276–288, 2002.

[5] M. Chujo and M. Sato, "On Japanese and Formosan species of the genus *Curtos* Motschulsky," *Memoirs of the Faculty of Education of Kagawa University, Part II*, vol. 192, pp. 59–65, 1970.

[6] M.-L. Jeng, P.-S. Yang, M. Sato, J. Lai, and J.-C. Chang, "The genus *Curtos* (Coleoptera, Lampyridae, Luciolinae) of Taiwan and Japan," *Japanese Journal of Systematic Entomology*, vol. 4, pp. 331–347, 1998.

[7] N. Ohba and Y. Goto, "Geographical variation on the morphology and behavior of *Curtos costipennis* and *C. okinawana* (Coleoptera: Lampyridae) in the Southwestern Islands," *Science Report of the Yokosuka City Museum*, vol. 41, pp. 1–14, 1993.

[8] D. O. Clary and D. R. Wolstenholme, "The mitochondrial DNA molecule of *Drosophila yakuba*: nucleotide sequence, gene organization, and genetic code," *Journal of Molecular Evolution*, vol. 22, no. 3, pp. 252–271, 1985.

[9] M. Muraji, N. Arakaki, S. Ohno, and Y. Hirai, "Genetic variation of the green chafer, *Anomala albopilosa* (Hope) (Coleoptera: Scarabaeidae), in the Ryukyu Islands of Japan detected by mitochondrial DNA sequences," *Applied Entomology and Zoology*, vol. 43, no. 2, pp. 299–306, 2008.

[10] S. Kumar, K. Tamura, and M. Nei, "MEGA3: integrated software for Molecular Evolutionary Genetics Analysis and sequence alignment," *Briefings in Bioinformatics*, vol. 5, no. 2, pp. 150–163, 2004.

[11] D. L. Swofford, *PAUP*. Phylogenetic Analysis Using Parsimony (*and other methods). Version 4*, Sinauer Associates, Sunderland, Mass, USA, 2003.

[12] T. E. Clarke, D. B. Levin, D. H. Kavanaugh, and T. E. Reimchen, "Rapid evolution in the *Nebria gregaria* group (coleoptera: Carabidae) and the paleogeography of the Queen Charlotte islands," *Evolution*, vol. 55, no. 7, pp. 1408–1418, 2001.

[13] C. Juan, P. Oromi, and G. M. Hewitt, "Phylogeny of the genus *Hegeter* (Tenebrionidae, Coleoptera) and its colonization of the Canary Islands deduced from cytochrome oxidase I mitochondrial DNA sequence," *Heredity*, vol. 76, no. 4, pp. 392–403, 1996.

[14] A. P. Vogler and R. Desalle, "Phylogeographic patterns in coastal North American tiger beetles (*Cicindela dorsalis* Say) inferred from mitochondrial DNA sequences," *Evolution*, vol. 47, pp. 1192–1202, 1993.

[15] T. Kawana, "Neotectonics of the Ryukyu Arc," in *The Formation of the Ryukyu Arc and Migration of Biota*, M. Kimura, Ed., pp. 59–83, Okinawa Times, Okinawa, Japan, 2002.

[16] K. Yamamoto, Y. Iryu, T. Sato, S. Chiyonobu, K. Sagae, and E. Abe, "Responses of coral reefs to increased amplitude of sea-level changes at the Mid-Pleistocene Climate Transition," *Palaeogeography, Palaeoclimatology, Palaeoecology*, vol. 241, no. 1, pp. 160–175, 2006.

[17] S. Shokita, T. Naruse, and H. Fujii, "*Geothelphusa miyakoensis*, a new species of freshwater crab (Crustacea: Decapoda: Brachyura: Potamidae) from Miyako Island, Southern Ryukyus, Japan," *Raffles Bulletin of Zoology*, vol. 50, no. 2, pp. 443–448, 2002.

[18] S. Ikehara, "Islands of valuable animals: fauna of the Ryukyu Archipelago," in *Nature in Japan, 8. Southern Islands*, K. Nakamura, H. Ujiie, S. Ikehara, H. Tagawa, and N. Hori, Eds., pp. 149–160, Iwanami-Shoten, Tokyo, Japan, 1996.

[19] WWF Japan, *Nansei Islands Biological Diversity Evaluation Project Report*, WWF Japan, Tokyo, Japan, 2010.

[20] T. Kiyoshi, "Differentiation of golden-ringed dragonfly *Anotogaster sieboldii* (Selys, 1854) (Cordulegastridae: Odonata) in the insular East Asia revealed by the mitochondrial gene genealogy with taxonomic implications," *Journal of Zoological Systematics and Evolutionary Research*, vol. 46, no. 2, pp. 105–109, 2008.

[21] M. Fujioka, *A List of Japanese Lamellicornia*, The Japanese Society of Scarabaeideans, Tokyo, Japan, 2001.

[22] T. Hosoya and K. Araya, "Molecular phylogeography of the genus *Neolucanus* in the Ryukyu Archipelago," *The Nature and Insect*, vol. 41, pp. 5–10, 2007.

[23] N. Ohba and Y. Goto, "Experimental mating in closely related species of Japanese fireflies," *Science Report of the Yokosuka City Museum*, vol. 38, pp. 1–5, 1990.

9

The Low Temperature Induced Physiological Responses of *Avena nuda* L., a Cold-Tolerant Plant Species

Wenying Liu,[1] Kenming Yu,[1] Tengfei He,[1] Feifei Li,[2] Dongxu Zhang,[1] and Jianxia Liu[1]

[1] *School of Life Science, Shanxi Datong University, Datong 037009, China*
[2] *School of Agriculture and Food Science, Zhejiang Agriculture and Forestry University, Hangzhou 311300, China*

Correspondence should be addressed to Wenying Liu; lwylwy5@163.com

Academic Editors: J. Huang and Z. Wang

The paperaim of the was to study the effect of low temperature stress on *Avena nuda* L. seedlings. Cold stress leads to many changes of physiological indices, such as membrane permeability, free proline content, malondialdehyde (MDA) content, and chlorophyll content. Cold stress also leads to changes of some protected enzymes such as peroxidase (POD), superoxide dismutase (SOD), and catalase (CAT). We have measured and compared these indices of seedling leaves under low temperature and normal temperature. The proline and MDA contents were increased compared with control; the chlorophyll content gradually decreased with the prolongation of low temperature stress. The activities of SOD, POD, and CAT were increased under low temperature. The study was designated to explore the physiological mechanism of cold tolerance in naked oats for the first time and also provided theoretical basis for cultivation and antibiotic breeding in *Avena nuda* L.

1. Introduction

Avena nuda L., also named naked oats, is originated and widely separated in north and high altitude region in China. It belongs to herb of gramineous plant with annual growing. Naked oats has great value in nutrition and medicine. It contains abundant proteins with 18 kinds of amino acid, and lots of unsaturated fatty acids. Naked oats likes to grow under cool weather and has tolerance to low temperature. Naked oats is a typical temperate crop adapted to cool climates.

Low temperature is a major abiotic stress that limits the growth, productivity, and geographical distribution of agricultural crops and can lead to significant crop loss [1, 2]. To cope with low temperature, plants have evolved a variety of efficient mechanisms that allow them to adapt to the adverse conditions [3, 4]. This adaptive process involves a number of biochemical and physiological changes, including increased levels of proline, soluble sugars, and MDA, as well as enzyme activities [5].

Understanding the mechanisms of low temperature adaptation is crucial to the development of cold-tolerant crops.

The study was designated to explore the physiological mechanism of cold tolerance in naked oats. The responses of the *Avena nuda* L. seedlings to low temperature stress were also evaluated by measuring electrolyte leakage (EL), chlorophyll content, and the concentration of MDA. We measured and compared these indices of seedlings leaves under low temperature and normal temperature. The study provided theoretical basis for cultivator and antibiotic breeding in *Avena nuda* L.

2. Materials and Methods

2.1. Materials and Cold Treatment. Naked oats cultivar Jinyan 14 (*Avena nuda* L.) was used in the experiment. Seeds were sterilized by incubation for 1 min in 75% ethanol and then washed thoroughly with sterile water. The seeds were germinated in soil in pots at 20°C under long-day conditions (16 h of cool white fluorescent light, photon flux of 70 umol m^{-2} s^{-1}).

Seedlings at the four-leaf stage were subjected to cold stress. Plants were divided into three groups, one group was under normal temperature as control, the other two were subjected to low temperature processing, and each grouphad the stress repeated three times. Seedlings of the control group were grown at 20°C continuously. For low temperature treatments, seedlings were transferred to a temperature of 1°C and −10°C in an artificial climate box under the same light and photoperiodic conditions for 7 days.

The leaves were sampled after 0, 1, 3, 5 and, 7 d of treatment for next measurement. The leaf samples were immediately frozen in liquid nitrogen and stored at −80°C until use. Three independent biological samples for each treatment were harvested, and each replicate contained 10 plants.

2.2. Determination of Relative Electrolyte Leakage. For electrolyte leakage measurement, protocol was used as described [6]. Briefly, 100 mg leaves were placed in 25 mL distilled water, shaken on a gyratory shaker (200 rpm) at room temperature for 2 h, and the initial conductivity (C1) was measured with a conductivity instrument. The samples were then boiled for 10 min to induce maximum leakage. After cooling down at room temperature, electrolyte conductivity (C2) was measured, and the relative electrical conductivity (C%) was calculated based on (C1/C2) × 100. All low temperature testing experiments were repeated three times. A paired t-test was used to determine the difference between the cold treatment and normal condition.

2.3. Determination of Chlorophyll Content. For estimation of total chlorophyll, protocol was followed as described [7]. About 100 mg of fine powder of leaf tissue was homogenized in 1 mL of 80% acetone and kept for 15 min at room temperature in dark. The crude extract was centrifuged for 20 min at 10,000 rpm at room temperature, and the resultant supernatant was used for assessing absorbance at 633 and 645 nm with a spectrophotometer. Total chlorophyll content was computed in terms of fresh weight (FW).

2.4. Determination of Proline Content. Proline concentrations in naked oats leaves were measured by the sulfosalicylic acid-acid ninhydrin method with slight modifications [8]. Around 100 mg of tissues were used and extracted in 5 mL of 3% sulphosalicylic acid at 95°C for 15 min. After filtration, 2 mL of supernatant was transferred to a new tube containing 2 mL of acetic acid and 2 mL of acidified ninhydrin reagent. After 30 min of incubation at 95°C, samples were kept at room temperature for 30 min and 5 mL of toluene was added to the tube with shaking at 150 rpm to extract red products. The absorbance of the toluene layer was determined at 532 nm using spectrophotometer.

2.5. Determination of Malondialdehyde (MDA) Content. Malondialdehyde (MDA) content in naked oats leaves was determined following the protocols as described [9]. Briefly, leaves were homogenized in 5 mL of 10% trichloroacetic acid containing 0.25% thiobarbituric acid. The mixture was incubated in water at 95°C for 30 min, and the reaction was stopped in an ice bath. The mixture was centrifuged at

10,000 g for 20 min, and the absorbance of the supernatant was measured at 450, 532, and 600 nm.

2.6. Determination of Peroxidase, Superoxide Dismutase, and Catalase Activity. Naked oats leaves (0.5 g) were ground thoroughly with a cold mortar and pestle in 50 mmol potassium phosphate buffer (pH 7.8) containing 1% polyvinylpyrrolidone. The homogenate was centrifuged at 15,000 g for 20 min at 4°C. The supernatant was crude enzyme extraction. The activities of peroxidase (POD), superoxide dismutase (SOD), and catalase (CAT) were measured using the protocols described [10].

3. Results and Discussion

3.1. Phenotypic Changes under Cold Stress. We investigated the phenotypic response to low temperature. During the cold treatment of 1°C, naked oats grew well as usual. Until 5 days later, the seedlings were always strong only except some leaf apexes began to get yellow. In the 7th day, most parts of seedling remained green as normal temperature as shown in Figure 1. The seedling got curl after 3-4 hours after exposure to −10°C cold stress. Some leaf began to get yellow and curled seriously in the third day, while the seedling grew slowly. Most leaves showed severe rolling and wilting in the 7th day.

No obvious differences have been found in growth and developed between normal temperature and at 1°C, indicating that the naked oats have cold tolerance at the low temperature of 1°C.

3.2. Changes of Relative Electrolyte Leakage under Cold Stress. Cold stress often causes damage to cell membranes, so, we tested the cell-membrane penetrability. Cell membrane penetrability was evaluated by the relative conductance of the cell membrane under cold stress [11, 12]. The electrolyte leakage test was performed to compare membrane integrity [13]. For such experiment, plants were subjected to low temperature of 1°C and −10°C. The relative electrolyte leakage of seedling leaves was increased greatly with cold treatment as shown in Figure 2. There was no significant difference under normal temperature; the average electrolyte leakage was about 9.7%. During the cold stress, the relative electrolyte leakage value gradually increased with the prolongation of low temperature stress. The electrolyte leakage of −10°C was significantly higher than that of 1°C growing plants. In the 7th day, elative electrolyte leakage increased to 66.0% under cold stress of −10°C, while 37.3% at 1°C and the normal condition still 11.0%.

When the plants were under low temperature stress, the structure of cellular membrane was damaged. The degree of cell membrane injury induced by cold stress can be reflected by intracellular electrolyte leakage rate. The relative conductance value is one of the effective indicators to indirectly evaluate plant response ability to low temperature stress [14]. The damage degree of cellular membrane was aggravated with the continuity of low temperature stress [12]. In the experiment, the electrolyte leakage in leaves of naked oats seedlings gradually increased under low temperature of 1°C

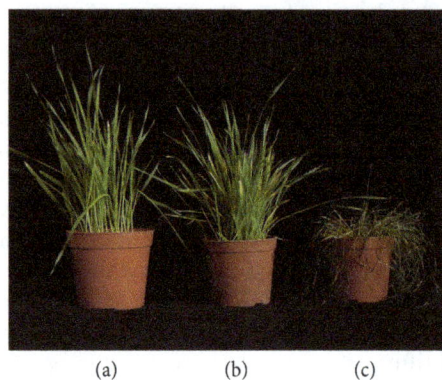

FIGURE 1: The seedlings of naked oats at normal temperature (a) and low temperature of 1°C (b) and −10°C (c) after 7 days.

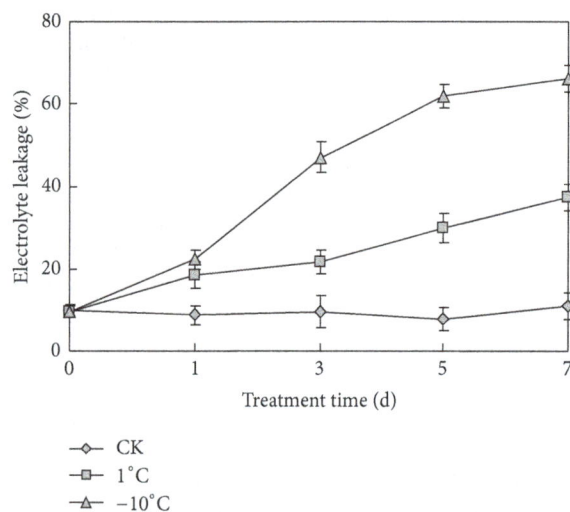

FIGURE 3: Chlorophyll content of seedling leaves under normal and low temperatures. CK represents normal growth condition. Each value is the average ± standard error (±SE) result of three independent measurements.

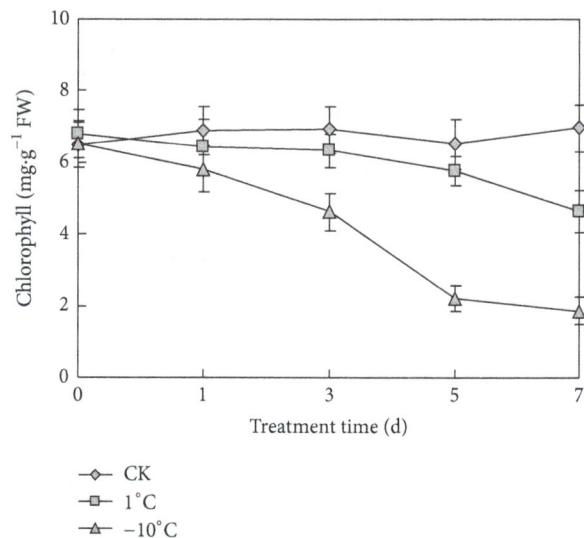

FIGURE 2: Relative electrolyte leakage of seedling leaves under normal and low temperatures. CK represents normal growth condition. Each value is the average ± standard error (±SE) result of three independent measurements.

stress, indicating that the damage of low temperature on cell membrane increased gradually. The electrolyte leakage increased more significantly at −10°C than 1°C, indicating that cell membrane was damaged seriously at −10°C than 1°C. At 1°C, the naked oats had some tolerance to protect the cell membraneavoiding cold damage.

3.3. Changes of Total Chlorophyll Content under Cold Stress.
Chlorophyll is an extremely important and critical biomolecule in photosynthesis with function of light absorbance and light energy transformation [15]. Low temperature stress can influent plant photosynthesis and decrease the utilization of light [16, 17]. We examined the content of chlorophyll under low temperature of 1°C and −10°C. Compared with the control, the chlorophyll content in seedling leaves under low temperature was lower than that under room temperature. As shown in Figure 3, the change range of chlorophyll content in leaves of naked oats was 6.4–6.9 mg/g

under room temperature. At 1°C, the chlorophyll content was decreased with the prolongation of low temperature stress. The chlorophyll decreased slightly from the 1st day to the 5th day, and the content was onlyslightly less than control, but in 7th day the chlorophyll decreased greatly to 4.6 mg/g, while at −10°C, the chlorophyll content decreased seriously. Especially in the 5th day, chlorophyll content decreased to 2.2 mg/g while 5.8 mg/g at 1°C and 6.5 mg/g at normal temperature. The results indicated that naked oats was damaged less at 1°C than at −10°C. Low temperature inhibits chlorophyll accumulations in actively growing leaves. Naked oats has some degree of cold tolerance at 1°C.

3.4. Changes of Free Proline Content under Cold Stress.
Proline is widely distributed in plants as protection material, which is an organic osmolyte [18]. It plays a vital role in maintaining osmotic balance and stabilizing cellular structures in plants. Many plants accumulate free proline in response to abiotic stress of low temperature. Increased free proline content protects the plant against the stress [19].

The effect of cold on content of proline was investigated. There was no significant difference in proline contents without cold stress as shown in Figure 4. An increase in proline content was observed upon exposure to cold stress. Compared with the control, the free proline content in seedling leaves under low temperature was obviously higher than that under room temperature. At 1°C, the proline content increased with the prolongation of low temperature stress. At −10°C, the proline content was increased in the first five days and reached the max of 601 μg/g in the 5th day, about 6 times of control. Then proline content decreased to 404 μg/g in the 7th day, which was still much higher than control. At −10°C, the plant accumulated more amounts of proline than at 1°C.

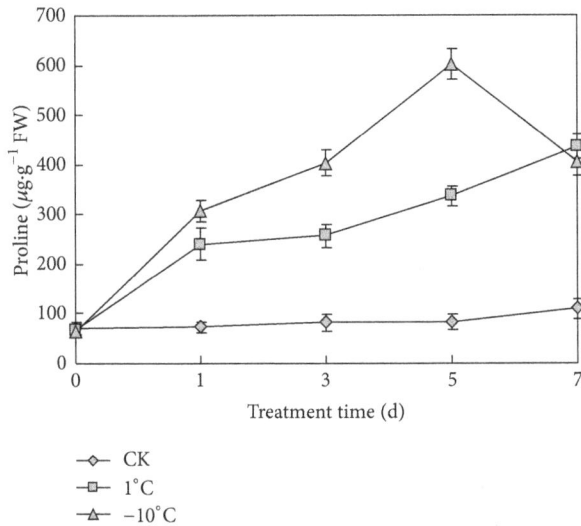

FIGURE 4: Free proline content of seedling leaves under normal and low temperatures. CK represents normal growth condition. Each value is the average ± standard error (±SE) result of three independent measurements.

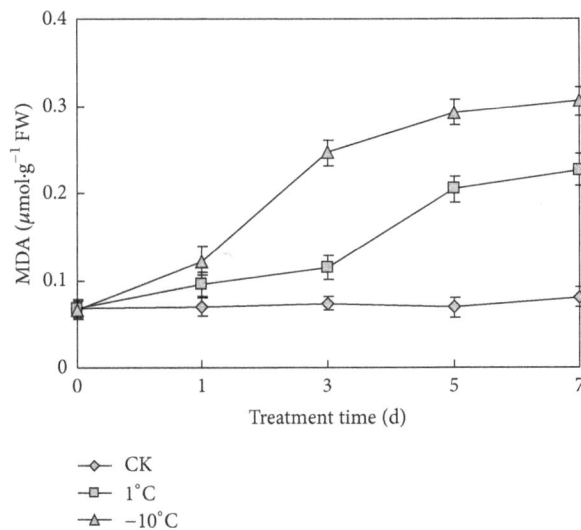

FIGURE 5: MDA content of seedling leaves under normal and low temperatures. CK represents normal growth condition. Each value is the average ± standard error (±SE) result of three independent measurements.

This indicated that the lower the temperature was, the more proline could be accumulated.

Proline plays a vital role in maintaining osmotic balance in plants. The accumulation of proline may function in preventing plants from being damaged by stress. The free proline acts as osmolytes to facilitate osmoregulation, thus protecting plants from dehydration resulting from cold stress by reducing water potential of plant cells [20]. In addition, proline can also function as a molecular chaperone to stabilize the structure of proteins as well as play a role in regulation of the antioxidant system [21, 22]. So increased free proline

content protects the plant against the stress. The study found greater accumulation of free proline under cold stress, which may partially account for the higher tolerance of plants to cold stress. Accumulation of proline to facilitate osmo-regulation is a common adaptive mechanism for tolerance of plants to abiotic stress. The results were consistent with previous studies that proline accumulated in leaves exposed to cold, salt, and other stresses [23].

3.5. Changes of Malondialdehyde (MDA) Content under Cold Stress. Cold stress often causes damage to cell membranes. Malondialdehyde (MDA) is an important indicator of membrane system injuries and cellular metabolism deterioration [24]. So, we further measured the cell membrane penetrability. The effects of MDA contents were investigated in the seedling of naked oats. In our experiment, naked oats had a significantly higher MDA level under low temperature stress compared to the control level. As shown in Figure 5, the MDA concentrations increased with the prolongation of low temperature stress. In the seventh day, the MDA content reached to the max nearly 3 times of control at 1°C and about 4 times at −10°C. The increase in MDA content at −10°C was significantly higher than 1°C all the time.

MDA has been well recognized as a parameter reflecting damage by cold stress. Cell membrane systems are the primary sites of freezing injury in plants [25]. Plants subjected to low temperatures frequently suffer membrane damage, which can be evaluated by relative electrolyte leakage and MDA production. MDA is considered to be the final product of lipid peroxidation in the plant cell membrane [26]. MDA is also an important intermediate in ROS scavenging, and a high level of MDA is toxic to plant cells. In this experiment, in the first day the MDA only increased $0.095 \, \mu mol \cdot g^{-1}$ at 1°C, while the MDA concentrations increased to $0.122 \, \mu mol \cdot g^{-1}$ at −10°C. In the third day, MDA increased slightly to $0.115 \, \mu mol \cdot g^{-1}$ at 1°C, while MDA increased rapidly to $0.247 \, \mu mol \cdot g^{-1}$ at −10°C, which was 4 times of control. The increase of MDA at −10°C was greater than at 1°C. These results suggested that cell membrane was little damaged at 1°C at the beginning of cold stress, but was hurt seriously at −10°C. It may have contributed to the different phenotypes under cold stress. Prolonged treatment finally led to a great cell damaged and MDA accumulation. The MDA content increased slowly in first three days at 1°C, but increased rapidly to $0.20 \, \mu mol \cdot g^{-1}$ in the fifth day. It indicated that cell membranes of seedlings were injured by cold seriously in 3–5 days at 1°C.

3.6. Changes of Antioxidant Enzymes under Cold Stress. Cold stress induces the accumulation of reactive oxygen species (ROS) such as superoxide, hydrogen peroxide, and hydroxyradicals [27, 28]. The elevated concentrations of ROS can damage cellular structures and macromolecules, leading to cell death [29, 30]. Plants under abiotic stress have evolved a defense system against oxidative stress by increasing the activity of ROS-scavenging enzyme. ROS can be scavenged by superoxide dismutase (SOD), peroxidase (POD), and catalase (CAT) [31]. SOD plays an important role in eliminating ROS induced by cold [32]. POD and CAT can scavenge H_2O_2.

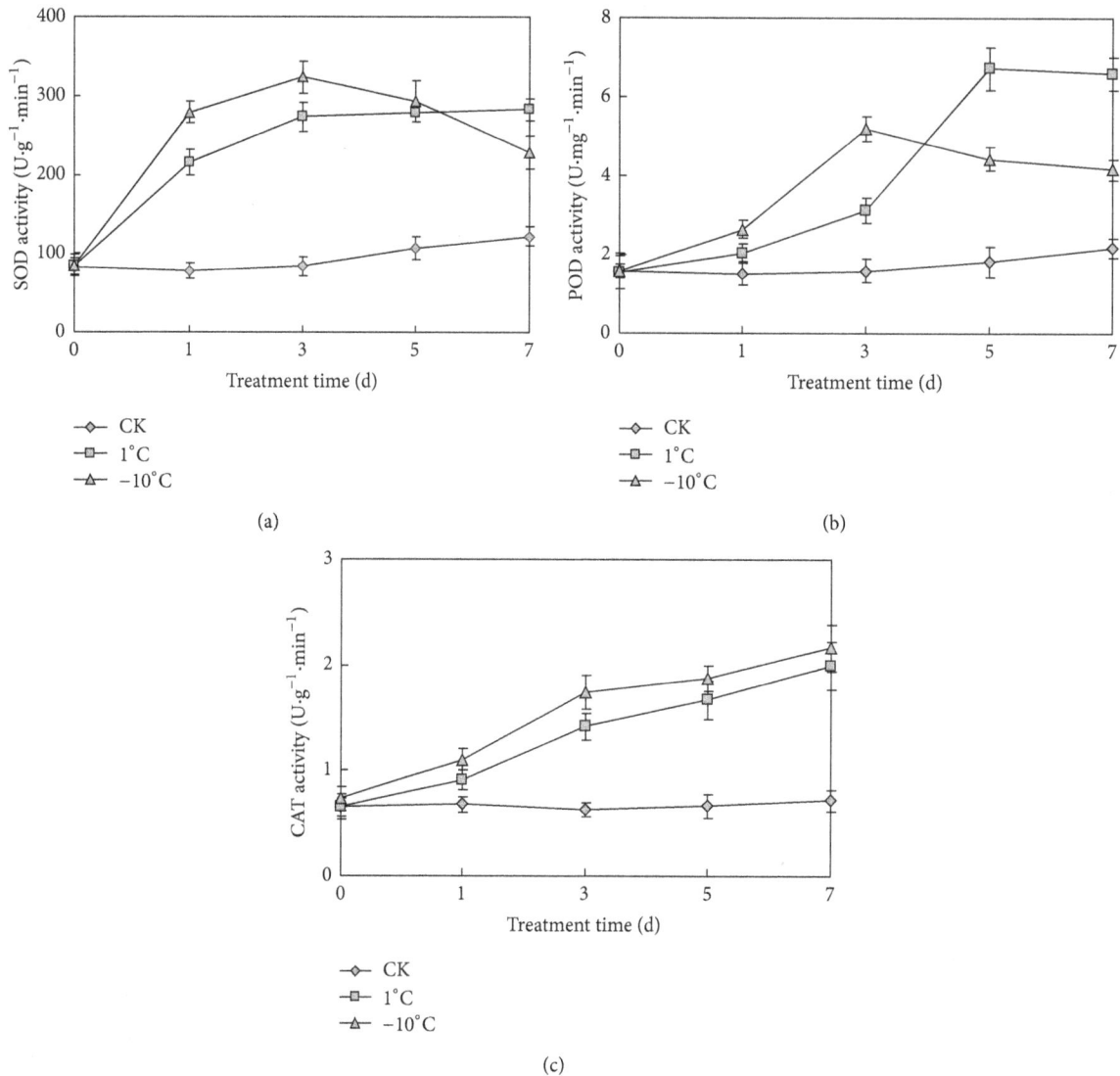

FIGURE 6: Activities of SOD (a), POD (b), and CAT (c) of seedling leaves under normal and low temperatures. CK represents normal growth condition. Each value is the average ± standard error (±SE) result of three independent measurements.

SOD, POD, and CAT are important enzymes protecting membrane system. Many kinds of antioxidant enzyme activities have been changed under low stress. Changes of enzyme activities are relvant to cold tolerance.

We measured the activities of these enzymes under normal and cold stress. Compared with normal, the SOD activity was higher under the cold treatment, as shown in Figure 6(a). At 1°C, the SOD activity was increased rapidly in the first three days and remained unchanged in later days. At −10°C, the SOD activity was increased first and decreased later. In the third day, the SOD activity was increased to the max value 323 U·g^{-1}·min^{-1}, which was nearly 4 times of normal temperature. Then, the SOD activity was decreased rapidly to a low value of 229 U·g^{-1}·min^{-1} in the seventh day. At this time, the seedlings were damaged seriously by low temperature. Compared with normal, the SOD activity was higher than normal under the cold treatment. Increased SOD

activity contributed for cold tolerance of naked oats under cold stress. Especially in the first day, the SOD activities had increasedgreatly, indicated that naked oats is resisting and adapting to low temperature, and contributed to reduce the damage by cold. The SOD activity decreased later at −10°C, which may be due to the greater damage by temperature of −10°C affected the synthesis of SOD in plant.

POD activities were higher under low temperature than normaltemperature as shown in Figure 6(b). At 1°C, POD activity was slowly increased in the first three days, rapidly increased in the third to fifth days, and reached the max of 6.7 U·min^{-1}·mg^{-1} in the fifth day, more than 4 times of control. Then POD activity slightly decreased to 6.6 U·mg^{-1}·min^{-1} in the 7th day, which was almost nearly the max value. At −10°C, POD activity was rapidly increased in the first three days and reached the max of 5.2 U·mg^{-1}·min^{-1} in the third day, more than 3 times of control. Then POD activity decreased in the third to seventh days. The POD activity decreased to $4.16\,U\cdot mg^{-1}\cdot min^{-1}$ in the 7th day, which was still 2.5 timesmore than that of control. PODs are a large family of enzymes that typically catalyze peroxide. The increased POD activities under low temperature had improved cold tolerance in some degree. POD activities decreased greatly in later days, indicating that low temperature had affected POD enzyme. It maybe due to low temperature affected RNA transcription and translation, reducing the synthesis of POD. At the same time proline content also increased under low temperature, which can degrade peroxidase. It can also decrease the POD activity.

Compared with normal, the CAT activity was higher under the cold treatment than normal temperature as shown in Figure 6(c). At $1^{\circ}C$, the CAT activity was increased with the prolongation of low temperature stress. CAT activity was $0.66\,U\cdot g^{-1}\cdot min^{-1}$ before of cold treatment and increased to $1.99\,U\cdot g^{-1}\cdot min^{-1}$ after the seven days cold treatment. At $-10^{\circ}C$, the CAT activity was increased rapidly in the first three days and increased slowly in later days. In the third day, the CAT activity was increased to the $1.74\,U\cdot g^{-1}\cdot min^{-1}$, which was nearly 2.5 times of normal temperature. Then, the CAT activity was increased to $2.16\,U\cdot g^{-1}\cdot min^{-1}$ in the seventh day. Compared with normal, the CAT activity was higher than normal under the cold treatment. Increased CAT activity contributed to cold tolerance of naked oats under cold stress.

In this experiment, the activities of these enzymes were increased in different degrees under cold stress. The results imply that higher SOD, POD, and CAT activities enhanced the capacity for scavenging ROS and contributed to enhanced tolerance of plant to cold stress. The result was similar to the SOD changes in other cold treatment plants.

4. Conclusions

Avena nuda L. is an important crop and has tolerance to low temperature. In the long time of evolution, it should have some cold resistant mechanism. The cold-stress leads to complex cellular and biochemical changes such as altered membrane composition and accumulation of proline, as well as the activities of antioxidant enzymes. In the present study, we have, for the first time, investigated the physiological changes in naked oats under low temperature. As results showed, low temperature stress caused a certain degree of physiological impairment in naked oats leaves. Here, we showed that cold increased the content of proline, MDA, electrolyte leakage, SOD, CAT, and POD activities. The content of chlorophyll was decreased. The study provided theoretical basis for cultivation and antibiotic breeding in Avena nuda L.

Acknowledgments

This work was supported by the Shanxi Datong University Research Starting Fund for Doctor no. 2009-B-16. The seeds of naked oats were supplied by Assistant Researcher Li Gang in High Latitude Crops Instituteof Shanxi Academy of Agriculture Sciences.

References

[1] C. Guy, F. Kaplan, J. Kopka, J. Selbig, and D. K. Hincha, "Metabolomics of temperature stress," *Physiologia Plantarum*, vol. 132, no. 2, pp. 220–235, 2008.

[2] H. J. Bohnert, D. E. Nelson, and R. G. Jensen, "Adaptations to environmental stresses," *Plant Cell*, vol. 7, no. 7, pp. 1099–1111, 1995.

[3] R. Bressan, H. Bohnert, and J.-K. Zhu, "Perspective: Abiotic stress tolerance: from gene discovery in model organisms to crop improvement," *Molecular Plant*, vol. 2, no. 1, pp. 1–2, 2009.

[4] W. Y. Liu, Y. F. Zhang, and D. X. Zhang, "Study on cold resistant genes in plants," *Journal of Shanxi Datong Unirersity*, vol. 28, no. 6, pp. 52–55, 2012.

[5] A. S. Holaday, W. Martindale, R. Aired, A. L. Brooks, and R. C. Leegood, "Changes in activities of enzymes of carbon metabolism in leaves during exposure of plants to low temperature," *Plant Physiology*, vol. 98, no. 3, pp. 1105–1114, 1992.

[6] M. Bajji, P. Bertin, S. Lutts, and J.-M. Kinet, "Evaluation of drought resistance-related traits in durum wheat somaclonal lines selected in vitro," *Australian Journal of Experimental Agriculture*, vol. 44, no. 1, pp. 27–35, 2004.

[7] D. Arnon, "Copper enzymes in isolated chloroplasts. Polyphenoloxidase in Beta vulgaris," *Plant Physioligacal*, vol. 24, no. 1, pp. 1–15, 1949.

[8] L. S. Bates, R. P. Waldren, and I. D. Teare, "Rapid determination of free proline for water-stress studies," *Plant and Soil*, vol. 39, no. 1, pp. 205–207, 1973.

[9] S.-Y. Song, Y. Chen, J. Chen, X.-Y. Dai, and W.-H. Zhang, "Physiological mechanisms underlying *OsNAC5*-dependent tolerance of rice plants to abiotic stress," *Planta*, vol. 234, no. 2, pp. 331–345, 2011.

[10] B.-H. Miao, X.-G. Han, and W.-H. Zhang, "The ameliorative effect of silicon on soybean seedlings grown in potassium-deficient medium," *Annals of Botany*, vol. 105, no. 6, pp. 967–973, 2010.

[11] P. Rohde, D. K. Hincha, and A. G. Heyer, "Heterosis in the freezing tolerance of crosses between two Arabidopsis thaliana accessions (Columbia-0 and C24) that show differences in non-acclimated and acclimated freezing tolerance," *Plant Journal*, vol. 38, no. 5, pp. 790–799, 2004.

[12] P. L. Steponkus, D. V. Lynch, and M. Uemura, "The influence of cold-acclimation on the lipid-composition and cryobehavior of the plasma-membrane of isolated rye protoplasts," *Philosophical Transactions B*, vol. 326, no. 1237, pp. 571–583, 1990.

[13] M. B. Murray, J. N. Cape, and D. Fowler, "Quantification of frost damage in plant-tissues by rates of electrolyte leakage," *New Phytologist*, vol. 113, no. 3, pp. 307–311, 1989.

[14] P. L. Steponkus, "Role of the plasma membrane in freezing-injury and cold acclimation," *Annual Review of Plant Physiology*, vol. 35, pp. 543–584, 1984.

[15] W. Xu, D. T. Rosenow, and H. T. Nguyen, "Stay green trait in grain sorghum: relationship between visual rating and leaf chlorophyll concentration," *Plant Breeding*, vol. 119, no. 4, pp. 365–367, 2000.

[16] J. C. Glaszmann, R. N. Kaw, and G. S. Khush, "Genetic divergence among cold tolerant rices (*Oryza sativa* L.)," *Euphytica*, vol. 45, no. 2, pp. 95–104, 1990.

[17] J. Wu, J. Lightner, N. Warwick, and J. Browse, "Low-temperature damage and subsequent recovery of fab1 mutant arabidopsis exposed to $2^{\circ}C$," *Plant Physiology*, vol. 113, no. 2, pp. 347–356, 1997.

[18] N. Verbruggen and C. Hermans, "Proline accumulation in plants: a review," *Amino Acids*, vol. 35, no. 4, pp. 753–759, 2008.

[19] Y. Igarashi, Y. Yoshiba, Y. Sanada, K. Yamaguchi-Shinozaki, K. Wada, and K. Shinozaki, "Characterization of the gene for Δ1-pyrroline-5-carboxylate synthetase and correlation between the expression of the gene and salt tolerance in *Oryza sativa* L.," *Plant Molecular Biology*, vol. 33, no. 5, pp. 857–865, 1997.

[20] G. Székely, E. Ábrahám, Á. Cséplo et al., "Duplicated *P5CS* genes of arabidopsis play distinct roles in stress regulation and developmental control of proline biosynthesis," *Plant Journal*, vol. 53, no. 1, pp. 11–28, 2008.

[21] P. Armengaud, L. Thiery, N. Buhot, G. G.-D. March, and A. Savouré, "Transcriptional regulation of proline biosynthesis in Medicago truncatula reveals developmental and environmental specific features," *Physiologia Plantarum*, vol. 120, no. 3, pp. 442–450, 2004.

[22] Z. Xin and J. Browse, "*eskimo1* mutants of Arabidopsis are constitutively freezing-tolerant," *Proceedings of the National Academy of Sciences of the United States of America*, vol. 95, no. 13, pp. 7799–7804, 1998.

[23] W.-Y. Liu, M.-M. Wang, J. Huang, H.-J. Tang, H.-X. Lan, and H.-S. Zhang, "The *OsDHODH1* gene is involved in salt and drought tolerance in rice," *Journal of Integrative Plant Biology*, vol. 51, no. 9, pp. 825–833, 2009.

[24] W. J. Fan, M. Zhang, H. X. Zhang, and P. Zhang, "Improved tolerance to various abiotic stresses in transgenic sweet potato (*Ipomoea batatas*) expressing spinach betaine aldehyde dehydrogenase," *PLoS One*, vol. 7, no. 5, Article ID e37344, 2012.

[25] R. A. J. Hodgson and J. K. Raison, "Lipid peroxidation and superoxide dismutase activity in relation to photoinhibition induced by chilling in moderate light," *Planta*, vol. 185, no. 2, pp. 215–219, 1991.

[26] D. M. Hodges, J. M. DeLong, C. F. Forney, and R. K. Prange, "Improving the thiobarbituric acid-reactive-substances assay for estimating lipid peroxidation in plant tissues containing anthocyanin and other interfering compounds," *Planta*, vol. 207, no. 4, pp. 604–611, 1999.

[27] R. Mittler, "Oxidative stress, antioxidants and stress tolerance," *Trends in Plant Science*, vol. 7, no. 9, pp. 405–410, 2002.

[28] N. Suzuki and R. Mittler, "Reactive oxygen species and temperature stresses: a delicate balance between signaling and destruction," *Physiologia Plantarum*, vol. 126, no. 1, pp. 45–51, 2006.

[29] K. Apel and H. Hirt, "Reactive oxygen species: metabolism, oxidative stress, and signal transduction," *Annual Review of Plant Biology*, vol. 55, pp. 373–399, 2004.

[30] J. Krasensky and C. Jonak, "Drought, salt, and temperature stress-induced metabolic rearrangements and regulatory networks," *Journal of Experimental Botany*, vol. 63, no. 4, pp. 1593–1608, 2012.

[31] G. Miller, N. Suzuki, S. Ciftci-Yilmaz, and R. Mittler, "Reactive oxygen species homeostasis and signalling during drought and salinity stresses," *Plant, Cell and Environment*, vol. 33, no. 4, pp. 453–467, 2010.

[32] R. G. Alscher, N. Erturk, and L. S. Heath, "Role of superoxide dismutases (SODs) in controlling oxidative stress in plants," *Journal of Experimental Botany*, vol. 53, no. 372, pp. 1331–1341, 2002.

10

GmNAC5, a NAC Transcription Factor, Is a Transient Response Regulator Induced by Abiotic Stress in Soybean

Hangxia Jin, Guangli Xu, Qingchang Meng, Fang Huang, and Deyue Yu

State Key Laboratory of Crop Genetics and Germplasm Enhancement, National Center for Soybean Improvement, Nanjing Agricultural University, Nanjing 210095, China

Correspondence should be addressed to Fang Huang; fhuang@njau.edu.cn and Deyue Yu; dyyu@njau.edu.cn

Academic Editors: J. Huang and Z. Wang

GmNAC5 is a member of NAM subfamily belonging to NAC transcription factors in soybean (*Glycine max* (L.) Merr.). Studies on NAC transcription factors have shown that this family functioned in the regulation of shoot apical meristem (SAM), hormone signalling, and stress responses. In this study, we examined the expression levels of *GmNAC5*. *GmNAC5* was highly expressed in the roots and immature seeds, especially strongly in immature seeds of 40 days after flowering. In addition, we found that *GmNAC5* was induced by mechanical wounding, high salinity, and cold treatments but was not induced by abscisic acid (ABA). The subcellular localization assay suggested that GmNAC5 was targeted at nucleus. Together, it was suggested that GmNAC5 might be involved in seed development and abiotic stress responses in soybean.

1. Introduction

Environmental stresses such as drought, salinity, and cold are major factors that significantly limit agricultural productivity. NAC transcription factors play essential roles in response to various abiotic stresses [1]. The N-terminal region of NAC proteins contains a highly conserved NAC domain, which can be divided into five subdomains based on sequence similarities and may function as DNA-binding region. The C-terminal regions of NAC proteins, which exhibit the trans-activation activity, are highly divergent in both sequence and length [2–4]. This family of transcription factors is involved in a lot of plant developmental processes, including shoot apical meristem formation [5], hormone signaling [2, 6], regulation of cell division and cell expansion [7], control of secondary wall formation [8–10], and responses to various stresses [11–14].

The NAC family consists of several subfamilies [15]. The NAM subfamily is the best studied NAC subfamily. *CUC1* and *CUC2*, encoding NAM subfamily proteins, are a pair of functionally redundant genes, expressed in *Arabidopsis* meristem and organ primordia boundary [1, 16]. The cotyledons of the transgenic seedlings overexpressing *CUC1* (*35S::CUC1*) regularly had two basal lobes, small and round epidermal cells between the sinuses, and adventitious SAMs on the adaxial surface of this region [17]. It has been reported that CUC2 is essential for dissecting the leaves of a wide range of lobed/serrated *Arabidopsis* lines. Inactivation of *CUC3* leads to a partial suppression of the serrations, indicating a role for this gene in leaf shaping. Morphometric analysis of leaf development and genetic analysis provide evidences for different temporal contributions of CUC2 and CUC3 [18]. The *CUP* played an important role in the lateral organ boundary forming snapdragon. Cupuliformis mutants are defective in shoot apical meristem formation, but cup plants overcome this early barrier to development to reach maturity. *CUP* encodes a NAM protein, homologous to the petunia NAM and *Arabidopsis* CUC proteins. The phenotype of *cup* mutants differs from the phenotype of *NAM* and *CUC1 CUC2* in that dramatic organ fusion is observed throughout development [19]. Phloem transport of *CmNACP* mRNA was proved directly by heterograft studies between pumpkin and cucumber plants, in which *CmNACP* transcripts were shown to accumulate in cucumber scion phloem and apical tissues [20]. Petunia NAM proteins were mainly expressed in the meristem and primordia boundaries, which might be required by embryo and flower pattern formation [5]. For abiotic stress, it was observed that *Arabidopsis AtNAC2* expression was induced by salt stress and this induction was reduced in magnitude in the transgenic *Arabidopsis* plants overexpressing tobacco ethylene receptor gene *NTHK1*. *AtNAC2* was localized in the nucleus and had transcriptional activation activity. It can form a homodimer in yeast. *AtNAC2* was highly expressed in roots and flowers but less expressed in other organs examined. In addition to the salt induction, *AtNAC2* can be induced by abscisic acid (ABA), ACC, and NAA [21]. These showed that the NAM subfamily members not only play a regulatory role in plant development but also participate in stress responses. *GmNAC5*, which is a member of NAM subfamily belonging to NAC transcription factor in soybean, was cloned and analysed [22]. In order to further study the physiological and biochemical processes that *GmNAC5* gene may be involved in, the soybean organ expression patterns of the gene and the relationship between *GmNAC5* gene and abiotic stress were examined.

2. Materials and Methods

2.1. Plant Materials.
Soybean cv. Ludou 10th was used in this study. Plants were field-grown under normal conditions in Nanjing Agricultural University. Vegetable tissues such as roots, stems, and leaves were collected from 4-week-old seedlings, while floral buds at R1 stage [23], young pods at R3 stage, and developing seeds from 15 to 50 days after flowering (DAF) were collected and frozen immediately in liquid nitrogen and stored at −80°C until use.

2.2. RNA Isolation, cDNA Synthesis, and Quantitative Real-Time PCR.
Total RNA was extracted using a Total RNA Plant Extraction Kit (Tiangen, Beijing, China), according to the manufacturer's protocol. First-strand cDNA was synthesized using the TaKaRa PrimeScript 1st strand cDNA Synthesis Kit (TaKaRa, Dalian, China), according to the manufacturer's instructions. Quantitative real-time polymerase chain reaction (qRT-PCR) was conducted using the SYBR Green Real-Time PCR Master Mix (TOYOBO, Osaka, Japan) on an ABI7500 Real-Time PCR System (Applied Biosystems, Foster City, CA, USA). Gene expression was quantified using the comparative method Ct: $2^{-\Delta\Delta Ct}$ method as previously described [24].

2.3. Stress Treatments.
The soybean seedlings cultured with sand were moved to Hoagland nutrient solution, when growing to two true leaves. After the first cluster of fronds grew, the plants were applied with stress treatments with three replicates. For hormone treatments, the seedlings were treated with 100 μM JA and 100 μM of ABA, respectively. For salt stress, the seedlings were treated with 200 mM NaCl. For dehydration stress, the seedlings were placed on filter paper, respectively. For cold stress, the seedlings were placed in 4°C light incubator. For mechanical wounding, the seedling leaves were cut into pieces with a sharp and clean scissor. After each treatment, the leaves were harvested and frozen in liquid nitrogen immediately.

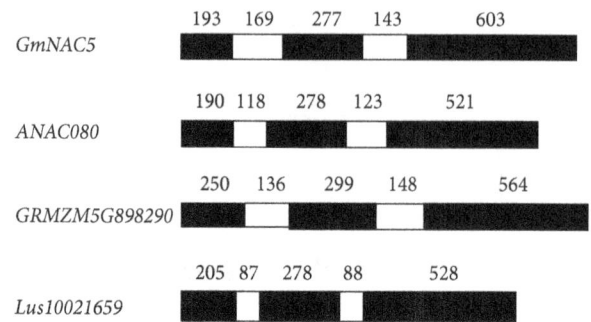

FIGURE 1: Schematic diagram of gene structures of *GmNAC5* and its homologous genes in *Arabidopsis thaliana* (*ANAC08*), *Zea mays* (*GRMZM5G898290*), and *Linum usitatissimum* (*Lus10021659*). The black and white boxes indicate exons and introns, respectively.

2.4. Subcellular Localization of NAC Proteins.
The full-length cDNA of *GmNAC5* was cloned in pBI121-GFP vector, in frame fusing with GFP reporter gene and producing the plasmid pBI-GmNAC5-GFP. After transient expression of the fusion plasmid in onion epidermal cells, the cells were observed under florescence microscope.

3. Results

3.1. Genomic Structure of GmNAC5.
NAC transcription factors have been considered one of the largest families of transcription factors so far discovered in the plant genomes. *GmNAC5* encodes a NAC transcription factor belonging to the NAM subfamily. It was found that the exon-intron structures were conserved among *GmNAC5* homologous genes in three common species, including *Arabidopsis thaliana*, *Zea mays*, and *Linum usitatissimum* (Figure 1).

3.2. Subcellular Localization.
GmNAC5 encoding product is presumed to act as a transcription factor. If transcription factors achieve the precise adjustment of the target genes, this specific transcription factor should be located in the nucleus. Interestingly, GmNAC5 lacks the traditional nuclear localization signal (NLS); even some researchers have found that some NAC domain proteins have the nuclear localization signals [16, 25, 26]. To clarify whether soybean NAC protein GmNAC5 is located in the nucleus, the subcellular localization assay was performed (Figure 2). Despite the transient expression in the onion epidermal cells, it was observed that the GmNAC5-GFP fusion protein was located predominantly in the nucleus whereas GFP alone was localized throughout the cells (Figure 2(b)).

3.3. Tissue-Specific Expression of GmNAC5.
In order to analyze the physiological and biochemical processes that *GmNAC5* gene may involve, qRT-PCR approach was used to analyze *GmNAC5* gene expression in soybean in different tissues and organs. *GmNAC5* was mainly expressed in the roots and seeds in soybean development and weakly expressed in the other organs (Figure 3). *GmNAC5* has the lowest expression level in the stems, but the highest expression level

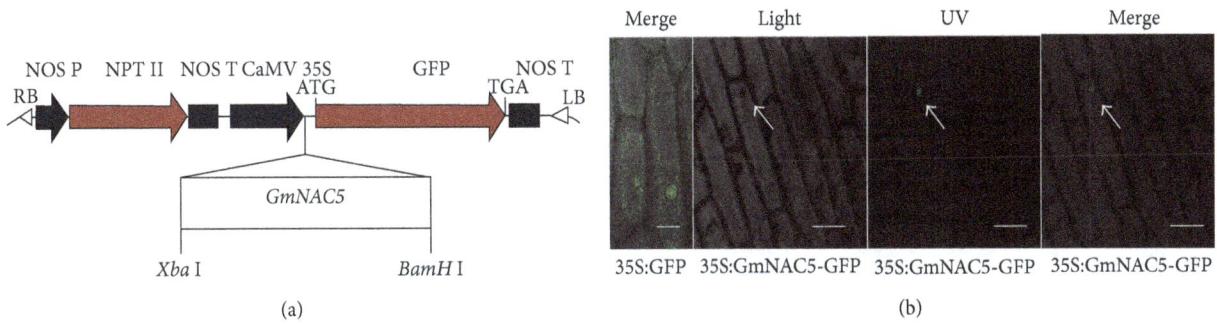

FIGURE 2: Subcellular localization of GmNAC5. (a) The structure of 35S:GmNAC5-GFP vector. (b) Subcellular localization of GmNAC5-GFP fusion protein. The arrow indicates the location of the nucleus. Bars: 40 μm in 35S:GFP; 80 μm in 35S:GmNAC5-GFP.

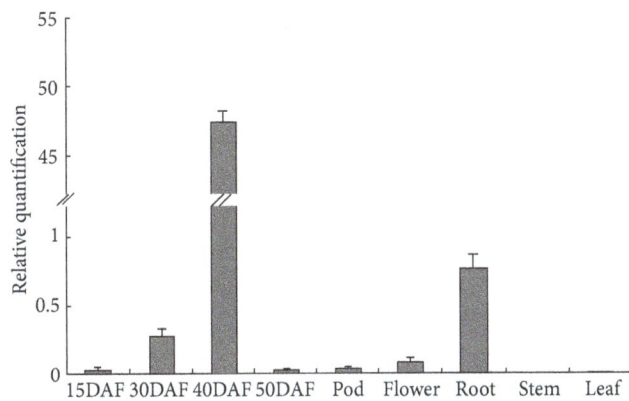

FIGURE 3: Real-time RT-PCR analysis of *GmNAC5* expression in various soybean tissues. DAF: days after flowering.

in soybean seeds of 40 days after flowering. The difference in *GmNAC5* expression level of each period in soybean seed development is obvious. The highest expression level was found 40 days after flowering (DAF), but only weak expression in the seeds of 15 days and 50 days after flowering, which indicates that the *GmNAC5* may participate in the middle stage of soybean seed development. We found that *GmNAC5* has strong expression in roots, but expression levels in stems, leaves, and pods are weak.

3.4. Expression of GmNAC5 in Soybean under Various Stresses. *GmNAC5* was weakly expressed in leaves in soybean under normal growth condition. The real-time qRT-PCR was performed to detect the expression of *GmNAC5* in soybean under various stresses (Figure 4). For jasmonic acid treatment, *GmNAC5* was significantly induced after 3 h of JA treatment (Figure 4(a)). For mechanical wounding, expression of *GmNAC5* was sharply induced after 1 h of treatment (Figure 4(b)). For NaCl treatment, *GmNAC5* expression was markedly upregulated by 8-fold after 3 h of treatment and then decreased (Figure 4(c)). Under drought treatment, expression of *GmNAC5* showed a weak increase and then declined (Figure 4(d)). For cold stress, it was found that *GmNAC5* expression was gradually increased and reached the maximum after 12 h of treatment (Figure 4(e)). In order to

reveal whether stress responsive expression of *GmNAC5* was involved in ABA pathway, we studied expression of *GmNAC5* under ABA treatment (Figure 4(f)). The qRT-PCR assay suggested that expression of *GmNAC5* was not markedly affected by ABA, suggesting that GmNAC5 may participate in ABA-independent signaling pathway in soybean under abiotic stresses.

4. Concluding Remarks

It has been documented that the plant-specific NAC (for NAM, ATAF1, 2, and CUC2) transcription factors play an important role in plant development and stress responses [27]. GmNAC5 belongs to the NAM subgroup and is most closely related to CUC1, CUC2, and NAM, which are involved in developmental events, maintenance of shoot meristem, and cotyledons separation [28]. In this study, we observed some new clues involved in the functions of GmNAC5. Tissue-specific expression analysis indicated that *GmNAC5* was highly expressed in immature seeds at 40 DAF and in the roots, suggesting the involvements of GmNAC5 in seed development and root development. It was also found that transcripts of *Arabidopsis* AtNAC2 were accumulated at the late stages of seed development [29].

It is also possible that higher expression of *GmNAC5* in soybean roots is associated with abiotic stress responses. *Arabidopsis* AtNAC2 expression was highly in roots and induced by salt stress [21]. Further studies suggested that AtNAC2 functioned downstream of ethylene and auxin signaling pathways and regulated lateral root development under salt stress. Expression of *GmNAC5* was significantly induced by multiple abiotic stresses but not by ABA, suggesting that GmNAC5 may be involved in ABA-independent stress responses in soybean under abiotic stresses. It was previously reported that NAC transcription factor involves the control of plant senescence and transient expression of *GmNAC5* in tobacco leaves induced senescence and necrosis, suggesting that GmNAC5 may play a role in the regulation of stress promoted senescence. Through microarray analysis, it was found that *Arabidopsis* AtNAC2 regulated many senescence-related genes and the majority of them are also regulated by salt stress, a major promoter of plant senescence [29]. Whether GmNAC5 plays a regulatory role in stress regulated

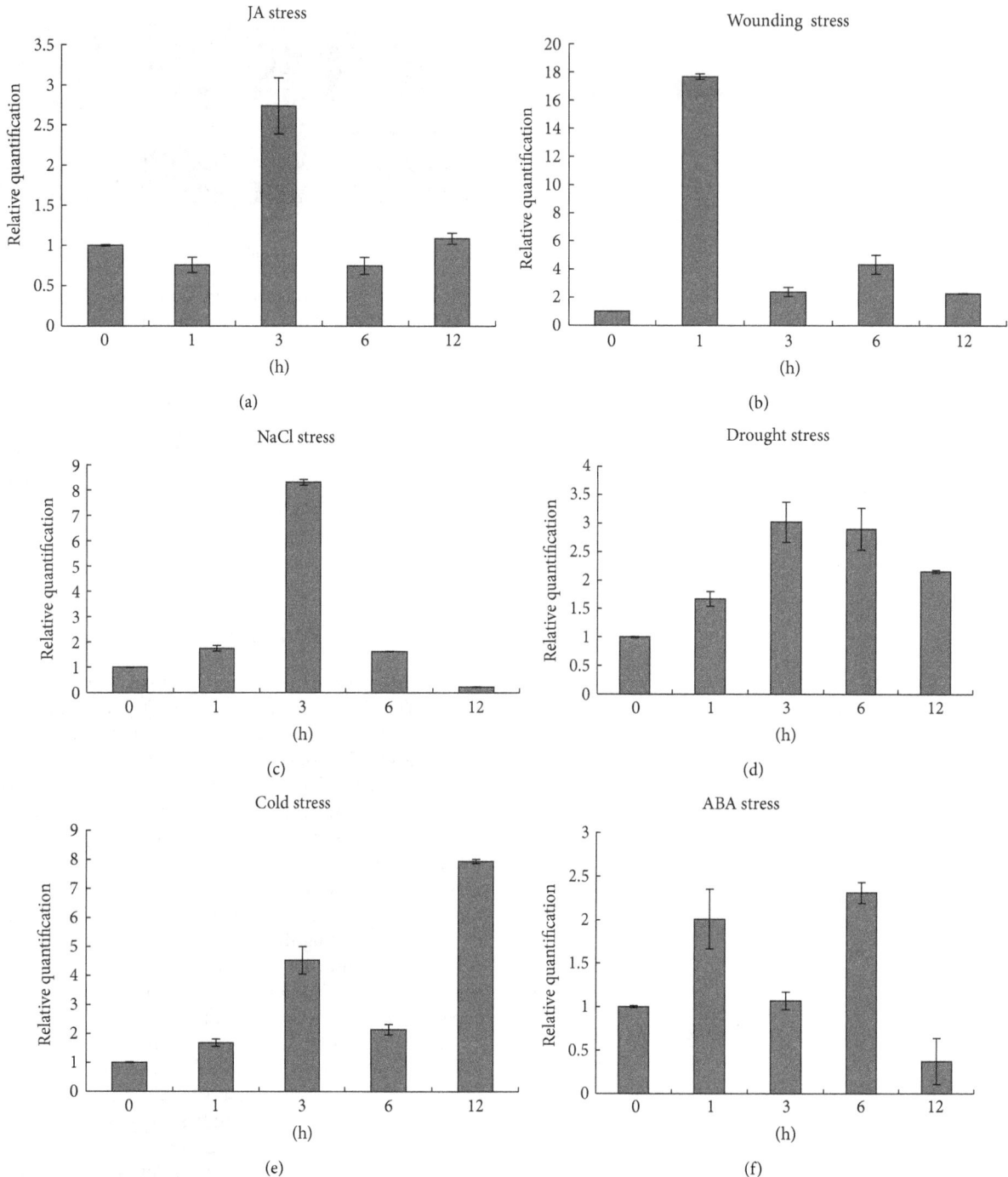

FIGURE 4: Expression of *GmNAC5* in soybean seedlings under various stresses. The soybean seedlings were stressed with 100 μM JA (a), wounding (b), 200 mM NaCl (c), drought (d), 4°C (e), and 100 μM ABA (f).

root development or stress promoted senescence still needs to be further analyzed.

Conflict of Interests

The authors declare no conflict of interests. The authors do not have a direct financial relation with the commercial identity mentioned in the current paper.

Authors' Contribution

Hangxia Jin and Guangli Xu contributed equally to this work.

Acknowledgments

This work was supported in part by the National Basic Research Program of China (973 Program) (2010CB125906) and the National Natural Science Foundation of China (30800692, 31171573, and 31201230).

References

[1] X. Mao, H. Zhang, X. Qian, A. Li, G. Zhao, and R. Jing, "TaNAC2, a NAC-type wheat transcription factor conferring enhanced multiple abiotic stress tolerances in Arabidopsis," *Journal of Experimental Botany*, vol. 63, no. 8, pp. 2933–2946, 2012.

[2] Q. Xie, G. Frugis, D. Colgan, and N.-H. Chua, "Arabidopsis NAC1 transduces auxin signal downstream of TIR1 to promote lateral root development," *Genes and Development*, vol. 14, no. 23, pp. 3024–3036, 2000.

[3] S. Takada, K.-I. Hibara, T. Ishida, and M. Tasaka, "The CUP-SHAPED *COTYLEDON1* gene of Arabidopsis regulates shoot apical meristem formation," *Development*, vol. 128, no. 7, pp. 1127–1135, 2001.

[4] M. Duval, T.-F. Hsieh, S. Y. Kim, and T. L. Thomas, "Molecular characterization of AtNAM: a member of the Arabidopsis NAC domain superfamily," *Plant Molecular Biology*, vol. 50, no. 2, pp. 237–248, 2002.

[5] E. Souer, A. Van Houwelingen, D. Kloos, J. Mol, and R. Koes, "The no apical Meristem gene of petunia is required for pattern formation in embryos and flowers and is expressed at meristem and primordia boundaries," *Cell*, vol. 85, no. 2, pp. 159–170, 1996.

[6] M. Fujita, Y. Fujita, K. Maruyama et al., "A dehydration-induced NAC protein, RD26, is involved in a novel ABA-dependent stress-signaling pathway," *Plant Journal*, vol. 39, no. 6, pp. 863–876, 2004.

[7] R. W. M. Sablowski and E. M. Meyerowitz, "A homolog of NO APICAL MERISTEM is an immediate target of the floral homeotic genes *APETALA3/PISTILLATA*," *Cell*, vol. 92, no. 1, pp. 93–103, 1998.

[8] A. Iwase, A. Hideno, K. Watanabe, N. Mitsuda, and M. Ohme-Takagi, "A chimeric NST repressor has the potential to improve glucose productivity from plant cell walls," *Journal of Biotechnology*, vol. 142, no. 3-4, pp. 279–284, 2009.

[9] N. Mitsuda, A. Iwase, H. Yamamoto et al., "NAC transcription factors, NST1 and NST3, are key regulators of the formation of secondary walls in woody tissues of Arabidopsis," *Plant Cell*, vol. 19, no. 1, pp. 270–280, 2007.

[10] N. Mitsuda and M. Ohme-Takagi, "NAC transcription factors NST1 and NST3 regulate pod shattering in a partially redundant manner by promoting secondary wall formation after the establishment of tissue identity," *Plant Journal*, vol. 56, no. 5, pp. 768–778, 2008.

[11] F. T. S. Nogueira, P. S. Schlögl, S. R. Camargo et al., "SsNAC23, a member of the NAC domain protein family, is associated with cold, herbivory and water stress in sugarcane," *Plant Science*, vol. 169, no. 1, pp. 93–106, 2005.

[12] T. Ohnishi, S. Sugahara, T. Yamada et al., "OsNAC6, a member of the NAC gene family, is induced by various stresses in rice," *Genes and Genetic Systems*, vol. 80, no. 2, pp. 135–139, 2005.

[13] H. Hu, M. Dai, J. Yao et al., "Overexpressing a NAM, ATAF, and CUC (NAC) transcription factor enhances drought resistance and salt tolerance in rice," *Proceedings of the National Academy of Sciences of the United States of America*, vol. 103, no. 35, pp. 12987–12992, 2006.

[14] X. Han, G. He, S. Zhao, C. Guo, and M. Lu, "Expression analysis of two NAC transcription factors *PtNAC068* and *PtNAC154* from poplar," *Plant Molecular Biology Reporter*, vol. 30, no. 2, pp. 370–378, 2012.

[15] H. Ooka, K. Satoh, K. Doi et al., "Comprehensive analysis of NAC family genes in *Oryza sativa* and *Arabidopsis thaliana*," *DNA Research*, vol. 10, no. 6, pp. 239–247, 2003.

[16] K.-I. Taoka, Y. Yanagimoto, Y. Daimon, K.-I. Hibara, M. Aida, and M. Tasaka, "The NAC domain mediates functional specificity of CUP-SHAPED COTYLEDON proteins," *Plant Journal*, vol. 40, no. 4, pp. 462–473, 2004.

[17] K.-I. Hibara, S. Takada, and M. Tasaka, "CUC1 gene activates the expression of SAM-related genes to induce adventitious shoot formation," *Plant Journal*, vol. 36, no. 5, pp. 687–696, 2003.

[18] A. Hasson, A. Plessis, T. Blein et al., "Evolution and diverse roles of the CUP-SHAPED COTYLEDON genes in Arabidopsis leaf development," *Plant Cell*, vol. 23, no. 1, pp. 54–68, 2011.

[19] I. Weir, J. Lu, H. Cook, B. Causier, Z. Schwarz-Sommer, and B. Davies, "Cupuliformis establishes lateral organ boundaries in *Antirrhinum*," *Development*, vol. 131, no. 4, pp. 915–922, 2004.

[20] R. Ruiz-Medrano, B. Xoconostle-Cázares, and W. J. Lucas, "Phloem long-distance transport of CmNACP mRNA: implications for supracellular regulation in plants," *Development*, vol. 126, no. 20, pp. 4405–4419, 1999.

[21] X.-J. He, R.-L. Mu, W.-H. Cao, Z.-G. Zhang, J.-S. Zhang, and S.-Y. Chen, "AtNAC2, a transcription factor downstream of ethylene and auxin signaling pathways, is involved in salt stress response and lateral root development," *Plant Journal*, vol. 44, no. 6, pp. 903–916, 2005.

[22] Q. Meng, C. Zhang, J. Gai, and D. Yu, "Molecular cloning, sequence characterization and tissue-specific expression of six NAC-like genes in soybean (*Glycine max* (L.) Merr.)," *Journal of Plant Physiology*, vol. 164, no. 8, pp. 1002–1012, 2007.

[23] W. R. Fehr, C. E. Caviness, D. T. Burmood, and J. S. Pennington, "Stage of development descriptions for soybeans, *Glycine max* (L.) Merrill," *Crop Science*, vol. 11, no. 6, pp. 929–931, 1971.

[24] F. Huang, Y. Chi, J. Gai, and D. Yu, "Identification of transcription factors predominantly expressed in soybean flowers and characterization of *GmSEP1* encoding a SEPALLATA1-like protein," *Gene*, vol. 438, no. 1-2, pp. 40–48, 2009.

[25] H. A. Ernst, A. N. Olsen, K. Skriver, S. Larsen, and L. Lo Leggio, "Structure of the conserved domain of ANAC, a member of the NAC family of transcription factors," *EMBO Reports*, vol. 5, no. 3, pp. 297–303, 2004.

[26] L.-S. P. Tran, K. Nakashima, Y. Sakuma et al., "Isolation and functional analysis of arabidopsis stress-inducible NAC transcription factors that bind to a drought-responsive *cis*-element in the early responsive to dehydration stress 1 promoter," *Plant Cell*, vol. 16, no. 9, pp. 2481–2498, 2004.

[27] H. Peng, H.-Y. Cheng, X.-W. Yu et al., "Characterization of a chickpea (*Cicer arietinum* L.) NAC family gene, CarNAC5, which is both developmentally- and stress-regulated," *Plant Physiology and Biochemistry*, vol. 47, no. 11-12, pp. 1037–1045, 2009.

[28] A. N. Olsen, H. A. Ernst, L. L. Leggio, and K. Skriver, "NAC transcription factors: structurally distinct, functionally diverse," *Trends in Plant Science*, vol. 10, no. 2, pp. 79–87, 2005.

[29] S. Balazadeh, H. Siddiqui, A. D. Allu et al., "A gene regulatory network controlled by the NAC transcription factor ANAC092/AtNAC2/ORE1 during salt-promoted senescence," *Plant Journal*, vol. 62, no. 2, pp. 250–264, 2010.

11

Developmental Principles: Fact or Fiction

A. J. Durston

Sylvius Laboratory, Institute of Biology, University of Leiden, Wassenaarseweg 72, 2333 BE Leiden, The Netherlands

Correspondence should be addressed to A. J. Durston, a.j.durston@biology.leidenuniv.nl

Academic Editor: David Tannahill

While still at school, most of us are deeply impressed by the underlying principles that so beautifully explain why the chemical elements are ordered as they are in the periodic table, and may wonder, with the theoretician Brian Goodwin, "*whether there might be equally powerful principles that account for the awe-inspiring diversity of body forms in the living realm*". We have considered the arguments for developmental principles, conclude that they do exist and have specifically identified features that may generate principles associated with Hox patterning of the main body axis in bilaterian metazoa in general and in the vertebrates in particular. We wonder whether this exercise serves any purpose. The features we discuss were already known to us as parts of developmental mechanisms and defining developmental principles (how, and at which level?) adds no insight. We also see little profit in the proposal by Goodwin that there are principles outside the emerging genetic mechanisms that need to be taken into account. The emerging developmental genetic hierarchies already reveal a wealth of interesting phenomena, whatever we choose to call them.

1. The Basic Idea

While still at school, most of us are deeply impressed by the underlying principles that so beautifully explain why the chemical elements are ordered as they are in the periodic table and may wonder, with the theoretician Brian Goodwin [1], "whether there might be equally powerful principles that account for the awe-inspiring diversity of body forms in the living realm." In fact, the question of how an organism acquires its structure and form during embryogenesis is one of the most intriguing and challenging questions in science. It is now becoming clear that there are indeed developmental principles. These principles define the developmental constraints that limit the life forms that can evolve. These constraints operate above and beyond the constraints imposed by Darwinian natural selection.

Many people have debated this point, for example, "Developmental constraints (defined as biases on the production of variant phenotypes or limitations on phenotypic variability caused by the structure, character, composition, or dynamics of the developmental system) undoubtedly play a significant role in evolution" [2], and "Darwin's theory of evolution by natural selection focuses on inheritance and survival without attempting to explain the forms organisms

take. The first part of Form and Transformation looks critically at the conceptual structure of Darwinism and describes the limitations of the theory of evolution. A theory of biological form is needed to understand the structure of organisms and their transformations." Forward to [3]. Not all agree: "we find ourselves perched on one tiny twig in the midst of a blossoming and flourishing tree of life and it is no accident, but the direct consequence of evolution by non-random selection" [4].

In fact, it is obvious that developmental principles or developmental constraints apply. The form of living organisms is generally predictable and limited. Animals cannot evolve to just anything. No metazoan, no matter how fast its movement, has yet evolved wheels. Pigs cannot fly.

2. Convergent Evolution

Another indicator about developmental constraints is the phenomenon of convergent evolution. Organisms in very different taxa can apparently independently evolve very similar structures, possibly suggesting that general principles apply. The best known example of this is the apparently independent evolution of the "camera eye" in vertebrates and cephalopod molluscs, while less complex members of

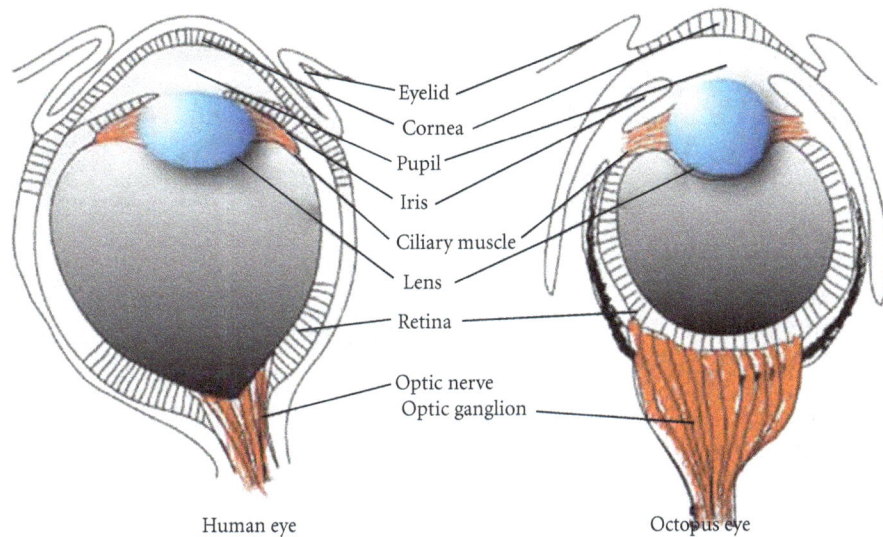

FIGURE 1: Human and octopus camera eyes have very similar morphology. Essentially the whole gene set needed to make this complex structure was already selected in the common ancestor of the bilaterians, long before camera eyes appeared. This poses a problem for "convergent evolution". See Ogura et al. [14].

the taxa concerned (nonvertebrate chordates and non-cephalopod molluscs lack this complex structure and have simpler eyes). There are many other examples of convergent evolution, including a remarkable number of parallels between cephalopods and vertebrates [7]. There are three obvious possible explanations for convergent evolution.

(1) The convergence is illusory. The gene set concerned is ancestral and evolved in a common ancestor but remained partly unused and cryptic until evolutionary manifestation of the phenotype. At first sight, this seems unlikely because it apparently requires evolution without natural selection—but the gene set concerned may have evolved for a different purpose. See below.

(2) The same mechanism evolved independently in two different taxa. This is inherently unlikely and is not considered in the literature.

(3) Two different mechanisms evolved in two different taxa but they deliver the same or a very similar phenotype. This is genuine convergent evolution.

The necessary information to distinguish between these alternatives is not available for most examples of convergent evolution although alternative 3 is generally the null hypothesis. It is known, however, for eye development that the basic machinery for making an eye is universal for the metazoa [8–10]. This includes a cascade downstream of Pax6 and genes for photoresponsive proteins that were probably originally imported into metazoan cells from synergistic photosynthetic bacteria. It has generally been assumed that convergent evolution of the complex camera eye in vertebrates and cephalopods is due to alternative 3 (true convergent evolution). We now thus know that the basic mechanism for making an eye is ancestral and common

to both taxa (alternative 1). The question arises whether the advanced special features of the camera eye evolved independently in vertebrates and cephalopods. Till recently, the consensus was yes [11–13]. However, a very interesting recent study by Ogura et al. [14] has challenged this view. These authors used bioinformatics to show that 729 out of 1052 (69.3%) of genes studied that were expressed in the octopus eye were also expressed in the human eye. In contrast, the expression similarity between human and octopus connective tissue was rather low. Ogura et al. also examined the availability of these eye genes in the sequenced genomes of a variety of other bilateria. The findings show that 1019 out of the 1052 genes (i.e., essentially all, presumably including the genes specifically needed for "camera" properties) had already existed (and were thus selected) in the common ancestor of bilateria, long before camera eyes appeared (Figure 1). This shows that there is a single ancestral gene set for the vertebrate and cephalopod camera eye. These findings possibly argue for alternative 1 and an ancestral mechanism as the driving force for this apparent example of convergent evolution. Obviously, the case is not complete. More needs to be done in studying different developmental stages in eye development and in understanding the developmental mechanisms involved, particularly those causing the known differences between vertebrate and cephalopod eyes. It is possible that while the relevant genes are conserved, the secret lies in their regulation and usage. We note that lens crystallins: typical "camera eye" genes, are a diverse group of proteins, some of which are either known to have other functions than in the eye lens or strongly resemble proteins that do [15].

In conclusion, genes expressed in the octopus eye belong to a gene set that was already selected in the ancestor of the first bilaterians. Some of these early selected genes were presumably concerned with specifying an ancestral eye.

Others, that specify the advanced features of the octopus camera eye, were not selected for this function because camera eyes were not yet available in early bilateria. These must be dual-purpose genes that were originally selected for functions that were operational in primitive bilateria and were secondarily recruited later for camera eye development. The camera eye lens crystallins appear to have exactly such a dual nature, suggesting that this view is correct. The lens crystallins in vertebrates and cephalopods are also generally different, indicating secondary recruitment of different members of a larger gene set in these two versions of the camera eye. In contrast, some camera eye genes are identical in cephalopods and vertebrates [16]. These findings indicate a complex mechanism for this example of "convergent evolution", with elements of explanations 1 and 3.

3. What Are Developmental Principles?

The nature of developmental principles has attracted discussion. "Do genes explain life? Can advances in evolutionary and molecular biology account for what we look like, how we behave, and why we die? In this powerful intervention into current biological thinking, Brian Goodwin argues that such genetic reductionism has important limits. Drawing on the sciences of complexity, the author shows how an understanding of the self-organizing patterns of networks is necessary for making sense of nature. Genes are important, but only as part of a process constrained by environment, physical laws, and the universal tendencies of complex adaptive systems. In a new preface for this edition, Goodwin reflects on the advances in both genetics and the sciences of complexity since the book's original publication" [17]. "While neo-Darwinism has considerable explanatory power, it is widely recognized as lacking a component dealing with individual development, or ontogeny. This lack is particularly conspicuous when attempting to explain the evolutionary origin of the thirty-five or so animal body plans, and of the developmental trajectories that generate them. This significant work examines both the origin of body plans in particular and the evolution of animal development in general. Wallace Arthur ranges widely in his treatment, covering topics as diverse as comparative developmental genetics, selection theory, and Vendian/Cambrian fossils. He places particular emphasis on gene duplication, changes in spatio-temporal gene-expression patterns, internal selection, coevolution of interacting genes, and coadaptation. The book will be of particular interest to students and researchers in evolutionary biology, genetics, paleontology, and developmental biology" [18]. In fact, although Goodwin and others are presumably right that dynamical stability, self-organisation, and physical laws are important and although this is a valid point of view, it is predictable that many developmental principles will be inherent in the developmental genetic machinery, that is, in the gene hierarchies that mediate development. Wallace Arthur is right. The theoretical approach to developmental principles can also, because of its lack of specific concrete detail, sometimes take on almost a religious tint, a fact that fails to inspire confidence and a point of view that evoked a strong response from

the Darwinists. "The human eye is so complex and works so precisely that surely, one might believe, its current shape and function must be the product of design. How could such an intricate object have come about by chance? Tackling this subject—in writing that the New York Times called "a masterpiece"—Richard Dawkins builds a carefully reasoned and lovingly illustrated argument for evolutionary adaptation as the mechanism for life on earth". See forward to [19].

4. An Example: The Hox Genes

An example of the evolution of a molecular mechanism and a possible source of some developmental principles is supplied by the well-studied Hox genes. What we see here is clusters of closely related genes for transcription factors that together mediate part of a developmental function. That function is patterning an embryonic axis and in all bilaterian metazoa, either one cluster of 10–13 Hox genes patterns part or all of the main body axis or 4 or 8 similar clusters act in parallel to do the same (in the vertebrates). A whole Hox cluster thus acts as a functional unit or metagene. No one Hox gene can pattern an embryonic axis but the genes in one cluster, acting together, can function in this [20, 21]. An important property that enables this is collinearity. The Hox genes in a cluster are expressed and act sequentially, from 3′ to 5′ to specify sequential levels in the body. This is called spatial collinearity and is evident in all bilateria. Hox clusters are thought to have evolved by tandem duplication of an ancestral ur-Hox gene and by sequential evolutionary modification of the duplicated genes. The individual Hox genes at the same homologous position in different clusters and different organisms thus have conserved properties. It is for example, possible to replace a *Drosophila*-Hox gene with a human Hox gene corresponding to the same cluster position. This functions correctly [22]. It is claimed in *Drosophila* that it is still possible to identify the present day Hox gene representing the evolutionary ground state-the ur-Hox gene [23].

Hox genes, their duplication, clustering, and spatial colinearity clearly provide universal developmental principles for axial patterning in the bilaterian metazoa. These seem to have evolved early. Among the bilaterian phyla, there are animals with clustered Hox genes and animals where the Hox cluster is in various stages of disintegration—from two pieces as in various *Drosophila* species [20] to scattered— "atomised" Hox genes, for example, in Oikopleura [24]. It is argued that clustering is ancestral and that scattered Hox genes have arisen by cluster disintegration. Interestingly, scattered Hox genes retain their spatial sequence of expression and action along the main body axis. The gene relationships that arose within the ancestral Hox cluster are thus preserved. No bilaterian metazoan has been detected with less than 10 Hox genes, so stages in the progress of Hox gene duplication and modification have not been preserved and this cannot be monitored.

There is a unique situation in the vertebrates. Here, Hox gene sizes, that are very large in invertebrates, have become small, so the Hox clusters are more compact. The

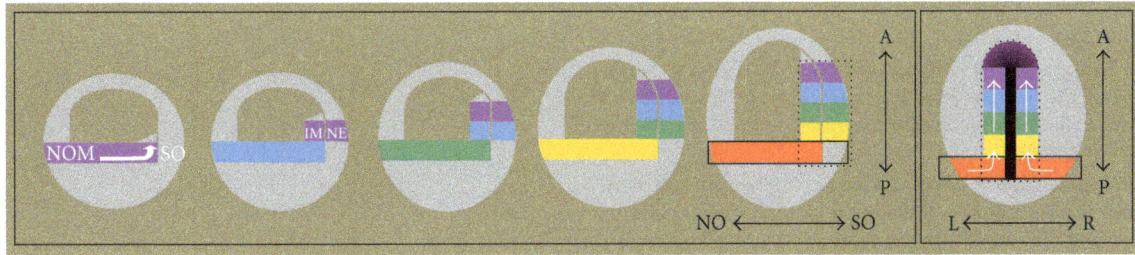

FIGURE 2: A developmental principle: time-space translation. The figure illustrates how temporally collinear Hox expression (in vertebrate gastrula mesoderm) is translated to a spatially collinear axial Hox pattern (in axial mesoderm and the neural plate). For a detailed explanation, see Durston et al [5, 6].

basic Hox cluster size has also been amplified from 10 to 13 genes, by amplification of the number of posterior AbdB orthologues (10–13). The genome has also been duplicated twice or more during vertebrate evolution, so most tetrapod vertebrates have 4 similar Hox clusters, following 2 genome duplications, and teleost fishes and a few other vertebrates have 8 Hox clusters, due to 3 genome duplications [20]. The multiple vertebrate Hox clusters are essentially copies of each other and share functions but each of the 4–8 copies misses different specific Hox genes. This is most extreme in the zebrafish Hox Db cluster, where all of the Hox genes are missing and only the Hox associated micro-RNA gene Mir10 persists [25].

The unique vertebrate Hox clusters are associated with unusual Hox collinearity principles. Vertebrate Hox genes unusually show temporal collinearity. For example, in early development, Hox genes are first expressed in a time sequence during gastrulation [26]. This time sequence antedates and is used to generate the spatially collinear axial pattern of Hox expression in the early embryo, by a time-space translation mechanism [5, 6] (Figure 2). temporal collinearity and time-space translation represent vertebrate developmental principles associated with Hox patterning of the main body axis. These are what Maynard Smith et al. [2] define as local developmental constraints, while spatial collinearity and Hox gene duplication-clustering are universal constraints.

5. Conclusions

Above, we have considered the arguments for developmental principles and have specifically identified some features that could generate principles associated with Hox patterning of the main body axis in bilaterian metazoa in general and in the vertebrates in particular. We wonder whether this exercise serves any purpose. The features we have discussed were known to us as parts of developmental mechanisms and defining developmental principles (how, and at which level?) adds no insight. We also see little profit in the proposal by B. C. Goodwin that there are principles outside the emerging genetic mechanisms that need to be taken into account. The emerging developmental genetic hierarchies already reveal a wealth of interesting phenomena, whatever we choose to call them.

References

[1] S. Harding, "Brian Goodwin: hugely influential and insightful biologist, philosopher and writer: obituary for BC Goodwin," *The Independent*, 2009.

[2] J. Maynard Smith et al., "Developmental constraints and evolution," *Quarterly Reviews of Biology*, vol. 60, no. 3, p. 265, 1985.

[3] G. Webster and B. Goodwin, *Form and Transformation: Generative and Relational Principles in Biology*, Cambridge University Press, 1996.

[4] R. Dawkins, *The Greatest Show on Earth*, Simon and Schuster, 2010.

[5] A. J. Durston, H. J. Jansen, and S. A. Wacker, "Review: time-space translation regulates trunk axial patterning in the early vertebrate embryo," *Genomics*, vol. 95, no. 5, pp. 250–255, 2010.

[6] A. J. Durston, H. J. Jansen, and S. A. Wacker, "Time-space translation: a developmental principle," *TheScientificWorldJournal*, vol. 10, pp. 2207–2214, 2010.

[7] S. Conway-Morris, "Consider the octopus," *EMBO Reports*, vol. 12, no. 3, p. 182, 2011.

[8] P. Callaerts, G. Halder, and W. J. Gehring, "Pax-6 in development and evolution," *Annual Review of Neuroscience*, vol. 20, pp. 483–532, 1997.

[9] M. Koyanagi, K. Takano, H. Tsukamoto, K. Ohtsu, F. Tokunaga, and A. Terakita, "Jellyfish vision starts with cAMP signaling mediated by opsin-Gs cascade," *Proceedings of the National Academy of Sciences of the United States of America*, vol. 105, no. 40, pp. 15576–15580, 2008.

[10] W. J. Gehring and K. Ikeo, "Pax 6: mastering eye morphogenesis and eye evolution," *Trends in Genetics*, vol. 15, no. 9, pp. 371–377, 1999.

[11] W. A. Harris, "Pax-6: where to be conserved is not conservative," *Proceedings of the National Academy of Sciences of the United States of America*, vol. 94, no. 6, pp. 2098–2100, 1997.

[12] D. J. Futuyma, *Evolutionary Biology*, Sinauer Associates, Sunderland, Mass, USA, 1997.

[13] R. C. Brusca and G. J. Brusca, *Invertebrates*, Sinauer Associates, Sunderland, Mass, USA, 1990.

[14] A. Ogura, K. Ikeo, and T. Gojobori, "Comparative analysis of gene expression for convergent evolution of camera eye

between octopus and human," *Genome Research*, vol. 14, no. 8, pp. 1555–1561, 2004.

[15] S. I. Tomarev, R. D. Zinovieva, and J. Piatigorsky, "Crystallins of the octopus lens. Recruitment from detoxification enzymes," *Journal of Biological Chemistry*, vol. 266, no. 35, pp. 24226–24231, 1991.

[16] M. A. Yoshida and A. Ogura, "Genetic mechanisms involved in the evolution of the cephalopod camera eye revealed by transcriptomic and developmental studies," *BMC Evolutionary Biology*, vol. 11, no. 1, article 180, 2011.

[17] B. C. Goodwin, *"How the Leopard Changed its Spots" The Evolution of Complexity*, Princeton University Press, 2001.

[18] W. Arthur, *The Origin of Animal Body Plans: A Study in Evolutionary Developmental Biology*, Cambridge University Press, 2000.

[19] R. Dawkins, *Climbing Mount Improbable*, W. W. Norton, 1997.

[20] D. Duboule, "The rise and fall of Hox gene clusters," *Development*, vol. 134, no. 14, pp. 2549–2560, 2007.

[21] G. Mainguy, J. Koster, J. Woltering, H. Jansen, and A. Durtson, "Extensive polycistronism and antisense transcription in the mammalian Hox cluster," *PLoS ONE*, vol. 2, no. 4, article e356, 2007.

[22] N. McGinnis, M. A. Kuziora, and W. McGinnis, "Human Hox-4.2 and Drosophila deformed encode similar regulatory specificities in Drosophila embryos and larvae," *Cell*, vol. 63, no. 5, pp. 969–976, 1990.

[23] W. J. Gehring, U. Kloter, and H. Suga, "Evolution of the Hox gene complex from an evolutionary ground state," *Current Topics in Developmental Biology*, vol. 88, pp. 35–61, 2009.

[24] H. C. Seo, R. B. Edvardsen, A. D. Maeland et al., "Hox cluster disintegration with persistent anteroposterior order of expression in *Oikopleura dioica*," *Nature*, vol. 430, no. 7004, pp. 67–71, 2004.

[25] J. M. Woltering and A. J. Durston, "The zebrafish hoxDb cluster has been reduced to a single microRNA," *Nature Genetics*, vol. 38, no. 6, pp. 601–602, 2006.

[26] S. A. Wacker, H. J. Jansen, C. L. McNulty, E. Houtzager, and A. J. Durston, "Timed interactions between the Hox expressing non-organiser mesoderm and the Spemann organiser generate positional information during vertebrate gastrulation," *Developmental Biology*, vol. 268, no. 1, pp. 207–219, 2004.

12

Differential Evolutionary Constraints in the Evolution of Chemoreceptors: A Murine and Human Case Study

Ricardo D'Oliveira Albanus,[1] **Rodrigo Juliani Siqueira Dalmolin,**[1]
José Luiz Rybarczyk-Filho,[2] **Mauro Antônio Alves Castro,**[1]
and José Cláudio Fonseca Moreira[1]

[1] *Departamento de Bioquímica, Universidade Federal do Rio Grande do Sul, Rua Ramiro Barcelos 2600,*
90040-180 Porto Alegre, RS, Brazil
[2] *Departamento de Física e Biofísica, Universidade Estadual Paulista, Distrito de Rubião Júnior, S/N, 18618-970 Botucatu, SP, Brazil*

Correspondence should be addressed to Ricardo D'Oliveira Albanus; ricardo.albanus@ufrgs.br

Academic Editors: M. Frank, B. Gantenbein-Ritter, B.-Y. Liao, and M. Shimoyama

Chemoreception is among the most important sensory modalities in animals. Organisms use the ability to perceive chemical compounds in all major ecological activities. Recent studies have allowed the characterization of chemoreceptor gene families. These genes present strikingly high variability in copy numbers and pseudogenization degrees among different species, but the mechanisms underlying their evolution are not fully understood. We have analyzed the functional networks of these genes, their orthologs distribution, and performed phylogenetic analyses in order to investigate their evolutionary dynamics. We have modeled the chemosensory networks and compared the evolutionary constraints of their genes in *Mus musculus*, *Homo sapiens*, and *Rattus norvegicus*. We have observed significant differences regarding the constraints on the orthologous groups and network topologies of chemoreceptors and signal transduction machinery. Our findings suggest that chemosensory receptor genes are less constrained than their signal transducing machinery, resulting in greater receptor diversity and conservation of information processing pathways. More importantly, we have observed significant differences among the receptors themselves, suggesting that olfactory and bitter taste receptors are more conserved than vomeronasal receptors.

1. Introduction

The ability to evaluate the environment has always been of vital importance to all organisms. In order to find food, detect dangers, and search for reproductive partners, a constant appraisal of the outside world must be made by any organism. Chemosensory reception is one such tool for this task, and it is present in all life forms. Over the last decade, several studies were conducted in order to characterize the different chemosensory receptors (CR) genes [1–4]. In vertebrates, they are coded by six major multigene families: the trace amine-associated receptors (TAAR) [5], the olfactory receptors (OR) [6], the type I and II vomeronasal receptors (V1R and V2R) [3, 4, 7], and type I and II taste receptors (T1R and T2R) [1, 2]. All proteins coded by these genes are G protein-coupled proteins [8].

Different from other environmental appraisal systems such as vision and hearing, which remained relatively stable once they were formed, chemosensory reception must be constantly tuned to an ever-changing environment of odors and toxins. This need for variability is reflected in the organization of the CR genes in the genome. In all studied species, it was found that these genes occur in great numbers, and there are considerable numbers of CR pseudogenes [4, 9–11], suggesting that they are prone to duplication and inactivation events. There are theories to explain the evolution of CR genes [9, 11–14], but several gaps regarding this subject still remain. For instance, there are no currently available data regarding the evolutionary dynamics of the chemosensory apparatus as a whole (i.e., the CR and its signal transducing machinery). Equally unclear are the differences in evolutionary dynamics among the CR families.

In this work, we have tackled the evolution of the mammalian CR gene families and their signal transducing machinery from a systems biology-oriented approach. We have analyzed the orthologs distribution of the chemosensory machinery, their functional networks topologies, and their phylogenetic diversity in *Mus musculus*, *Rattus norvegicus*, and *Homo sapiens*. We have found evidences that there are distinct evolutionary dynamics in the CR genes and the signal transducing apparatus. More importantly, we have observed significant differences among the CR gene families, suggesting distinct evolutionary dynamics for each receptor type.

2. Methods

2.1. Data Collection. In order to determine which receptors are involved in each sensory modality, we have gathered data from the Gene Ontology (GO) Consortium [15] regarding *Homo sapiens, Mus musculus,* and *Rattus norvegicus.* GO groups used were 0004984—MF Olfactory receptor activity, 0007608—BP Sensory perception of smell, 0008527—MF Taste receptor activity, 0050909—BP Sensory perception of taste, 0016503—MF Pheromone receptor activity, 0019236—BP Response to Pheromone. We have chosen these three species for our study due to robustness of their genomic/proteomic data available in databases. Studied genes were sorted in groups according to their receptor modality: olfactory receptors; taste receptors; and vomeronasal receptors. We made one further division of the GO taste group to separate taste receptors type 1 and 2 and study them separately because of their functional differences. TAAR genes were withdrawn from our analysis due to lack of data in the databases. Also due to lack of available data, we have combined the two vomeronasal families (V1R and V2R) and studied them as a single group (VN). We have sorted all genes in GO groups into two functional categories: the first consisted of genes coding the proteins directly involved in binding chemical stimuli (the chemosensory receptors *per se*), and the second consisted of the rest of the genes related to signal transduction machinery (STM).

Functional network parameters of proteins coded by CR genes were gathered using STRING database (String-DB), version 8.3 [16], using their corresponding ENSEMBL IDs. To assemble these IDs, a cross-search was performed between GO, String-DB, HUGO Gene Nomenclature Consortium [17], Mouse Genome Informatics [18], Rat Genome Database [19], and BioMart [20] databases. Genes that presented ID divergences among databases were manually curated or removed from our analysis. String-DB analyses were made with a 0.7 combined score and only interactions generated from experiments and databases were used. This is a medium to high stringency parameter.

2.2. Topology and Evolutionary Plasticity Analysis. Topologies of the receptors functional networks were analyzed by connectivity [$k(i)$] and clusterization [$c(i)$] indexes of their components. $k(i)$ index is calculated by the number of

neighbors that an i node has in a network, and $c(i)$ by the equation

$$c(i) = \frac{2n_i}{k_i(k_i-1)}, \qquad (1)$$

which represent general interactivity of i's neighbors, where n_i represents the number of their connections among each other. Evolutionary Plasticity Index (EPI) of the orthologous groups of these proteins was calculated by equation

$$EPI = 1 - \frac{H_\alpha}{\sqrt{D_\alpha}}, \qquad (2)$$

where H_α is the ortholog diversity in the eukaryotic tree, calculated using how many species the ortholog is found, and D_α is its abundance, calculated by the number of ortholog members found in each species [21]. Orthology data of these proteins was also gathered using String-DB. All statistical analyses were made using one-way ANOVA with Tukey's test. $k(i)$ and $c(i)$ indexes were compared by the Shannon diversity (S) of their distribution, using equation

$$S = -\sum p \ln p, \qquad (3)$$

where p is the probability of a value occurrence in any dataset. Entropy calculation was used in a complementary way in order to mathematically support or refute any observations in the connectivity and clusterization distribution behavior. In order to generate the graphical representations of the CR network, we have plotted String-DB interactions of all Gene Ontology groups proteins among each other using RedeR R package [22]. The list of all the genes analyzed in this work is presented in the Supplementary Material available online at http://dx.doi.org/10.1155/2014/696485.

2.3. Phylogenetic Analysis. Chemoreceptor genes sequences were gathered from the Chemosensory Receptor Database [23]. Alignments and trees were made with the MEGA 5.2 software [24], using, respectively, the Muscle alignment algorithm [25, 26] and the Tamura-Nei model [27]. Parameters used were the default for each algorithm. Branch reliability was calculated using bootstrap method. 100 bootstrap replications were performed for T1R, T2R, and VN and 50 replications for OR. For entropy analysis, we have calculated the Shannon diversity of the phylogenetic trees by subsetting each tree into n consecutives samples of w size, where n is the tree size and w is the maximum tree depth (i.e., number of levels). This was made to detect whether tree diversity was consistent throughout the entire tree radius. One-way ANOVA was used in order to compare these results. We have chosen w as the number of levels as a means for defining proportional windows for each tree.

3. Results

3.1. Differences between the Chemosensory Receptors and the Signal Transducing Machinery. We have calculated separately the Evolutionary Plasticity Index (EPI) [21] of the CR genes

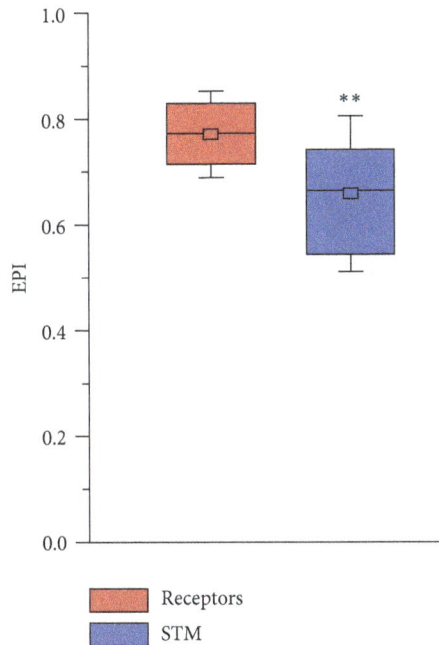

FIGURE 1: Mean EPI values of chemoreceptors (red) and signal transducing machinery (blue). The edges of the boxes indicate the upper and lower quartiles. The line at the center of each box indicates the median, the square represents the mean, and whiskers represent the standard deviation. Asterisks indicate statistically significant data ($P < 0.001$).

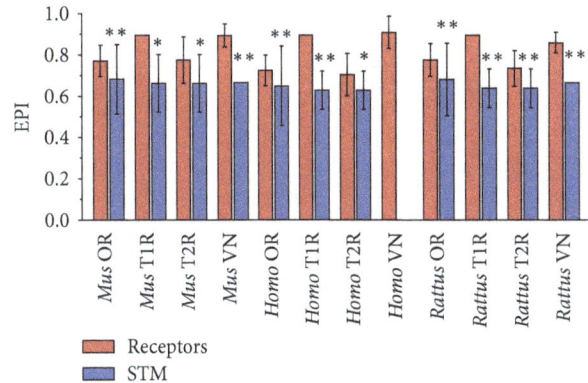

FIGURE 2: Mean EPI values of chemoreceptors families (red) and their respective signal transducing machineries (blue). Plasticity values are shown in the vertical axis and the different subgroups are listed on the horizontal axis. Whiskers represent the standard deviation. Statistically significant data are indicated by double ($P < 0.001$) and single ($P < 0.05$) asterisks.

and their signal transducing machinery (STM). We have observed that CR genes as a whole have significantly higher plasticity values than the STM (Figure 1), indicating that CR genes have a broader ortholog distribution than the STM, meaning that the latter was subject to less variation during the course of evolution. To further corroborate these findings, we have compared each CR family separately to its signal transducing machinery. We have found that, in all cases but one, the EPI of each CR family was significantly higher than its STM (Figure 2). The exception was the human vomeronasal (VN) genes, which lack their STM due to the loss of the TRPC2 channel [28, 29].

Next, we have compared the network topologies of each CR family and their STM. We have observed that most CR genes are functionally less connected than their STM. Most CR genes are connected only to their respective G proteins, indicating that they are located in the periphery of their functional networks (Figures 3 and 4). This assumption is further supported by analyzing the Shannon diversity of the connectivity and clusterization indexes. We have found that the STM has higher diversity values for these indexes ($P < 0.05$), suggesting that they occupy a broader range of niches in their network. Exceptions to this are some olfactory receptors, which presented higher connectivity and clusterization values among each other.

3.2. Differences among the Different Chemosensory Families. We have compared the EPI of the different CR families with themselves in order to identify differences in their orthologs

distribution. Due to lack of data regarding the V1R and V2R, we have considered these genes as a single group in our analysis (VN). Strikingly, we have observed that CR families can be sorted in two groups regarding their plasticity. The OR and T2R have significantly lower plasticity than the T1R and VN in the three mammals we have studied, indicating that they had evolved under different constraints in these species (Figure 5). To further assess these differences, we have reconstructed the phylogenetic relationships among each CR family. We have observed that the OR, T1, and T2 genes form branches preferentially with their orthologs in other species, whereas the VN genes branches with their inparalogs (Figure 6). These results are further supported by calculating the Shannon diversity index stepwise for each CR tree. We have found that the VN tree had significantly lower diversity values ($P < 10^{-16}$) than the other CR, suggesting that the VN genes are less conserved than the other CR. The original trees with bootstrap replications confidence values can be found in the Supplementary Material.

3.3. The Functional Organization of the CR Genes Network. Finally, we have reconstructed the CR genes network in order to visualize its functional organization. We can observe that even though they form completely separate functional clusters, all the CR families, with the exception of VN, share the same STM cluster (Figure 7). This indicates that the STM machinery is essentially the same in every CR cell type.

4. Discussion

Chemosensory perception is one of the most important systems for appraisal of the environment. It is of vital necessity to every organism that the chemical species detected by each chemoreceptor are tuned to tastes or odorants which bring meaningful information from the outside world. Unlike physical sensory modalities, whose stimuli nature is constant (e.g., light, sound), chemical perception may be

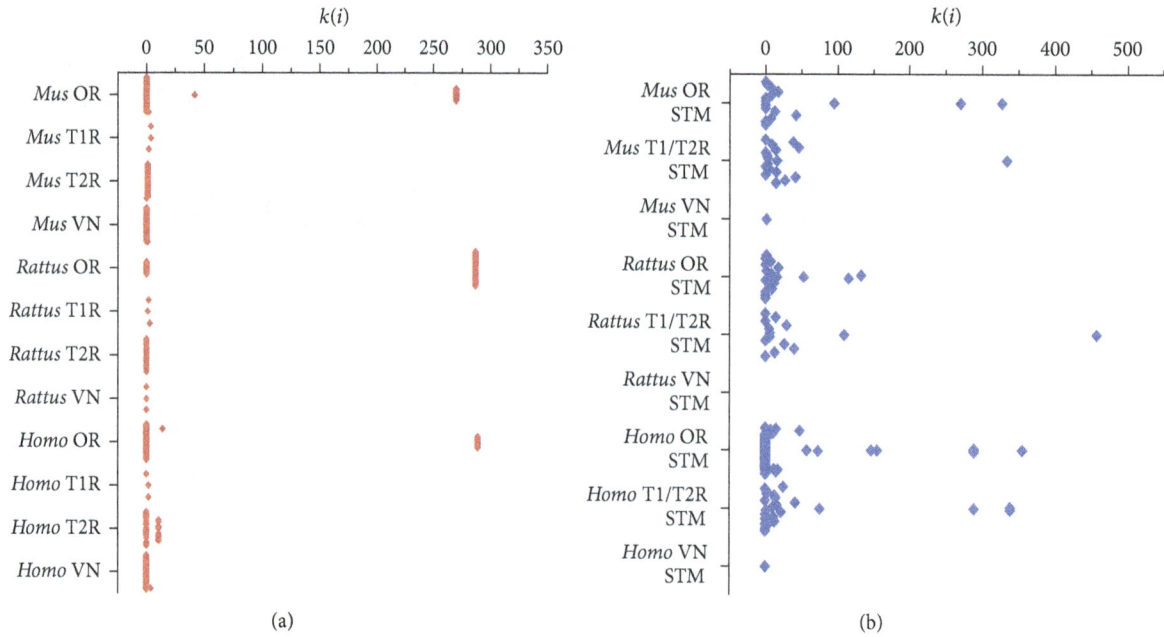

FIGURE 3: Connectivity values distribution for the chemoreceptors families (red) and their respective signal transducing machineries (blue). Values are shown in the vertical axis and the different subgroups are listed in the horizontal axis.

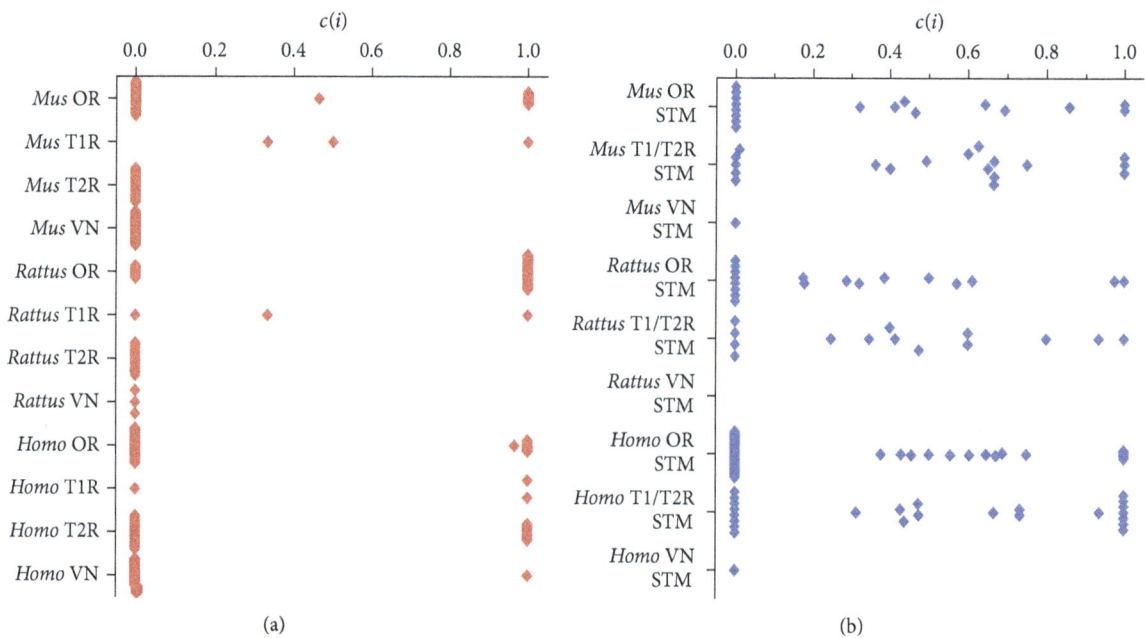

FIGURE 4: Clusterization values distribution for the chemoreceptors families (red) and their respective signal transducing machineries (blue). Values are shown in the vertical axis and the different subgroups are listed in the horizontal axis.

subject to radical changes in very short time windows. For instance, some plants are able to change their repertory of toxic secondary compounds in just a few generations [30], forcing herbivorous species that can potentially ingest these compounds to keep equally updated their ability for detecting these toxins. From an evolutionary point of view, this means that the genes coding these receptors must have a more relaxed behavior in order to accommodate novelties in the environment.

When comparing all CR genes to their STM, we have observed that CR have higher evolutionary plasticity values, suggesting that they were more subject to variation in the course of evolution than the STM. This indicates that the STM has remained relatively unchanged since its appearance,

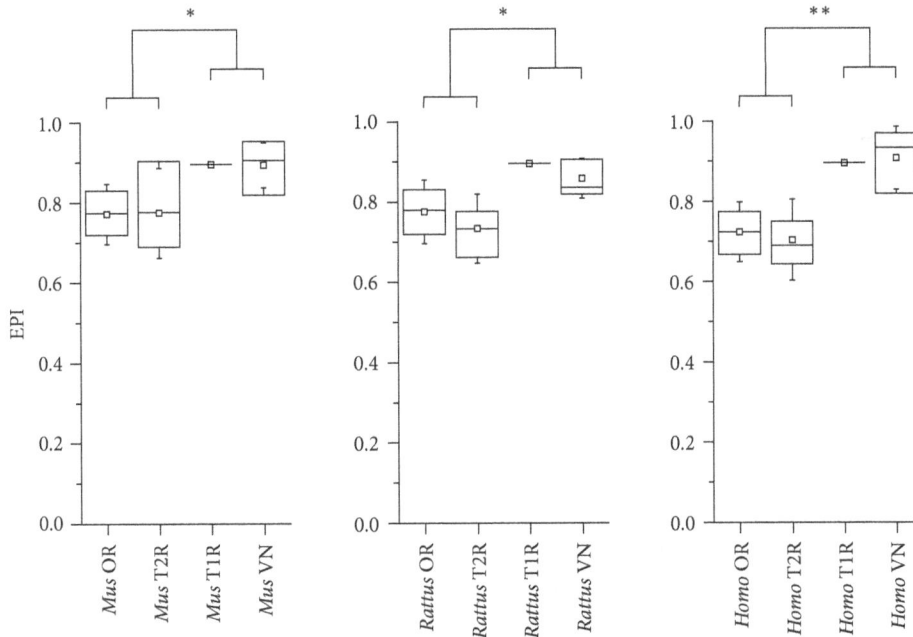

FIGURE 5: Mean EPI values of chemoreceptors families. Plasticity values are shown in the vertical axis and the different families are listed on the horizontal axis. The line at the center of each box indicates the median, the square represents the mean, and whiskers represent the standard deviation. Statistically significant data are indicated by double ($P < 0.001$) and single ($P < 0.05$) asterisks.

while the receptors themselves were free to experiment with the environment. By analyzing the network topology of the CR and STM, we have observed that CR occupy a peripheral position in their functional network. It has been proposed that proteins located in the periphery of their respective functional networks have elevated propensity to duplicate and undergo positive selection [31, 32]. This happens because poorly connected and loose clusters are able to more efficiently accommodate evolutionary novelties such as gene duplications, deletions, and changes of function, and thus they become the "evolutionary motors" of their biological networks [33–35]. D'Antonio and Ciccarelli have recently demonstrated evidences supporting this assertion [36]. In their paper, these authors have thoroughly analyzed network properties, sequences, and orthology data from *E. coli*, yeast, fly, and human. They observed that genes acquired during evolution encode less connected and less central proteins that are subject to more duplication events. Conversely, it has been observed in other types of signal transducing cascades that the receptors are more constrained than the intermediate elements of their networks [37–40]. These studies, however, were made with pathways such as those of insulin/TOR, which integrate information from inside the organism. As corporeal composition remained relatively the same throughout vertebrate evolution, intra- and extracellular components are not subject to radical variation, making necessary that internal signal transducing cascades must be more tightly constrained in order to consistently maintain their behavior. The environment, however, is constantly subject to changes, and the chemoreceptors cannot be too tightly constrained in

order to accommodate these fluctuations. Our data support that CR are a special case of signal transducing pathways that have loosely constrained receptors.

Our subsequent insight into CR evolution was made when comparing the receptors with themselves. We have observed striking evidences suggesting that the vomeronasal receptors are less constrained than the other CR families. First, their plasticity is significantly higher than the other CR, suggesting that this gene family was probably more subject to duplications and deletions than the other receptors. This is further supported by their phylogenetic tree, which is grouped by inparalogs rather than orthologs, suggesting that these genes have arisen from recent duplications and, therefore, are probably less constrained. This finding is similar to what Grus and Zhang observed when studying the dynamics of vomeronasal and olfactory receptors in vertebrate species [41]. Lastly, we have observed that their functional network is completely detached from the other receptors, making them the most peripheral CR. From an evolutionary point of view, one would be tempted to think that the VN code the least important CR in terms of individual survival. The T2R are responsible for detection of bitter tastes. In general, these tastes are typically associated with toxic nitrogenated compounds, such as alkaloids and amines [42]. The perception of these toxins is a major issue in the survival of any organism that has chances of ingesting them. Equally important to their survival is the detection of food, predators, and members of the same species by OR. Conversely, the VN genes likely give clues about potential reproductive partners by detecting genetic likeness and even

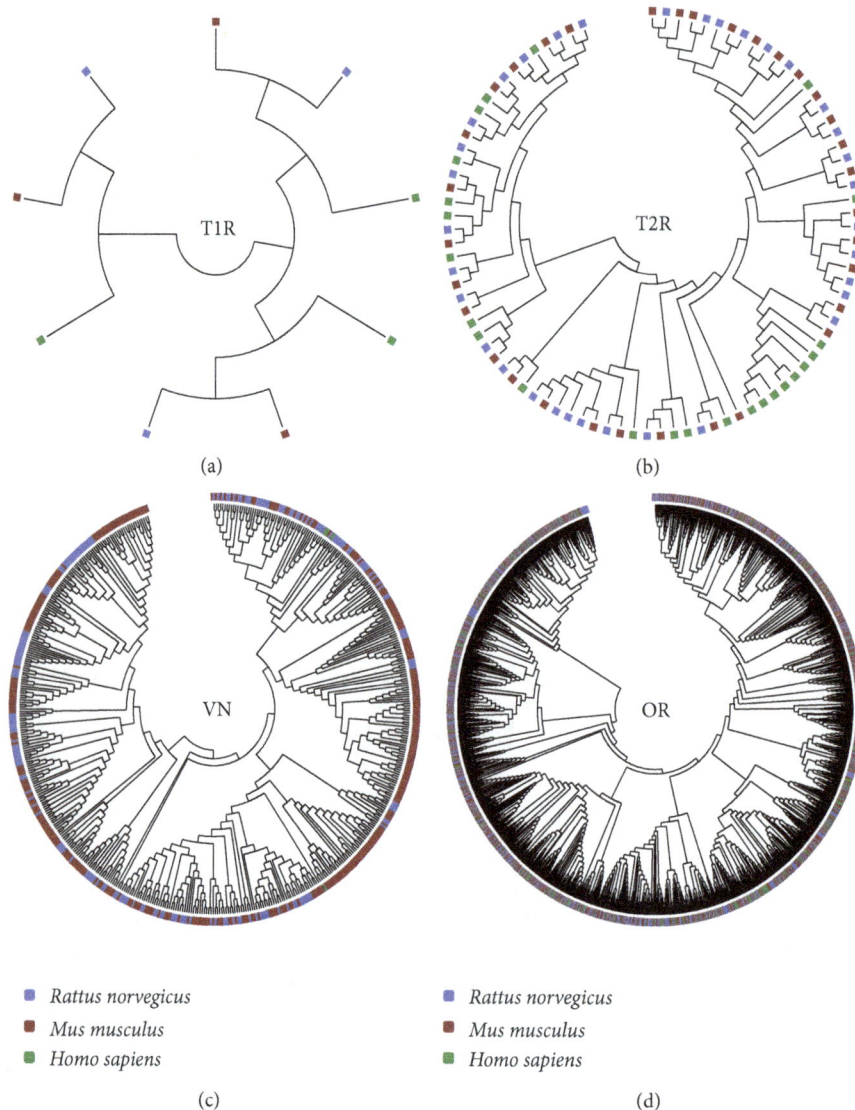

FIGURE 6: Reconstructed phylogenetic tree of the chemoreceptor families. Each square represents a CR gene. Blue, red, and green squares represent *Rattus norvegicus*, *Mus musculus*, and *Homo sapiens* genes, respectively. Phylogenetic trees were reconstructed with Tamura-Nei model. T1R: type I taste receptors; T2R: type II taste receptors; VN: vomeronasal receptors.

immune compatibility [43, 44]. These characteristics, albeit very important to long term adaptation and survival of the species as a whole, are not a major issue in direct survival of the individual.

An apparent contradiction in our analysis was the case of the T1R, which code sweet and *umami* receptors. From our phylogenetic analysis, these receptors are tightly constrained. All three species have the same number of these receptors, each branching more closely with its orthologs in other species rather than the others of the same species. This finding is supported by an earlier analysis that found the same pattern in all vertebrate species [45]. However, by their ortholog distribution, we have found high EPI values. These receptors are grouped in the KOG1056 group, which encompasses 1790 proteins in 52 species, with most varied functions (e.g., bride of sevenless, a homeotic gene). The high-plasticity values

of T1R family are owed to the comprehensive reach of this orthologous group, suggesting that these receptors are constrained members of a larger and more dynamic family of proteins. Albeit instigating, these assumptions can only be confirmed with further in-depth study of this interesting orthologous group.

From a systems perspective, we have found evidences that the CR evolved through duplication events that resulted in gain of function. We have observed that all CR families share the same STM cluster, suggesting that the latter is an older transducing core that was reused in several cell types. The CR, on the other hand, are specific and only expressed in their appropriate cell type. The only CR family that diverges fromz this behavior is the vomeronasal receptors, which were adapted to convey their signal directly to an ion-channel. This

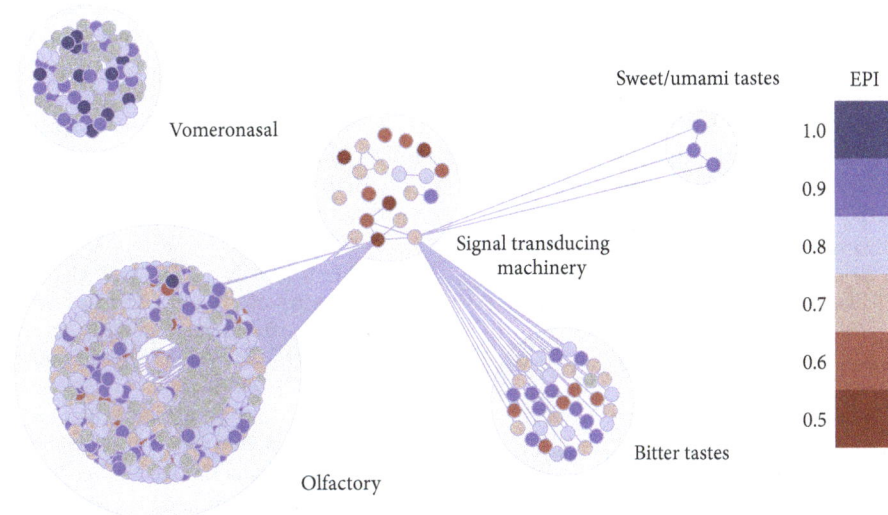

FIGURE 7: Graphical representation of *Mus musculus* chemosensory network. EPI values are plotted on each node by a color scale. Higher plasticity is indicated by bluish colors and lower plasticity by reddish colors. The other networks are not shown in this paper. Nodes represent protein coding genes and edges, functional interactions.

deviation may be the reason why these receptors are under different evolutionary constraints.

Our results suggest that genes coding chemoreceptors were subject to more variation in the course of evolution than those coding signal transducing machinery, reflecting their distinct functional roles in organisms. We have also found significant variation even among the different receptor modalities, suggesting, for the first time to our notice, that olfactory and bitter taste receptors are, albeit less constrained than the transduction machinery, more conserved than vomeronasal receptors. These differences are due to the distinct ecological roles played by the receptors, with the low-plasticity olfactory and bitter taste receptors taking major part in direct survival of the organism, whereas high-plasticity vomeronasal receptors contribute to overall adaptation of the species. Sweet/*umami* receptors cannot be analyzed by their orthologous distribution alone due to the large variability of their ortholog group, and further studies are needed in order to understand the selective pressures imposed on them. We believe that the chemoreceptor networks case is illustrative to demonstrate the generation of novelties through evolutionary tinkering. During the course of evolution, the chemosensory cells generated novel receptor clusters probably by duplicating older ones in order to perceive different sensory inputs. Even among these clusters, there is a great deal of evolutionary experimentation, so that the organisms can be kept up to date with their environment. The signal transduction machinery and other information pathways, however, remained essentially the same throughout generations.

Conflict of Interests

The authors declare that there is no conflict of interests regarding the publication of this paper.

References

[1] E. Adler, M. A. Hoon, K. L. Mueller, J. Chandrashekar, N. J. P. Ryba, and C. S. Zuker, "A novel family of mammalian taste receptors," *Cell*, vol. 100, no. 6, pp. 693–702, 2000.

[2] G. Nelson, M. A. Hoon, J. Chandrashekar, Y. Zhang, N. J. P. Ryba, and C. S. Zuker, "Mammalian sweet taste receptors," *Cell*, vol. 106, no. 3, pp. 381–390, 2001.

[3] I. Rodriguez, K. Del Punta, A. Rothman, T. Ishii, and P. Mombaerts, "Multiple new and isolated families within the mouse superfamily of V1r vomeronasal receptors," *Nature Neuroscience*, vol. 5, no. 2, pp. 134–140, 2002.

[4] J. M. Young, M. Kambere, B. J. Trask, and R. P. Lane, "Divergent V1R repertoires in five species: amplification in rodents, decimation in primates, and a surprisingly small repertoire in dogs," *Genome Research*, vol. 15, no. 4, pp. 231–240, 2005.

[5] Y. Hashiguchi and M. Nishida, "Evolution of trace amine-associated receptor (TAAR) gene family in vertebrates: Lineage-specific expansions and degradations of a second class of vertebrate chemosensory receptors expressed in the olfactory epithelium," *Molecular Biology and Evolution*, vol. 24, no. 9, pp. 2099–2107, 2007.

[6] L. Buck and R. Axel, "A novel multigene family may encode odorant receptors: A molecular basis for odor recognition," *Cell*, vol. 65, no. 1, pp. 175–187, 1991.

[7] J. M. Young, H. F. Massa, L. Hsu, and B. J. Trask, "Extreme variability among mammalian V1R gene families," *Genome Research*, vol. 20, no. 1, pp. 10–18, 2010.

[8] P. Shi and J. Zhang, "Extraordinary diversity of chemosensory receptor gene repertoires among vertebrates," *Results and Problems in Cell Differentiation*, vol. 47, pp. 1–23, 2009.

[9] Y. Niimura, "Evolutionary dynamics of olfactory receptor genes in chordates: interaction between environments and genomic contents," *Human genomics*, vol. 4, no. 2, pp. 107–118, 2009.

[10] M. Nei, Y. Niimura, and M. Nozawa, "The evolution of animal chemosensory receptor gene repertoires: Roles of chance and

necessity," *Nature Reviews Genetics*, vol. 9, no. 12, pp. 951–963, 2008.

[11] Y. Niimura and M. Nei, "Evolutionary dynamics of olfactory and other chemosensory receptor genes in vertebrates," *Journal of Human Genetics*, vol. 51, no. 6, pp. 505–517, 2006.

[12] J. M. Eirín-López, L. Rebordinos, A. P. Rooney, and J. Rozas, "The birth-and-death evolution of multigene families revisited," *Genome Dynamics*, vol. 7, pp. 170–196, 2012.

[13] D. Dong, G. He, S. Zhang, and Z. Zhang, "Evolution of olfactory receptor genes in primates dominated by birth-and-death process," *Genome Biology and Evolution*, vol. 1, pp. 258–264, 2009.

[14] Y. Niimura, "Evolutionary dynamics of olfactory receptor genes in chordates: interaction between environments and genomic contents," *Human genomics*, vol. 4, no. 2, pp. 107–118, 2009.

[15] M. Ashburner, C. A. Ball, J. A. Blake et al., "Gene ontology: tool for the unification of biology," *Nature Genetics*, vol. 25, no. 1, pp. 25–29, 2000.

[16] L. J. Jensen, M. Kuhn, M. Stark et al., "STRING 8—A global view on proteins and their functional interactions in 630 organisms," *Nucleic Acids Research*, vol. 37, no. 1, pp. D412–D416, 2009.

[17] R. L. Seal, S. M. Gordon, M. J. Lush, M. W. Wright, and E. A. Bruford, "Genenames.org: The HGNC resources in 2011," *Nucleic Acids Research*, vol. 39, no. 1, pp. D514–D519, 2011.

[18] J. A. Blake, C. J. Bult, J. A. Kadin, J. E. Richardson, and J. T. Eppig, "The mouse genome database (MGD): Premier model organism resource for mammalian genomics and genetics," *Nucleic Acids Research*, vol. 39, no. 1, pp. D842–D848, 2011.

[19] M. R. Dwinell, E. A. Worthey, M. Shimoyama et al., "The rat genome database 2009: variation, ontologies and pathways," *Nucleic Acids Research*, vol. 37, no. 1, pp. D744–D749, 2009.

[20] S. Haider, B. Ballester, D. Smedley, J. Zhang, P. Rice, and A. Kasprzyk, "BioMart central portal: unified access to biological data," *Nucleic Acids Research*, vol. 37, no. 2, pp. W23–W27, 2009.

[21] R. J. S. Dalmolin, M. A. A. Castro, J. L. Rybarczyk Filho, L. H. T. Souza, R. M. C. de Almeida, and J. C. F. Moreira, "Evolutionary plasticity determination by orthologous groups distribution," *Biology Direct*, vol. 6, article 22, 2011.

[22] M. A. A. Castro, X. Wang, M. N. C. Fletcher, K. B. Meyer, and F. Markowetz, "RedeR: R/Bioconductor package for representing modular structures, nested networks and multiple levels of hierarchical associations," *Genome Biology*, vol. 13, article R29, 2012.

[23] D. Dong, K. Jin, X. Wu, and Y. Zhong, "CRDB: Database of chemosensory receptor gene families in vertebrate," *PLoS ONE*, vol. 7, no. 2, Article ID e31540, 2012.

[24] K. Tamura, D. Peterson, N. Peterson, G. Stecher, M. Nei, and S. Kumar, "MEGA5: Molecular evolutionary genetics analysis using maximum likelihood, evolutionary distance, and maximum parsimony methods," *Molecular Biology and Evolution*, vol. 28, no. 10, pp. 2731–2739, 2011.

[25] R. C. Edgar, "MUSCLE: multiple sequence alignment with high accuracy and high throughput," *Nucleic Acids Research*, vol. 32, no. 5, pp. 1792–1797, 2004.

[26] R. C. Edgar, "MUSCLE: a multiple sequence alignment method with reduced time and space complexity," *BMC Bioinformatics*, vol. 5, article 113, 2004.

[27] K. Tamura and M. Nei, "Estimation of the number of nucleotide substitutions in the control region of mitochondrial DNA in humans and chimpanzees," *Molecular Biology and Evolution*, vol. 10, no. 3, pp. 512–526, 1993.

[28] W. T. Swaney and E. B. Keverne, "The evolution of pheromonal communication," *Behavioural Brain Research*, vol. 200, no. 2, pp. 239–247, 2009.

[29] B. Nilius and G. Owsianik, "The transient receptor potential family of ion channels," *Genome Biology*, vol. 12, no. 3, article 218, 2011.

[30] T. Hartmann, "From waste products to ecochemicals: fifty years research of plant secondary metabolism," *Phytochemistry*, vol. 68, no. 22-24, pp. 2831–2846, 2007.

[31] P. M. Kim, J. O. Korbel, and M. B. Gerstein, "Positive selection at the protein network periphery: Evaluation in terms of structural constraints and cellular context," *Proceedings of the National Academy of Sciences of the United States of America*, vol. 104, no. 51, pp. 20274–20279, 2007.

[32] T. Yamada and P. Bork, "Evolution of biomolecular networks lessons from metabolic and protein interactions," *Nature Reviews Molecular Cell Biology*, vol. 10, no. 11, pp. 791–803, 2009.

[33] X. Zhu, M. Gerstein, and M. Snyder, "Getting connected: Analysis and principles of biological networks," *Genes and Development*, vol. 21, no. 9, pp. 1010–1024, 2007.

[34] R. Albert, "Scale-free networks in cell biology," *Journal of Cell Science*, vol. 118, no. 21, pp. 4947–4957, 2005.

[35] Y. I. Wolfa, P. S. Novichkovb, G. P. Kareva, E. V. Koonina, and D. J. Lipmana, "The universal distribution of evolutionary rates of genes and distinct characteristics of eukaryotic genes of different apparent ages," *Proceedings of the National Academy of Sciences of the United States of America*, vol. 106, no. 18, pp. 7273–7280, 2009.

[36] M. D'Antonio and F. D. Ciccarelli, "Modification of gene duplicability during the evolution of protein interaction network," *PLoS Computational Biology*, vol. 7, no. 4, Article ID e1002029, 2011.

[37] D. Alvarez-Ponce, M. Aguadé, and J. Rozas, "Network-level molecular evolutionary analysis of the insulin/TOR signal transduction pathway across 12 Drosophila genomes," *Genome Research*, vol. 19, no. 2, pp. 234–242, 2009.

[38] X. Wu, X. Chi, P. Wang, D. Zheng, R. Ding, and Y. Li, "The evolutionary rate variation among genes of HOG-signaling pathway in yeast genomes," *Biology Direct*, vol. 5, article 46, 2010.

[39] R. Jovelin and P. C. Phillips, "Expression level drives the pattern of selective constraints along the insulin/tor signal transduction pathway in caenorhabditis," *Genome Biology and Evolution*, vol. 3, no. 1, pp. 715–722, 2011.

[40] D. Alvarez-Ponce, M. Aguadé, and J. Rozas, "Comparative genomics of the vertebrate insulin/TOR signal transduction pathway: A network-level analysis of selective pressures," *Genome Biology and Evolution*, vol. 3, no. 1, pp. 87–101, 2011.

[41] W. E. Grus and J. Zhang, "Distinct evolutionary patterns between chemoreceptors of 2 vertebrate olfactory systems and the differential tuning hypothesis," *Molecular Biology and Evolution*, vol. 25, no. 8, pp. 1593–1601, 2008.

[42] J. I. Glendinning, "Is the bitter rejection response always adaptive?" *Physiology and Behavior*, vol. 56, no. 6, pp. 1217–1227, 1994.

[43] P. A. Brennan and K. M. Kendrick, "Mammalian social odours: Attraction and individual recognition," *Philosophical Transactions of the Royal Society B: Biological Sciences*, vol. 361, no. 1476, pp. 2061–2078, 2006.

[44] P. Chamero, T. Leinders-Zufall, and F. Zufall, "From genes to social communication: molecular sensing by the vomeronasal organ," *Trends in Neurosciences*, vol. 35, no. 10, pp. 597–606, 2012.

[45] P. Shi and J. Zhang, "Contrasting modes of evolution between vertebrate sweet/umami receptor genes and bitter receptor genes," *Molecular Biology and Evolution*, vol. 23, no. 2, pp. 292–300, 2006.

13

Origins and Evolution of WUSCHEL-Related Homeobox Protein Family in Plant Kingdom

Gaibin Lian,[1] **Zhiwen Ding,**[1,2,3] **Qin Wang,**[2] **Dabing Zhang,**[1] **and Jie Xu**[1]

[1] School of Life Sciences and Biotechnology, Shanghai Jiao Tong University, Shanghai 200240, China
[2] Department of Physics, Shanghai Jiao Tong University, Shanghai 200240, China
[3] Institutes of Biomedical Sciences, Fudan University, Shanghai 200032, China

Correspondence should be addressed to Jie Xu; xujie3000@gmail.com

Academic Editors: B. Gantenbein-Ritter and B. Lieb

WUSCHEL-related homeobox (WOX) is a large group of transcription factors specifically found in plants. WOX members contain the conserved homeodomain essential for plant development by regulating cell division and differentiation. However, the evolutionary relationship of WOX members in plant kingdom remains to be elucidated. In this study, we searched 350 WOX members from 50 species in plant kingdom. Linkage analysis of WOX protein sequences demonstrated that amino acid residues 141–145 and 153–160 located in the homeodomain are possibly associated with the function of WOXs during the evolution. These 350 members were grouped into 3 clades: the first clade represents the conservative WOXs from the lower plant algae to higher plants; the second clade has the members from vascular plant species; the third clade has the members only from spermatophyte species. Furthermore, among the members of *Arabidopsis thaliana* and *Oryza sativa*, we observed ubiquitous expression of genes in the first clade and the diversified expression pattern of *WOX* genes in distinct organs in the second clade and the third clade. This work provides insight into the origin and evolutionary process of WOXs, facilitating their functional investigations in the future.

1. Introduction

Homeobox genes encode transcript factors containing a classic DNA binding domain (called homeodomain) with about 60 amino acids (aa), which forms three helixes in space. The homeobox gene was first identified in *Drosophila* [1, 2]. Subsequently, more homeobox members have been reported in most eukaryotes [3]. WOX (WUSCHEL-related homeobox) is the member of ZIP superfamily belonging to homeobox proteins family [4].

In *Arabidopsis thaliana*, WUSCHEL (*WUS*) is essential in maintaining shoot apical meristem (SAM); *WUS* mutants display aborted SAM maintenance during embryonic and later developmental stages [5, 6]. The expression of *AtWOX2* is detectable in the egg cell and zygote, and *AtWOX2* functions in zygotic apical cell development [7]. Transcripts of *AtWOX3* are observed in the peripheral area of SAM, and *AtWOX3* is implicated in forming horizontal regions of vegetative and floral organs [8]. *AtWOX5*, a close homolog of

AtWUS, is mainly expressed in the quiescent centre and plays a role in maintaining stem cells of root apical meristem (RAM) [9]. *AtWOX6* regulates ovule development [10], and *AtWOX9* maintains cell division and inhibits SAM differentiation [11]. In rice (*Oryza sativa*), *OsWOX5* is involved in the specification and maintenance of the RAM stem cells and its expression is specifically detectable in the quiescent center of the root [12]. *OsWOX11* is expressed in emerging crown roots and later in cell division regions of the root meristem and is involved in the activation of crown root emergence and growth [13].

In *Arabidopsis*, 15 *WOX* genes are grouped into 3 clades: group 1 containing *WOX1–WOX7* and *WUS*; group 2 containing *WOX8*, *WOX9*, *WOX11*, and *WOX12*; group 3 containing *WOX10*, *WOX13*, and *WOX14* [7, 14]. *WOX*s from monocots including rice, maize (*Zea mays*), *Brachypodium* (*Brachypodium distachyon*), and *Sorghum* (*Sorghum bicolor*) are divided into 3–5 clades [15–17]. Nardmann et al. (2009) analyzed the evolution of WOX family in two basal

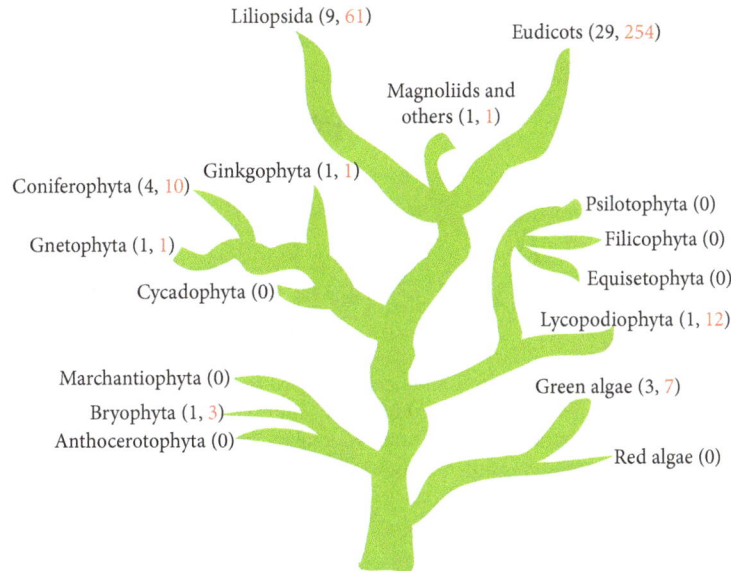

FIGURE 1: Summary of investigated plant species and their phylogenetic relationship. There were 350 WOX sequences of 50 species in 21 families from green algae to angiosperms. The numbers in black mean the amount of species and the numbers in red mean the amount protein sequences of WOX family. The species include *Arabidopsis lyrata, Arabidopsis thaliana, Arachis hypogaea, Artemisia annua, Brassica napus, Brassica rapa, Carica papaya, Citrus sinensis, Cucumis sativus, Glycine max, Gossypium hirsutum, Helianthus annuus, Lotus japonicus, Malus x domestica, Manihot esculenta, Medicago truncatula, Mimulus guttatus, Nicotiana tabacun, Petunia x hybrida, Phaseolus coccineus, Populus trichocarpa, Prunus persica, Raphanus sativus, Ricinus communis, Solanum lycopersicum, Solanum tuberosum, Theobroma cacao, Vigna unguiculata, Vitis vinifera, Brachypodium distachyon, Hordeum vulgare, Lycoris longituba, Oryza sativa, Panicum virgatum, Saccharum officinarum, Sorghum bicolor, Triticum aestivum, Zea mays, Amborella trichopoda, Ginkgo biloba, Gnetum gnemon, Picea abies, Picea sitchensis, Pinus sylvestris, Pinus taeda, Selaginella moellendorffii, Physcomitrella patens, Micromonas, Ostreococcus tauri,* and *Ostreococcus lucimarinus*.

angiosperms (*Amborella trichopoda* and *Nymphaea jamesoniana*) and three gymnosperms (*Pinus sylvestris, Ginkgo biloba,* and *Gnetum gnemon*) and observed common ancestors of WOX members in monocots and dicots [18]. Deveaux et al. (2008) proposed that WOX13 subgroup probably represents the oldest clade among WOX families [19]. More recently, we divided WOX members of *Arabidopsis,* poplar (*Populus trichocarpa*), rice, maize, and *Sorghum* into nine subgroups of three clades and revealed multiplication/duplication of WOX families in these five plants during evolutionary process [14]. However, the origins and evolutionary history of WOX family in plant kingdom remain unclear.

In this study, we systematically performed the analysis of the origin and evolutionary history of 350 WOX sequences from 50 plants species. In addition, the conserved and the possible active amino acid residues located in the homeodomain were revealed. Moreover, we analyzed the expression and methylation of *WOX* genes in *Arabidopsis* and rice and discussed their possible roles.

2. Results

2.1. Identification of WOX Sequences in Plants. To obtain sequences of WOX family in plant kingdom, we used the full-length sequence of *AtWUS* (*At2g17950*) [7] as a TBLASTN query against the available genome sequences (see Supplementary Table 1 in Supplementary Materials available

online at http://dx.doi.org/10.1155/2013/534140) including the unicellular green algae (3 species), the bryophyte (1 species), the Lycopodiophyta (1 species), the Gnetophyta (1 species), the coniferophyta (4 species), the Ginkgophyta (1 species), and flowering plants (39 species). Generally, candidate sequences containing the characteristic homeodomain with higher similarity were identified. Further, the homologous and conserved sequences were manually reconstructed by repeated sequence alignment. A total of 367 putative WOX family proteins with E values below $<1e-5$ were identified from the databases of TAIR (The *Arabidopsis* Information Resource database), RGAP (the Rice Genome Annotation Project database), JGI (genomic databases in Joint Genome Institute Eukaryotic Genomics), and PlantTFDB (the Plant Transcription Factor Database). Redundant members were removed according to Sol Genomics Network (SGN) (http://solgenomics.net/), UniProt (http://www.uniprot.org/), SMART (http://smart.embl-heidelberg.de/), and Pfam (http://pfam.sanger.ac.uk/), and eventually 350 WOX proteins from 50 species in 21 families ranging from green algae to angiosperms were obtained, that is, 3 species (*Micromonas pusilla, Ostreococcus tauri,* and *Ostreococcus lucimarinus*) in Chlorophyta, 1 species (*Physcomitrella patens*) in Bryophyta, 1 species (*Selaginella moellendorffii*) in Lycopodiophyta, 6 species in gymnosperms, 29 species in eudicots, and 9 species in monocots (Figure 1 and Supplementary Table 1). 244 of the WOX sequences have not been defined in previous publications, for example, WOX15 (AT5g46010)

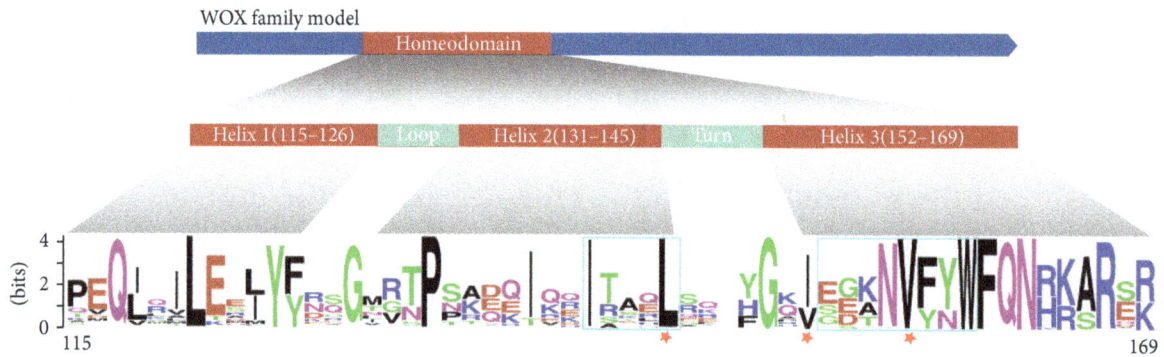

FIGURE 2: Alignment of the WOX homeodomain sequences. Canonical conserved residues analyzed by WebLogo (Web-based sequence logo generating application; Weblogo.berkeley.edu). The homeodomain of WOX family contained the helix-loop-helix-turn-helix structure [12]. The highly conserved residues are revealed by the alignment of homeodomains, the three highest conserved residues were marked using asterisk, and the residues within the two boxed motifs within the homeodomain marked using azure have close positional correlation which is revealed in Figure 3.

from *Arabidopsis*. Interestingly, we observed that several different genes encode the same WOX proteins; that is, *LOC_Os11g01130* and *LOC_Os12g01120* encode rice OsWOX3 (Supplementary Table 1). Additionally, the lower plants such as green algae and moss have fewer *WOX* genes than those of higher plants: 2 members found in *Ostreococcus*, 3 members in *Physcomitrella patens*, and above ten members in most of the higher plants. This suggested that the WOX members expanded as the evolution of plants.

Previously, various names were used in WOXs [8, 14–16, 18–21]; we used the WOX names described by Zhang et al. (2010) [14] and the LOCUS number (RGAP, http://rice.plant-biology.msu.edu/; JGI, http://genome.jgi-psf.org/) to avoid confusion.

2.2. Multiple Sequences Alignment and Analysis of Conserved Residues/Domains. To examine sequence features of these WOX family proteins, we conducted multiple sequence alignment of 350 WOXs (Figure 2 and Supplementary Figure 1). Generally, these 350 WOXs have the conserved amino acids among the homeodomain, and the average size of the homeodomain is 60 aa, and all the homeodomains contain the helix-loop-helix-turn-helix structure [12]. Previously, it has been reported that the homeodomain of WOXs contains 17 conserved amino acids [1, 14]. In this study, we observed 9 additional conserved residues, including E (122) and F (126) in helix 1; G (129) in loop; T (132), I (138), and T (142) in helix 2; N (156), Y (159), and A (166) in helix 3, among these WOX members (Figure 2).

To understand the possible relationship between amino acid residues of WOXs and the function during the evolutionary change, the physicochemical value of amino acid sites was calculated using the modified version of an algorithm CRASP [22]. Physicochemical value reflects a significant correlation between the protein sequence and function possibly because of structural and functional constraints or results from evolutionary history and stochastic events [22]. The physicochemical analysis results showed that the pairwise positions from 141 (I) to 145 (L) correlated with the pairwise

positions from 153 (E) to 160 (W) within the homeodomain in WOX family (Figure 3 and Supplementary Figure 2). In particular, the positive correlation coefficient between 153 (E) and 155 (K) was 0.813, and the positive correlation coefficient between 141 (I) and 157 (V) was 0.743, and there was a negative correlation between 158 (Y) and 159 (N) with a correlation coefficient of 0.770. This result suggested that these amino acids may be required for the function of the homeodomain during the evolutionary change.

2.3. Phylogenetic Analysis of WOXs. To understand the evolutionary change of WOXs, we conducted phylogenetic analyses using the full-length sequences of all 350 sequences. Although some bootstrap values for interior branches were low because of the large number of sequences included [23], a relatively well-supported phylogenetic tree was obtained (Figure 4 and Supplementary Figure 3). The phylogenetic tree constructed using the full length of WOX sequences was nearly identical to that by the WOX homeodomain, and we thus only show the phylogenetic tree conducted using the full-length sequences (Figure 4 and Supplementary Figure 3). These 350 WOX members were divided into three clades, and the first clade (also called the ancient clade) contained 98 WOXs in 47 species from lower plants to seed plants, including 7 WOXs from green algae, 3 WOXs from bryophyta, 6 WOXs from Lycopodiophyta, 1 WOX from Gnetophyta, 3 WOXs from coniferophyta, 1 WOX from Ginkgophyta, 1 WOX from *Amborella trichopoda*, 13 WOXs from Liliopsida, and 63 WOXs from eudicots, which are homologous to *Arabidopsis* WOX10, WOX13, and WOX14.

The second clade (also called the intermediate clade) consisted of 86 WOXs homologous to *Arabidopsis* WOX8, WOX9, WOX11, and WOX12, which are only from 28 vascular plant species, that is, 6 from Lycopodiophyta, 5 from coniferophyta, 23 from Liliopsida, and 52 WOXs from eudicots. These members in the intermediate clade were further divided into two subgroups, designed WOX8/9 and WOX11/12. WOX8/9 contained 42 members and WOX11/12

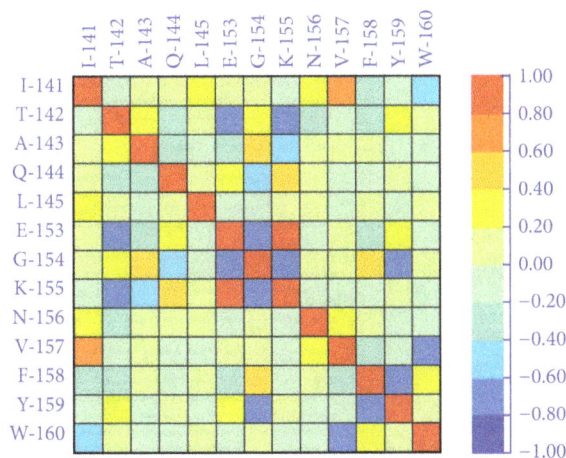

FIGURE 3: Pairwise positional correlation estimation of WOX sequences. The pairwise positional correlation estimation was analyzed by CRASP (http://www.bionet.nsc.ru/en). Two regions within the homeodomain, one is from 141 (I) to 145 (L) and the other is from 153 (E) to 160 (W), were closely correlated. The color refers correlation coefficiency of the amino acid at a certain position.

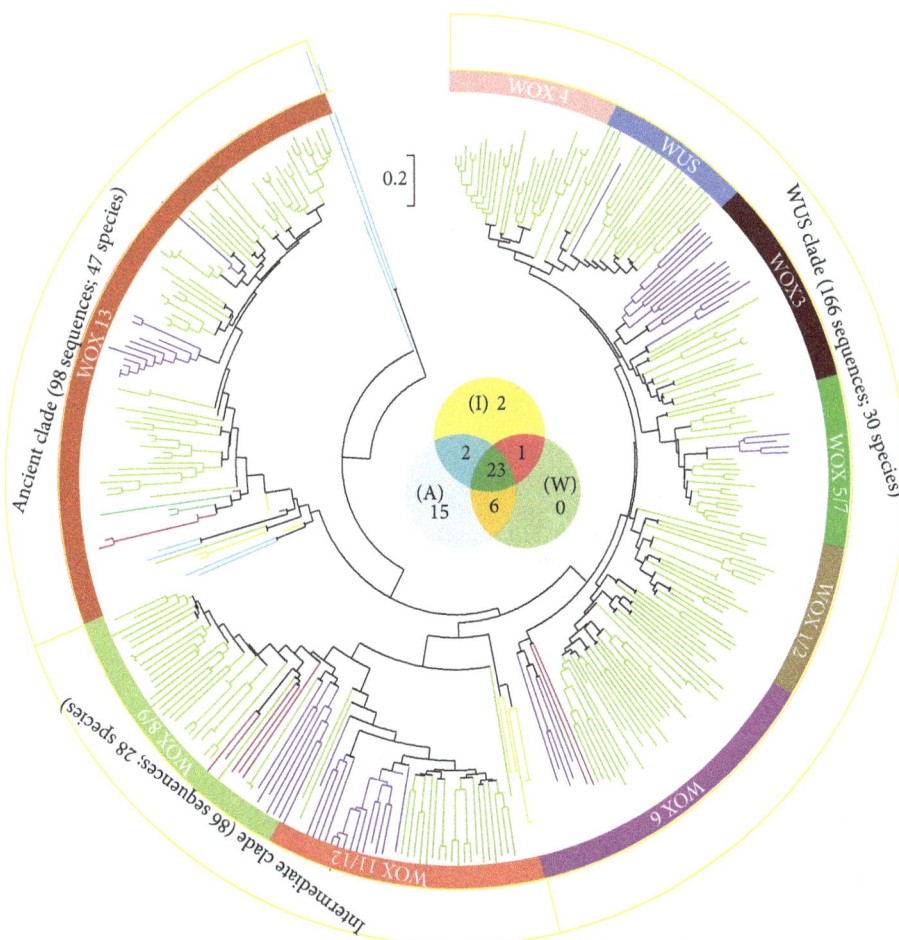

FIGURE 4: Phylogenetic analysis of WOX family. A simplified version of the neighbor joining (NJ) tree, with 350 sequences of proteins from 50 species from green algae to angiosperms. The tree was divided into three clades. The three clades were the WUS clade, containing 166 sequences, 30 species; the intermediate clade, containing 86 sequences, 28 species; the ancient clade, containing 98 sequences, 47 species. The full phylogeny is shown in Supplementary Figure 3. In the inset, (A), (I), and (W) refer to the ancient clade, the intermediate clade, and the WUS clade. Two species (*P. staeda* and *S. moellendorffii*) have members in both ancient clade and intermediate clade; one species (*P. hybieda*) has members in both intermediate clade and WUS clade; six species (*B. rapa, Gossypium hirsutum, P. sitchensis, R. sativus, Solanum tuberosum,* and *Vigna unguiculata*) have members in both ancient clade and WUS clade; twenty-three species have members belonging to three clades. The species names are shown in Supplementary Table 2.

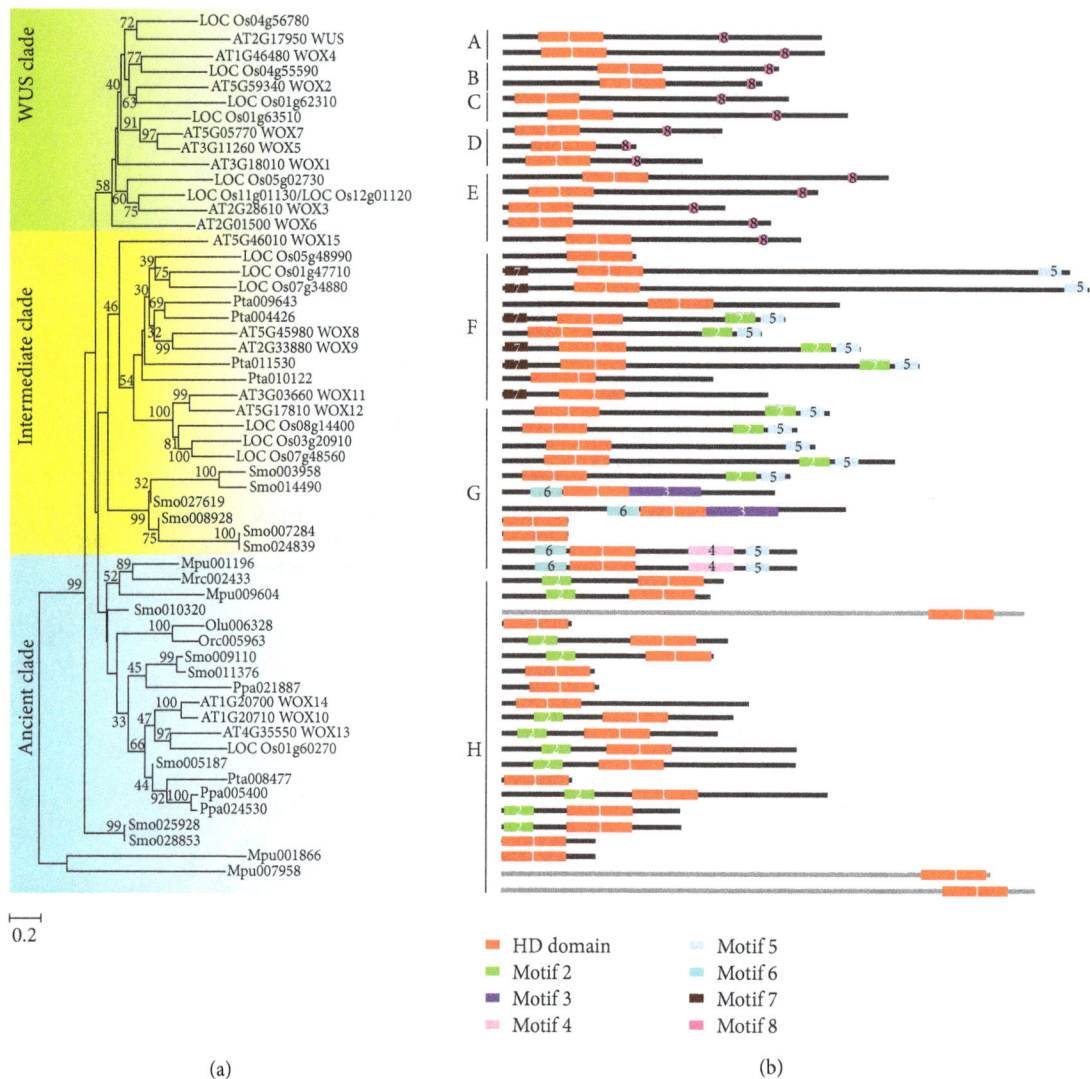

(a)   (b)

FIGURE 5: Phylogenetic tree of WOX family and motif analysis in model plants. (a) The phylogenetic tree (NJ) was constructed by MAGE 4 (Kumar et al., 2004) using 56 sequences from green algae (*Micromonas, Ostreococcus tauri, and Ostreococcus lucimarinus*), *Physcomitrella patens* (Ppa), *Selaginella moellendorffii* (Smo), *Pinus taeda* (Pta), *Oryza sativa* (Os), and *Arabidopsis thaliana* (At). WUS clade, intermediate clade, and ancient clade are color-coded green, orange, and blue, respectively. The WUS clade contains members only from angiosperms; the intermediate clade contains members only from vascular plants; the ancient clade contains members from lower plants to higher plants. (b) The conserved motifs among the members are highlighted in colored boxes with an arranged number and the sequences of the motifs are listed in Supplementary Table 3.

with 44 members. The third clade (also called the WUS Clade) contained 166 WOXs including *Arabidopsis* WUS which are only from 30 seed plants. These 166 members had 2 WOXs from coniferophyta, 25 WOXs from Liliopsida, and 139 WOXs from eudicots.

2.4. Motifs of WOXs. To further understand how WOX family evolved in plants and the protein sequence change, we selected 56 WOX members from model plants including *Arabidopsis*, rice, loblolly (*Pinus taeda*), *S. moellendorffii*, moss (*P. patens*), *Ostreococcus*, and *Micromonas* for phylogenetic analysis (Figure 5(a)). Consistently, these WOX members were also divided into three clades: the ancient clade, the intermediate clade, and the WUS clade. The ancient clade contained the members from green algae, moss, and vascular

plants; the intermediate clade contained members from vascular plants; the WUS clade only contained the members from seed plants, confirming the evolutionary relationship of WOXs.

Furthermore, we observed that most of the members in the same clade shared one or more common motifs besides the homeodomain. The multiple EM for motif elicitation tool (MEME, http://meme.nbcr.net/meme/) and Surveyed Conserved Motif Alignment Diagram and the Associating Dendrogram Database (SALAD database, http://salad/en/) were used to identify the similar motifs among WOXs. In addition to the homeodomain, a total of 7 motifs were observed in these members from the 56 WOX members (Figure 5(b) and Supplementary Table 3), and most of these motifs have not yet been characterized. The proteins in the same cluster display the same or similar domain organization, suggesting the reliable phylogenetic analysis [24]. In the ancient clade, three WOX members from green algae, one WOX member from moss, and all WOXs members from *S. moellendorffii* were observed to only have one domain, that is, the homeodomain, suggesting that they may represent the ancestral form of WOXs. Furthermore, other members of ancient clade from green algae, moss, loblolly, *Arabidopsis*, and rice were observed to contain another motif (number 2 motif) at the N-terminus of the WOX sequences, indicating that these WOXs might gain additional motifs after the divergence from the ancestor. Compared with the ancient clade, the motif distribution of members in the intermediate clade seemed more diversified and seven motifs were observed, that is, except the homeodomain, two motifs were located at the N-terminus of the WOX sequences while four were at the C-terminus of the WOX sequences (Figure 5(b)). No. 2 motif located at the N-terminus of the WOX sequences in the ancient clade was also observed at the C-terminus in eight members of seed plants (loblolly, *Arabidopsis*, and rice), suggesting the conserved role of this motif during the evolution. Interestingly, No. 5 motif is close to No. 2 motif among the intermediate clade members except that three rice members (LOC_Os01g47710, LOC_Os05g48990, and LOC_Os07g34880) which didn't have No. 2 motif. Furthermore, two members (Smo008928 and Smo027619) of WOX family from *S. moellendorffii* only had the homeodomain; however, the other numbers in *S. moellendorffii* had more than one motif. In seed plants, six members of the intermediate clade had an extra new motif (No. 7 motif) at the N-terminus of the WOX sequences. All the members of WUS clade contained two motifs: the homeodomain and WUS box (No. 8 motif) (Figure 5(b)). This observation suggested that formation of motif is associated with the subfunctionalization and neofunctionalization of WOX members.

To understand the relationship between WOX function and evolutionary events, we analyze the three highly conserved residues in the homeodomain: L (145), I (152), and V (157) (Figures 2 and 6). 3D-structure prediction of the homeodomain showed that these three amino acids were located in the interior of the homeodomain (Figure 6), implying that these three residues may perform key roles. Moreover, the angles in 3D structure formed by the three residues of ancient clade proteins (WOX13), intermediate clade proteins (WOX 11), and WUS clade proteins (WUS) were 79.32 degree, 122.62 degree, and 110.29 degree, respectively. The result suggested that the homeodomain 3D structures from different clades have differences even though they share similar primary structures. It is obvious that the angle in ancient clade was smaller than those of the intermediate clade and the WUS clade. It may be from the functional change of WOX family during the evolution.

2.5. Evolution of WOX Family.
The observation that lower plants only have the WOX members from the ancient clade and that the members from the WUS clade were only observed in higher plants (Figures 4 and 5; Supplementary

Figure 3 and Supplementary Table 2) suggested that the ancient clade represents the ancient WOX members, and the members in the intermediate clade and the WUS clade formed subsequently by gene duplication and diversification from the ancient members during the evolutionary history. Statistically, the average number of WOXs per species in the ancient clade, the intermediate clade, and the WUS clade is 2.09, 3.07, and 5.53, respectively.

The presence of the homeodomain of the WOX proteins from in extant eukaryotes from the algae to flowering plants supported the previous hypothesis that this DNA-binding domain might be originated before the divergence of the eukaryotes [25]. The phylogenetic analysis suggested that there was at least one WOX member as the last common ancestor among the green algae and land plants (Figures 4 and 5). To better understand how WOX family has evolved in plants, we analyzed the MRCA (most recent common ancestor) of *O. tauri, P. patens, S. moellendorffi, P. taeda*, rice, and *Arabidopsis* and deduced that WOX family originated from the ancient clade and the members in the ancient clade evolved independently among plant species. The WOX members of green algae, Bryophyta, Gnetophyta, Ginkgophyta, and *Amborella trichopoda* obtained by our query conditions were divided into the ancient clade, and members of intermediate clade and the WUS clade were not observed in nonvascular plants, confirming the ancient and conserved role of the ancient clade.

The phylogenetic tree showed that the first expansion of members in the intermediate clade from the ancient clade ancestor occurred in plants from ferns to higher plants (Figure 7). Subsequently, the ancestor of intermediate clade might have undergone a duplication and formed two subgroups: WOX8/9 and WOX11/12 in vascular plants (Figures 4, 5, and 7). Furthermore, all WOX members in the WUS clade except the subgroup WOX6 containing WOX members from coniferophyta were only observed in flowering plants, suggesting that the WUS clade plays a key role during the evolution of higher plants, and the WOX6 subgroup may represent the oldest subgroup in the WUS clade. Additionally, the WOX1/2 subgroup appeared to be generated from the WOX6 subgroup, and the WOX5/7 subgroup might be originated from the WOX1/2 subgroup.

2.6. Expression Analyses of WOXs in Arabidopsis and Rice and the Predictive DNA Methylated Region.
To better understand the duplication event in *Arabidopsis* and rice, we constructed a phylogenetic tree of 29 WOXs from these two species (Figure 8(a)). The phylogenetic tree contained three clades: the ancient clade, the intermediate clade, and the WUS clade. We observed relatively low bootstrap values in interior branches of the WUS clade, which is consistent with previous reports [14, 19], suggesting that WUS clade has more diversified members. In the WOX3 subgroup of the WUS clade, there were three rice genes and one from *Arabidopsis*, and their phylogenetic role suggested that WOX3 subgroup duplicated before and after the divergence of rice, or the homolog(s) of LOC_Os05g02730 in *Arabidopsis* was (were) lost during evolution. In the WOX5/7 subgroup of the WUS

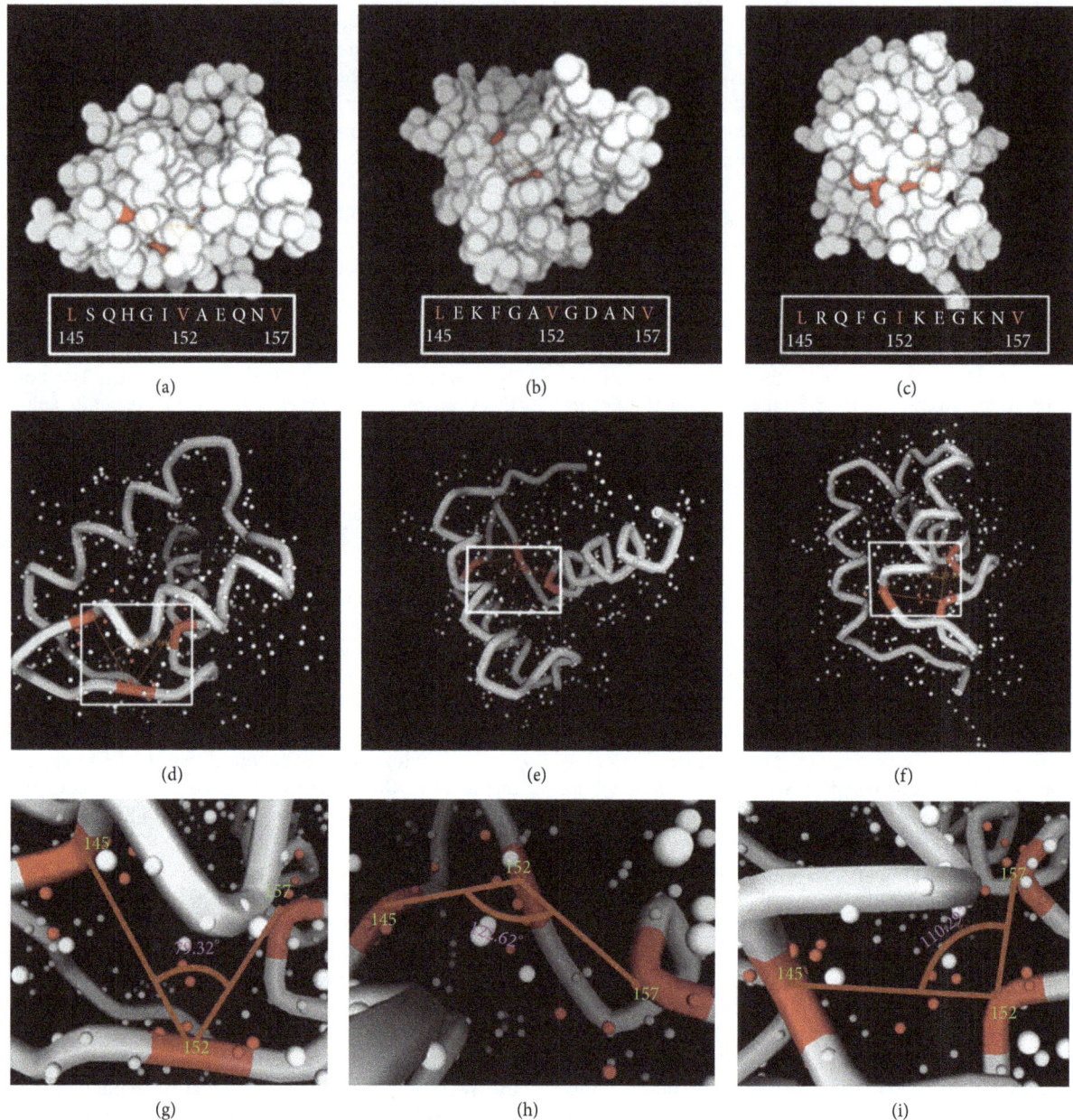

FIGURE 6: 3D structures of the homeodomain in different clades of WOXs. All the WOX proteins had 3 conserved residues in the homeodomain: 145 (L), 152 (V/I), and 157 (V). The 3 residues formed an angle in (a), (d), and (g) 3D structures of ancient clade proteins (WOX13): (b), (e), and (h) 3D structures of intermediate clade proteins (WOX11); (c), (f), and (i) 3D structures of WUS clade proteins (WUS); 79.32 deg, 122.62 deg, and 110.29 deg, respectively.

clade, *Arabidopsis* contained two members and one in rice, suggesting that rice might lose one member during evolution or this subgroup in *Arabidopsis* duplicated recently. In the WOX1/6 subgroup of the WUS clade, no homologs were found in rice which is consistent with a previous study by Zhang et al. (2010), suggesting that WOX1/6 subgroup in rice was lost. The subgroups, WUS, WOX4, and WOX2, contain only one individual member with high bootstrap values from *Arabidopsis* and rice, respectively, implying that these members may play a conserved and crucial role. In the WOX11/12 subgroup of the intermediate clade, members of rice formed one branch and members of *Arabidopsis* formed another one, suggesting that both rice and *Arabidopsis* underwent one duplication after the divergence of them. In the ancient clade, three members of *Arabidopsis* and one member of rice were grouped into two separated branches with high bootstrap values. One branch contained AtWOX13 and LOC_Os1g60270, and the other contained AtWOX10 and

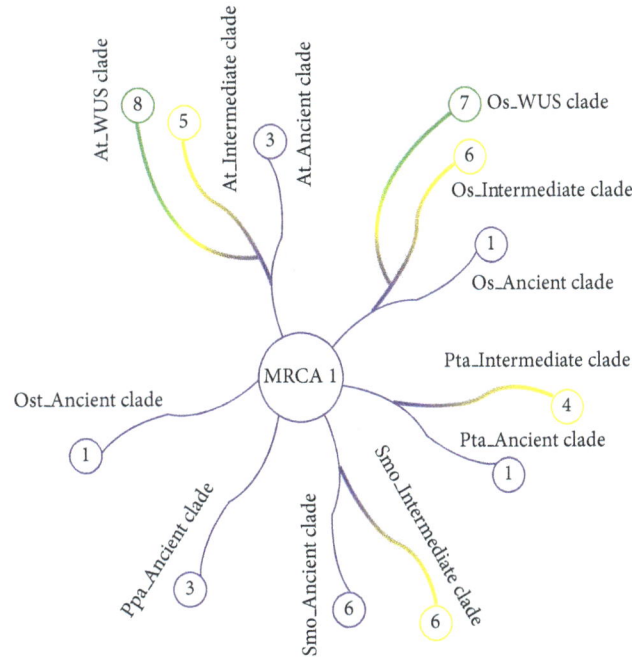

FIGURE 7: Evolutionary change of the number of WOX proteins in model plants. The numbers in circles represent the numbers of genes in extant species. Ost: *Ostreococcus tauri*, Ppa: *Physcomitrella patens*, Smo: *Selaginella moellendorffii*, Pta: *Pinus taeda*, Os: rice, and At: *Arabidopsis thaliana*.

AtWOX14, suggesting that AtWOX13 and LOC_Os1g60270 may play a conserved role, and AtWOX10 and AtWOX14 were generated by duplication.

To understand the function of WOXs, we investigated the expression pattern of *AtWOXs* and *OsWOXs* and the methylation information of their promoter regions using the available dataset from AtGenExpress (http://www.weigel-world.org/resources/microarray/AtGenExpress), RiceXPro (http://ricexpro.dna.affrc.go.jp), and SIGnAL (Salk Institute Genomic Analysis Laboratory; http://signal.salk.edu/) (Figure 8(b)). Consistent with previous observation [14], wide expression of *WOX* genes from *Arabidopsis* and rice was detectable in roots, stems, leaves, flowers, and seeds, suggesting that these WOXs play regulatory roles at various developmental events. Furthermore, some *WOX* homologs showed conserved expression pattern; for instance, members of the ancient clade *AtWOX13* and *AtWOX14* and *OsWOX13* (*LOC_Os01g60270*) were highly expressed in different organs (Figure 8(b)). In the intermediate clade, *WOX8* and *WOX9* exhibited detectable expression signals in seeds, and *WOX9* also in flowers; *WOX15* (*AT5g46010*), *OsWOX9c* (*LOC_Os05g48990*), and *OsWOX9a* (*LOC_Os01g47710*) had lower expression levels in various tissues except flower, where *OsWOX9c* (*LOC_Os05g48990*) was expressed higher than other tissues (Figure 8(b)). Moreover, the WUS clade members exhibited higher expression in flowers (Figure 8(b)), suggesting that these members play an important role in the development of flowers, consistently with previous observation of activation role of WUS in floral pat-terning. Furthermore, the expression of *OsWOX4*

(*LOC_Os4g55590*), *WOX2*, and *OsWOX2* (*LOC_Os1g62310*) was highly detectable in seeds.

DNA methylation is closely associated with the transcriptional regulation of gene expression [26]. Recent studies showed that the expression of *WUS* is regulated by DNA methylation, and there are three characteristic epigenetic marks of DNA methylation, that is, CpG motif within the *WUS* genomic sequences [27]. The sequences of one-kilobase (kb) promoter fragment and the genomic DNA region of *Arabidopsis* and rice *WOXs* were analyzed and CpG islands were observed in promoter regions of 10 *Arabidopsis* WOXs and 18 CpG islands within the promoter regions of 13 rice *WOXs*. In addition, 7 CpG islands were seen in the homeodomain of *Arabidopsis* WOXs, and 12 in the homeodomain of rice *WOXs* (Figure 8(c)). This observation suggests that *WOXs* may share epigenetic methylation modification modulating their expression during evolution.

3. Discussions

3.1. WOXs May Originate in Green Algae. Evolution created a tremendous variation in organ shapes within the plant kingdom. Plant diverse morphologies are associated with the activity of stem cells, which are regulated by WOX genes such as WUS and WOX5 in model eudicot *Arabidopsis* for maintaining stem cell in the shoot and the root, respectively [28]. In this study, we revealed 244 previously undefined WOX sequences. Our phylogenetic analysis using the 350 WOXs family members from 50 plant species supports that

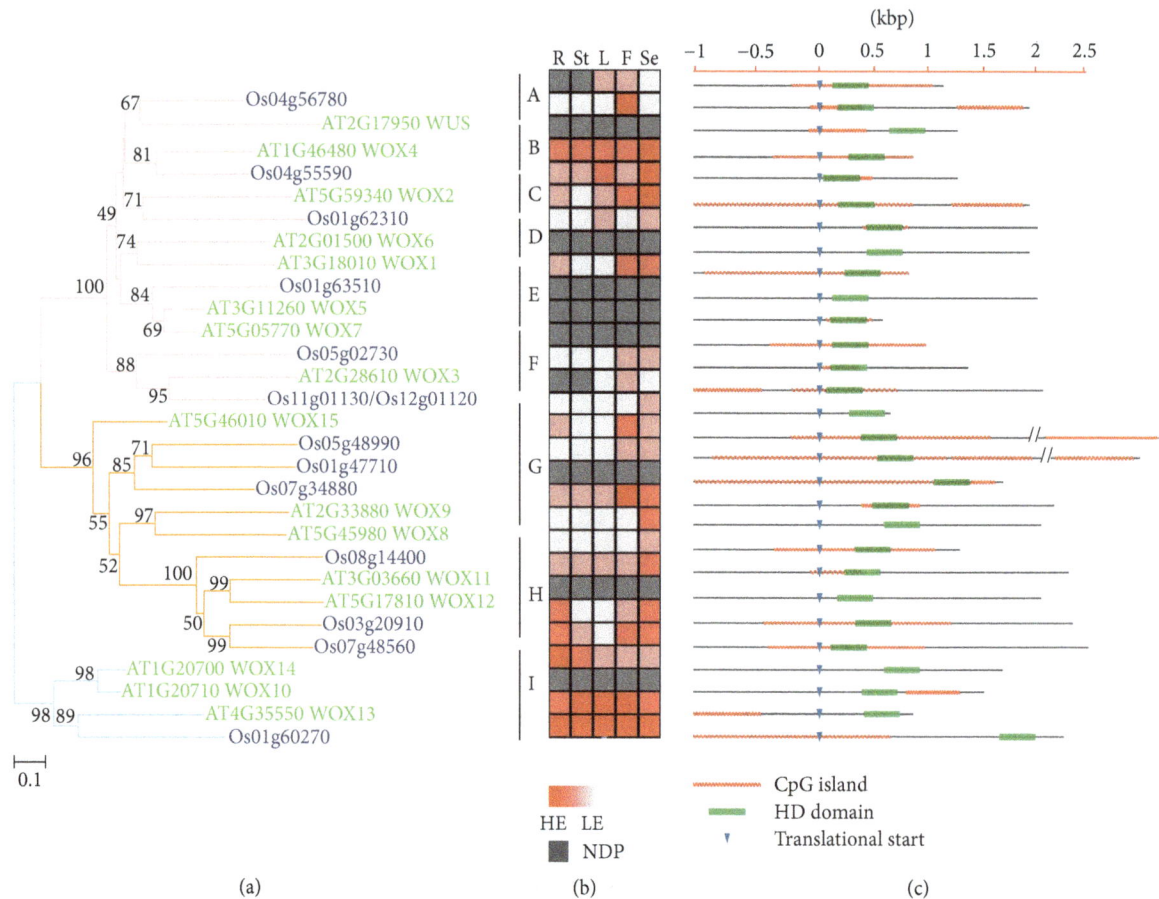

FIGURE 8: Phylogenetic and expression analysis of WOX family in *Arabidopsis* and rice. (a) The phylogenetic tree (NJ) of 29 WOX members from *Arabidopsis* and rice constructed by MAGE 4 (Kumar et al., 2004). A, B, C, D, E, F, G, H, and I refer to subgroups WUS, WOX4, WOX2, WOX1/6, WOX5/7, WOX3, WOX8/9, WOX11/12, and WOX13, respectively. (b) Expression pattern analysis; the average values were chosen among the expression values published in AtGenExpress. (http://www.weigelworld.org/resources/microarray/AtGenExpress), SIGnAL (http://signal.salk.edu/), and RiceXPro (http://ricexpro.dna.affrc.go.jp). R: root; St: stem; L: leaf; F: flower; Se: seed; HE: high expression; LE: low expression; NDP: no data published. (c) The CpG islands among the members are highlighted in red lines.

WOX gene family has a monophyletic origin [8, 14–16, 18–21]. Previous evolutionary analyses of WOX family genes using limited available genome sequences [18, 19, 29, 30] proposed that the green alga WOX genes may represent the earliest WOXs. We collected WOX family sequences using 3 green algal species: *Micromonas pusilla*, *Ostreococcus lucimarinus*, and *Ostreococcus tauri*, and comprehensive analysis supports the notion that WOX proteins in green alga represent the oldest members in WOX family. Supportively, we did not find out any WOX family gene in the genome of *Cyanidioschyzon merolae*, which belongs to the red algae group and is supposed to be earlier than green alga during evolution, even though we can not exclude the possibility that the red alga species lost WOXs during the evolution.

In addition, our phylogenetic analysis revealed that the ancient clade is the most ancient one and the WUS clade represents the latest members, which is consistent with previous analysis of WOX family in *Arabidopsis* [7] as well as other phylogenetic analyses [8, 14, 19–21]. Consistently, the

subclade encompassing WOX13 is considered the oldest one [19] and WUS/WOX5 as the modern one (Figure 9) [18].

3.2. The Homeodomain Region Plays a Key Role in Plant Development. The homeodomain can recognize sequence-specific targets in a precise spatial and temporal pattern, and helix 3 plays an important role in this process [1]. We did linkage analysis on WOX amino acid sequences in the plant kingdom and showed the correlation between amino acids in the homeodomain region, suggesting the importance of these residues to the role of the homeodomain in WOX family. Particularly, we observed that all the WOX proteins have three highly conserved residues in the homeodomain: L (145), I (152), and V (157), and the homeodomain 3D structures have differences in different clade, suggesting the reliability of the phylogenetic analysis. Moreover, we observed the putative methylated regions of the promoter and the homeodomain-encoding sequences, suggesting that the homeodomain may

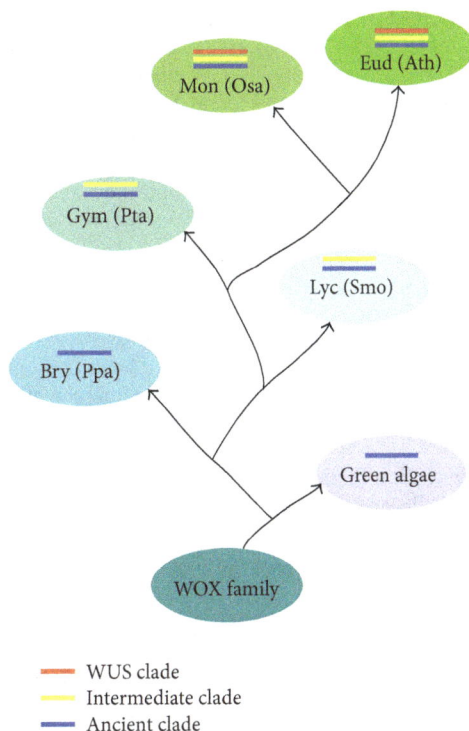

FIGURE 9: A proposed model of the evolutionary history of plant WOX family. The schematic tree represents the proposed evolutionary history of WOX family. Different color represents different clade. The WOX family might originate from green algae and expand in Lycopodiophyta with the appearance of the intermediate clade, and along with the emergence of seed plants, the WOX family generated WUS clade from the intermediate clade. Bry: Bryophyta, Lyc: Lycopodiophyta, Gym: Gymnosperm, Mon: Monocots, Eud: Eudicots, Ppa: *Physcomitrella patens*, Smo: *Selaginella moellendorffii*, Pta: *Pinus taeda*, Osa: *Oryza sativa*, and Ath: *Arabidopsis thaliana*.

be modulated by epigenetic marks and contributing to the expression control of WOXs.

4. Methods

4.1. Search of WOX Members. WOX family proteins were retrieved by TBLASTN using the following databases: the National Center for Biotechnology Information (NCBI) database, The *Arabidopsis* Information Resource (TAIR) database, the Rice Genome Annotation Project (RGAP) database, genomic databases in Joint Genome Institute (JGI) Eukaryotic Genomics, and the Plant Transcription Factor Database (PlantTFDB). We used the full-length sequence of AtWUS as a query sequence for TBLASTN. The E value of all the sequences we obtained was below $1e-5$. The structure and function of all the sequences were also checked to remove the redundant and non-WOX family sequences using the Sol Genomics Network (SGN) database, the UniProt, the SMART, and Sanger database, respectively.

4.2. Multiple Sequence Alignments. Multiple sequence alignments were carried out by using MUSCLE 3.6 with the default parameter setting. In order to obtain a better alignment, we adjusted manually the results based on the location of

the corresponding amino acids in the WOX motif using GeneDoc (version 2.6.002) software.

4.3. Construction of Phylogenetic Tree. A phylogenetic tree using neighbor joining method was constructed with the aligned WOX protein family sequences by MEGA (version 3.0;). NJ analyses were done using the following parameters: poisson correction methods, pairwise deletion of gaps, and bootstrap (1000 replicates; random seed).

4.4. Expression Analysis of AtWOXs and OsWOXs. Expression pattern data of AtWOXs and OsWOXs were obtained from the following databases: AtGenExpress Visualization Tool (AVT) and RiceXPro, respectively. The average values were calculated among the expression values of the organs.

4.5. Prediction of Methylated Region of AtWOXs and OsWOXs. Methyl Primer Express Software was used with the following parameter setting: minimum length of island is 300 bp, C + Gs/Total bases >50%, and CpG observed/CpG expected >0.6.

4.6. Analysis of Pairwise Positional Correlations. Analysis of pairwise positional correlations was obtained from the

Correlation Analysis of Protein Sequences (CRASP) database with the default parameter setting.

Conflict of Interests

The authors declare that there is no conflict of interests regarding the publication of this paper.

Authors' Contribution

Gaibin Lian and Zhiwen Ding contributed equally to this work.

Acknowledgments

The authors would like to acknowledge Novel Bioinformatics Co., Ltd. for the bioinformatics supports kindly provided by Jie Zong and Dai Chen. This work was supported by funds from the National Natural Science Foundation of China (31110103915, 31000593, and 31370026).

References

[1] W. J. Gehring, M. Affolter, and T. Bürglin, "Homeodomain proteins," *Annual Review of Biochemistry*, vol. 63, pp. 487–526, 1994.

[2] E. van der Graaff, T. Laux, and S. A. Rensing, "The WUS homeobox-containing (WOX) protein family," *Genome Biology*, vol. 10, no. 12, article 248, 2009.

[3] R. Derelle, P. Lopez, H. L. Guyader, and M. Manuel, "Homeodomain proteins belong to the ancestral molecular toolkit of eukaryotes," *Evolution & Development*, vol. 9, no. 3, pp. 212–219, 2007.

[4] G. Bharathan, B.-J. Janssen, E. A. Kellogg, and N. Sinha, "Did homeodomain proteins duplicate before the origin of angiosperms, fungi, and metazoa?" *Proceedings of the National Academy of Sciences of the United States of America*, vol. 94, no. 25, pp. 13749–13753, 1997.

[5] K. F. X. Mayer, H. Schoof, A. Haecker, M. Lenhard, G. Jürgens, and T. Laux, "Role of WUSCHEL in regulating stem cell fate in the Arabidopsis shoot meristem," *Cell*, vol. 95, no. 6, pp. 805–815, 1998.

[6] T. Laux, K. F. X. Mayer, J. Berger, and G. Jürgens, "The WUSCHEL gene is required for shoot and floral meristem integrity in Arabidopsis," *Development*, vol. 122, no. 1, pp. 87–96, 1996.

[7] A. Haecker, R. Groß-Hardt, B. Geiges et al., "Expression dynamics of WOX genes mark cell fate decisions during early embryonic patterning in Arabidopsis thaliana," *Development*, vol. 131, no. 3, pp. 657–668, 2004.

[8] R. Shimizu, J. Ji, E. Kelsey, K. Ohtsu, P. S. Schnable, and M. J. Scanlon, "Tissue specificity and evolution of meristematic WOX3 function," *Plant Physiology*, vol. 149, no. 2, pp. 841–850, 2009.

[9] A. K. Sarkar, M. Luijten, S. Miyashima et al., "Conserved factors regulate signalling in Arabidopsis thaliana shoot and root stem cell organizers," *Nature*, vol. 446, no. 7137, pp. 811–814, 2007.

[10] S. O. Park, Z. Zheng, D. G. Oppenheimer, and B. A. Hauser, "The PRETTY FEW SEEDS2 gene encodes an Arabidopsis homeodomain protein that regulates ovule development," *Development*, vol. 132, no. 4, pp. 841–849, 2005.

[11] X. Wu, T. Dabi, and D. Weigel, "Requirement of homeobox gene STIMPY/WOX9 for Arabidopsis meristem growth and maintenance," *Current Biology*, vol. 15, no. 5, pp. 436–440, 2005.

[12] N. Kamiya, H. Nagasaki, A. Morikami, Y. Sato, and M. Matsuoka, "Isolation and characterization of a rice WUSCHEL-type homeobox gene that is specifically expressed in the central cells of a quiescent center in the root apical meristem," *The Plant Journal*, vol. 35, no. 4, pp. 429–441, 2003.

[13] Y. Zhao, Y. Hu, M. Dai, L. Huang, and D.-X. Zhou, "The WUSCHEL-Related homeobox gene WOX11 is required to activate shoot-borne crown root development in rice," *Plant Cell*, vol. 21, no. 3, pp. 736–748, 2009.

[14] X. Zhang, J. Zong, J. Liu, J. Yin, and D. Zhang, "Genome-wide analysis of WOX gene family in rice, sorghum, maize, arabidopsis and poplar," *Journal of Integrative Plant Biology*, vol. 52, no. 11, pp. 1016–1026, 2010.

[15] J. Nardmann and W. Werr, "The shoot stem cell niche in angiosperms: expression patterns of WUS orthologues in rice and maize imply major modifications in the course of mono- and dicot evolution," *Molecular Biology and Evolution*, vol. 23, no. 12, pp. 2492–2504, 2006.

[16] J. Nardmann, R. Zimmermann, D. Durantini, E. Kranz, and W. Werr, "WOX gene phylogeny in poaceae: a comparative approach addressing leaf and embryo development," *Molecular Biology and Evolution*, vol. 24, no. 11, pp. 2474–2484, 2007.

[17] M. Dai, Y. Hu, Y. Zhao, H. Liu, and D.-X. Zhou, "A WUSCHEL-LIKE HOMEOBOX gene represses a YABBY gene expression required for rice leaf development," *Plant Physiology*, vol. 144, no. 1, pp. 380–390, 2007.

[18] J. Nardmann, P. Reisewitz, and W. Werr, "Discrete shoot and root stem cell-promoting WUS/WOX5 functions are an evolutionary innovation of angiosperms," *Molecular Biology and Evolution*, vol. 26, no. 8, pp. 1745–1755, 2009.

[19] Y. Deveaux, C. Toffano-Nioche, G. Claisse et al., "Genes of the most conserved WOX clade in plants affect root and flower development in Arabidopsis," *BMC Evolutionary Biology*, vol. 8, no. 1, article 291, 2008.

[20] X. Wu, J. Chory, and D. Weigel, "Combinations of WOX activities regulate tissue proliferation during Arabidopsis embryonic development," *Developmental Biology*, vol. 309, no. 2, pp. 306–316, 2007.

[21] M. Vandenbussche, A. Horstman, J. Zethof, R. Koes, A. S. Rijpkema, and T. Gerats, "Differential recruitment of WOX transcription factors for lateral development and organ fusion in Petunia and Arabidopsis," *Plant Cell*, vol. 21, no. 8, pp. 2269–2283, 2009.

[22] D. A. Afonnikov and N. A. Kolchanov, "CRASP: a program for analysis of coordinated substitutions in multiple alignments of protein sequences," *Nucleic Acids Research*, vol. 32, pp. W64–W68, 2004.

[23] G. Xu, H. Ma, M. Nei, and H. Kong, "Evolution of F-box genes in plants: different modes of sequence divergence and their relationships with functional diversification," *Proceedings of the National Academy of Sciences of the United States of America*, vol. 106, no. 3, pp. 835–840, 2009.

[24] X. Li, X. Duan, H. Jiang et al., "Genome-wide analysis of basic/helix-loop-helix transcription factor family in rice and Arabidopsis," *Plant Physiology*, vol. 141, no. 4, pp. 1167–1184, 2006.

[25] F. D. Ariel, P. A. Manavella, C. A. Dezar, and R. L. Chan, "The true story of the HD-Zip family," *Trends in Plant Science*, vol. 12, no. 9, pp. 419–426, 2007.

[26] A. Serman, M. Vlahović, L. Serman, and F. Bulić-Jakus, "DNA methylation as a regulatory mechanism for gene expression in mammals," *Collegium Antropologicum*, vol. 30, no. 3, pp. 665–671, 2006.

[27] W. Li, H. Liu, Z. J. Cheng et al., "Dna methylation and histone modifications regulate de novo shoot regeneration in arabidopsis by modulating wuschel expression and auxin signaling," *PLoS Genetics*, vol. 7, no. 8, Article ID e1002243, 2011.

[28] A. K. Sarkar, M. Luijten, S. Miyashima et al., "Conserved factors regulate signalling in Arabidopsis thaliana shoot and root stem cell organizers," *Nature*, vol. 446, no. 7137, pp. 811–814, 2007.

[29] M. Ikeda, N. Mitsuda, and M. Ohme-Takagi, "Arabidopsis wuschel is a bifunctional transcription factor that acts as a repressor in stem cell regulation and as an activator in floral patterning," *Plant Cell*, vol. 21, no. 11, pp. 3493–3505, 2009.

[30] K. Mukherjee, L. Brocchieri, and T. R. Bürglin, "A comprehensive classification and evolutionary analysis of plant homeobox genes," *Molecular Biology and Evolution*, vol. 26, no. 12, pp. 2775–2794, 2009.

14

Effects of Zoledronic Acid on Physiologic Bone Remodeling of Condylar Part of TMJ: A Radiologic and Histomorphometric Examination in Rabbits

Ufuk Tatli,[1] **Yakup Üstün,**[2] **Mehmet Kürkçü,**[1] **and Mehmet Emre Benlidayı**[1]

[1] *Department of Oral and Maxillofacial Surgery, Faculty of Dentistry, Çukurova University, Saricam-Balcali, 01330 Adana, Turkey*
[2] *Private Practice in Oral and Maxillofacial Surgery, 01120 Adana, Turkey*

Correspondence should be addressed to Ufuk Tatli; dr.ufuktatli@gmail.com

Academic Editors: G. Nocca, C. Rossa Jr., and I. Tomas

Objective. The purpose of the present study is to evaluate the effects of systemically administered zoledronic acid (ZA) on the physiological bone remodeling and the microarchitectural parameters of the condylar part of TMJ in a rabbit model. *Study Design.* Thirty skeletally mature male New Zealand white rabbits were randomly divided into two groups. The experimental group was administered an intravenous, single dose of 0.1 mg/kg ZA diluted with 15 mL of saline in a 15-minute perfusion with an infusion pump. The control group was administered only saline infusion for 15 minutes. All rabbits were sacrificed on the 21st postoperative day. Radiodensitometric and histomorphometric examinations were performed on the harvested mandibular condyles. The data were analyzed statistically. *Results.* Radiodensitometric findings showed that ZA treatment resulted in a significant increase in the mineralization of mandibular condyle. This result was supported by the histomorphometric findings. *Conclusion.* The present study has revealed that a temporary delay in the physiological bone remodeling using single dose of ZA increases bone mineral content and makes the microarchitecture of the mandibular condyle more compact. These effects may be regarded as base data and considered in numerous clinical situations including TMJ.

1. Introduction

The bony components of the temporomandibular joint (TMJ) are the articular fossa and articular eminence of temporal bone and mandibular condyle. Under normal physiologic conditions, a balance exists in synovial joints between tissue breakdown and repair. When the balance is disturbed by a mechanical, biomechanical, or inflammatory insult, the internal cartilaginous remodeling system may fail, resulting in accelerated tissue breakdown and articular bone resorption [1]. In the resorption phase, catabolic activities preponderate over anabolic responses resulting in radiographically visible degenerative changes such as flattening, sclerosis, or osteophyte in the articular bony areas [2]. As such, the amount of bone tissue could theoretically be bolstered by increasing anabolism or decreasing catabolism or both. Since the condylar bone is a load-bearing part of TMJ, remodeling process of the condyle is important in preventing microdamage

accumulation as a consequence of repetitive loading during jaw moment and clenching [3, 4]. In recent years there has been increased interest in the effects of antiresorptive therapies on trabecular architecture. Suppression of bone turnover using antiresorptive agents such as bisphosphonates (BPs) prevents bone loss but may also increase tissue mineralization [5].

BPs are a group of synthetic analogs of inorganic pyrophosphate, an endogenous regulator of bone mineralization [6]. BPs are well-recognized inhibitors of osteoclastic activity and have widely been used in the clinical treatment of various systemic metabolic bone diseases. Current indications include Paget's disease [7], hypercalcemia of malignancy [8], postmenopausal osteoporosis [9], fibrous dysplasia [10], osteogenesis imperfect [11], osteoarthritis [12], and rheumatoid arthritis [13]. Zoledronic acid (ZA), a new generation of intravenous BPs, has exhibited the greatest affinity for bone mineral with the longest retention [6]. Nowadays a novel

Effects of Zoledronic Acid on Physiologic Bone Remodeling of Condylar Part of TMJ: A Radiologic and Histomorphometric Examination in Rabbits

115

effect of BPs on bone healing has been defined. Researchers showed that single dose of ZA in rabbits improved bone healing during distraction osteogenesis [14], osseointegration period of dental implants [15], and fracture healing [16] in maxillofacial area.

Published information is lacking on the physiologic trabecular bone remodeling (TBR) in the mandibular condyle, as well as the effects of BP therapy on this condylar TBR [3]. Physiologic bone remodeling and architecture and density of the condylar subchondral bone are continuously constructed to withstand the mechanical forces and to accommodate the stress on the fibrocartilage [17]. Thus, understanding of changes occurring in physiologic bone remodeling of mandibular condyle after BP administration is crucial in the future development of treatment modalities of degenerative TMJ diseases causing condylar bone resorption.

With this background, the purpose of the present study is to evaluate the effects of systemically administered ZA on the physiological bone remodeling and the microarchitectural parameters of the condylar part of TMJ in a rabbit model using radiodensitometric and histomorphometric methods.

2. Materials and Methods

The ethical review committee of Çukurova University Medical Scientific Research Center approved the study. The experimental procedures and care of animals were in accordance with the European Convention for the Protection of Vertebrate Animals used for Experimental Scientific Purposes. A total of 30 skeletally mature, male New Zealand, white rabbits, weighing from 2.8 to 3.4 kg (mean 3.15 ± 0.25), were included in the study. The rabbits were randomly divided into two groups. The experimental group received a single intravenous infusion of 0.1 mg/kg ZA (Zometa; Novartis, Istanbul, Turkey) diluted with 15 mL of saline in a 15-minute perfusion with an infusion pump. The control group received a saline infusion only for 15 minutes. All the rabbits received the drug under general anesthesia, obtained by intramuscular injection of 35 mg/kg ketamine (Ketalar; Pfizer, Istanbul, Turkey) and 3 mg/kg xylazine (Rompun; Bayer, Istanbul, Turkey). Then, the rabbits were kept in separate cages. The food and water intake and weight of the rabbits were recorded daily.

No surgical intervention was performed in the rabbits to see the isolated BP effect on the physiological bone remodeling of the mandibular condyle without cofactors (steroids, TMJ surgery, etc.). Twenty-one days after the ZA infusion, all the rabbits were killed by an intravenous injection of 100 mg/kg sodium pentobarbitone (Pental; IE Ulagay, Istanbul, Turkey), and the mandibles were dissected subperiosteally. The mandibles were split at the midline. Thus, two hemimandibles including condyles were obtained from each rabbit (Figure 1). The condyles were resected from the subcondylar region and the samples were wrapped in saline-soaked gauze and stored at −20°C until the examinations.

2.1. Radiographic Examination. Digital radiographs of all the condyles were taken from the lateral aspect, with an aluminum step wedge attached to the sensor of the digital radiography device (RVG, Trophy Radiologie, Vincennes,

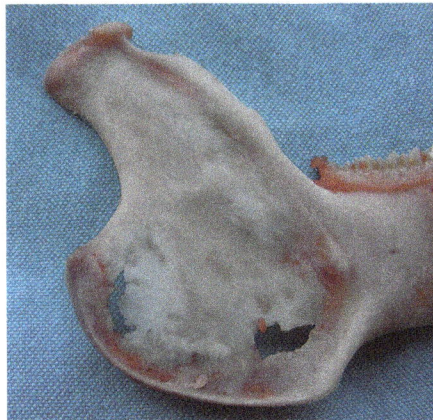

FIGURE 1: Subperiosteally dissected hemimandible of the rabbit including condyle.

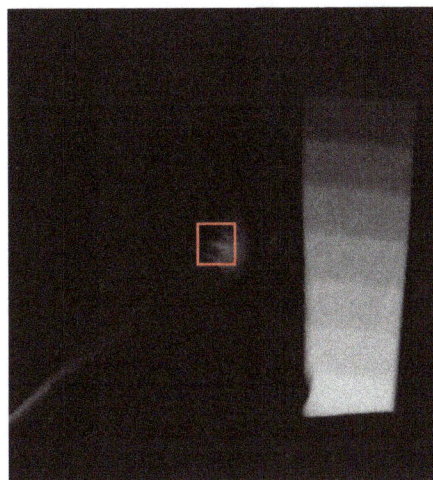

FIGURE 2: Radiographic image of rabbit condyle and aluminum step wedge from control group.

France). The aluminum step wedge consisted of 10 steps, with a thickness of 1 to 10 mm. The same aluminum step wedge was used for all radiographs. The X-ray unit (Philips Densomat, Eindhoven, The Netherlands) was set at 65 kVp, 300 mA, and 0.16 ms. The X-ray cone was directed perpendicularly to the sensor from a distance of 20 cm. The digital images were converted to "tiff" format using imaging software (Adobe Photoshop CS2; Adobe Systems, San Jose, CA, USA) and a standardized measurement area (2 × 2 mm) was outlined in the middle of the condylar bone (Figures 2 and 3). The bone density was measured using image analyzing software (ImageJ, version 1.33u; Wayne Rasband, National Institutes of Health, Bethesda, MD, USA). The gray level of each step of the aluminum step wedge was measured and used for calibration of the software. The aluminum-equivalent bone density of the condylar bone was measured. The results were expressed as millimeters of aluminum.

2.2. Histomorphometric Examination. Undecalcified sections of 30 intact samples from each group were prepared.

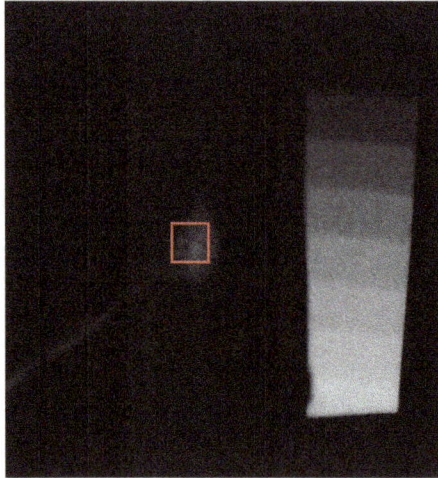

FIGURE 3: Radiographic image of rabbit condyle and aluminum step wedge from ZA-treated group.

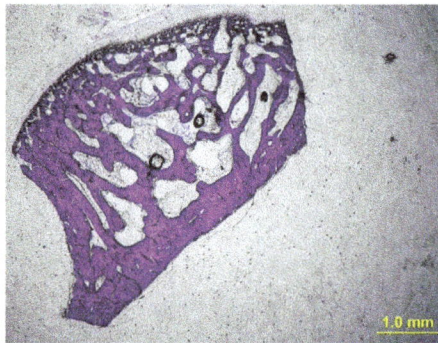

FIGURE 4: A 50 μm thick histologic section prepared for histomorphometric analysis from control group (toluidine blue stain, original magnification ×2).

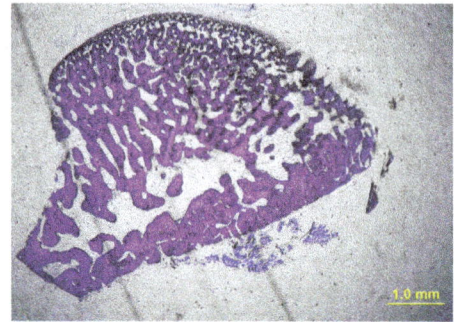

FIGURE 5: A 50 μm thick histologic section prepared for histomorphometric analysis from ZA-treated group (toluidine blue stain, original magnification ×2).

TABLE 1: Comparison of the densitometric data.

Group	Condyle (n)	Aluminum equivalent (mm) (mean ± SD)
Control	30	5.535 ± 2.754
ZA	30	9.676 ± 3.475

$P = 0.001$ (statistically significant).

TABLE 2: Comparison of the histomorphometric data.

Histomorphometric parameters	Control-condyle ($n = 30$)	ZA-condyle ($n = 30$)
Bone volume (%)	57.681 ± 15.95	75.483 ± 9.02*
Trabecular width (mcm)	54.188 ± 13.379	77.296 ± 18.352*
Trabecular thickness (mcm)	41.65 ± 11.547	58.017 ± 19.539*
Trabecular separation (mcm)	35.194 ± 12.025	16.721 ± 8.296*
Node-terminus ratio (NNd/NTm)	1.999 ± 1.399	3.159 ± 1.262*

*$P < 0.05$ (statistically significant).

The histomorphometric examination was performed as described by our previous study [16]. The specimens were fixed in 10% buffered formalin, dehydrated in increasing concentrations of ethanol of 70% to 99% for 10 days, and embedded in methylmethacrylate (Technovit 7200VLC; Heraeus Kulzer GmbH, Wehrheim, Germany). The 50 μm thick sagittal sections were prepared using an electric diamond saw and grinding system (Exakt; Exakt Vertriebs, Norderstedt, Germany) and stained with toluidine blue. Digital images of the sections were obtained using a digital camera (Camedia C4040; Olympus, Tokyo, Japan) attached to an Olympus BX50 microscope (Olympus) at a magnification rate of 2x (Figures 4 and 5). The images were transferred to a personal computer, and a standardized measurement area (2 × 2 mm) was outlined in the middle of the condylar bone. Bone volume, trabecular width, trabecular thickness, trabecular separation, and node/terminus ratio measurements were made using histomorphometry software (TAS, version 1.2.9; Steve Paxton, University of Leeds, Leeds, West Yorkshire, UK). The nomenclature and calculations for bone histomorphometry were applied in accordance with the report from the American Society for Bone and Mineral Research [18].

2.3. Statistical Analysis. Statistical analysis was performed with SPSS software, version 11.5 (SPSS, Chicago, IL). The data from the radiographic and histomorphometric evaluations were statistically analyzed using the unpaired t-test (Student's t-test). $P < 0.05$ was considered significant.

3. Results

The rabbits developed no complication during the study period. All the rabbits were considered for evaluation.

3.1. Radiographic Analysis. The mean aluminum thickness equivalent of the gray pixel value at the condylar area was 5.535 ± 2.754 mm of aluminum for the control group and 9.676 ± 3.475 mm of aluminum for the ZA-treated group (Table 1). The bone density was 1.74 times increased in the ZA-treated group and the difference between the two groups was statistically significant ($P = 0.001$).

3.2. Histomorphometric Analysis. The histomorphometric data were listed in Table 2. The differences in bone volume,

Effects of Zoledronic Acid on Physiologic Bone Remodeling of Condylar Part of TMJ: A Radiologic and
Histomorphometric Examination in Rabbits

117

trabecular width, trabecular thickness, trabecular separation, and node/terminus ratio between the two groups were statistically significant ($P = 0.001$, $P = 0.001$, $P = 0.009$, $P < 0.001$, and $P = 0.024$, resp.). In terms of bone microarchitecture, bone volume was 1.3 times, trabecular width was 1.42 times, trabecular thickness was 1.39 times, and node/terminus ratio was 1.58 times increased; on the contrary, trabecular separation was 0.47 times decreased in the ZA-treated condyles.

4. Discussion

Bone resorption occurs on the condylar part of TMJ in pathological conditions such as excessive trauma and inflammation [2]. The main goal of the treatment of degenerative and osteoarthritic changes of TMJ is to resolve the inflammatory resorptive activity at the articular region. BPs have become the primary therapy for treating diseases of unbalanced bone resorption [6]. In vivo bone turnover is determined by a delicate balance between osteoclastic bone resorption and osteoblastic bone formation. von Knoch et al. [19] suggest that BPs impact both sides of this balance: inhibit osteoclastic activity and have an anabolic effect on osteoblasts. Suppression of bone remodeling was demonstrated in dogs on BP therapy, with greater suppression at sites with higher levels of physiologic remodeling [20]. So, we hypothesized that high local bone turnover particular to the condylar bone may be shifted towards a positive balance by an adjunct BP therapy in the condylar part of the TMJ.

The underlying molecular mechanism in nitrogen-containing BPs, such as alendronate and risedronate, is the inhibition of enzymes in the mevalonate pathway of cholesterol synthesis that are essential for osteoclast activity and survival [21]. Consequently, BPs inactivate osteoclasts, which then undergo apoptosis, resulting in reduced bone resorption, lower bone turnover, and a positive bone balance [21]. Another pharmacologic action of BPs is the proliferation and maturation of osteoblasts [19]. Naidu et al. [22] reported that lower concentrations of BPs had a beneficial effect on osteoblast viability and function. Thus, reduced bone turnover allows more time for mineralization of existing bone, increasing the bone density [6]. BP treatment leads to the retention of trabeculae that act as a scaffold for more bone to be deposited on [23]. In the light of radiodensitometric analysis, the present study indicated a significantly greater amount of mineralized bone (1.74 times greater) in the ZA-treated group. In the literature, it was reported that the experimental TMJ arthritis resulted in low degree of mineralization compared to healthy condyles and was associated with morphological changes [24]. According to this background with the result of the present study, adjunct antiresorptive effects of ZA may theoretically support the treatment of TMJ arthritis by improving bone mineralization. Further studies involving samples with experimentally induced degenerative TMJ disorders are necessary in order to make more clear comments.

Microarchitecture is an important element of bone quality. Thus, the assessment of bone microarchitecture is crucial in evaluating the effects of adjunct antiresorptive drug therapies. Several methods are available to assess the bone architecture, particularly at the trabecular level, including histomorphometry, quantitative computed tomography, high-resolution computed tomography, volumetric quantitative computed tomography, and high-resolution magnetic resonance imaging [25, 26]. In the present study, the quantitative assessment of the condylar bone was performed using the histomorphometric method. Histomorphometric examination allows the measurement of the trabecular profiles and the count of their connections on two-dimensional sections. Recent observations seem to confirm that microstructural alterations are important determinants of bone strength, independently of bone density [27]. Trabecular separation has been defined as the distance between the edges of the trabeculae [18]. The ratio between the nodes and termini in a section is an index of the spatial connectivity in the trabecular network [27]. In the light of histomorphometric analysis, the present study showed that the administration of single dose ZA made the microarchitecture of the mandibular condyle more compact in rabbits. In an experimental study in dogs, Helm et al. [3] demonstrated that a total of 4 infusions of ZA administered monthly resulted in reduction in trabecular bone remodeling of mandibular condyle. However, the authors of the aforementioned study did not report significant difference in dogs between ZA-treated and control groups in terms of microarchitectural parameters. This might be due to different animal models.

The increase in the mechanical fixation of metallic biomaterials (metallic joint prosthesis, plate, and screws) in bone is considered an important factor in terms of treatment success. Tengvall et al. [28] demonstrated that surface treatment with BPs improved the mechanical fixation of stainless-steel screws. Consequently, BPs could also be used to improve the fixation of prosthetic joint replacement components in the surrounding bone. Further studies involving the samples with prosthetic joint replacements are necessary in order to make more clear comments about this phenomenon.

The levels of physiologic bone remodeling differ among types of bone, skeletal sites, and regions within skeletal sites, as well as with age [3]. The jawbones might be more affected than other parts of the skeletal system because of the increased bone remodeling that occurs around teeth in the alveolar region [29]. Mandibular condyle is an important growth center and also functions as an articular structure that resists compressive forces [3]. This might result in an excessive amount of BPs deposited in these mentioned regions. Therefore, the positive and negative effects of BPs in such regions must be well recognized in maxillofacial practice. The present study is the first investigation in which the effects of single dose of ZA on physiological bone remodeling of the condylar part of TMJ were evaluated.

BPs have a well-documented profile of possible side effects. An initial influenza-like illness has been documented with the first infusion of BPs. Renal failure has been noted in patients with cancer after repetitive high-dose infusions [30]. Recently, an association between BPs and osteonecrosis of the jaw was reported after oral surgical procedure or trauma [29]. Most of these complications have occurred in patients with cancer who have often received monthly high-dose BP infusions. To our knowledge, no data are available concerning the relationship between single-dose administration and the possible side effects of BPs.

In the present study, ZA was administered as a single dose of 0.1 mg/kg consistent with previous studies [14–16]. It has been proved that the plasma concentration of the drug gradually declines within 28 days [31]. Thus, a repeat dose of ZA could be administered 28 days after the initial single dose, if required. However, further studies are necessary to evaluate the effects of redosing on the physiological bone remodeling compared with the application of a single dose.

In conclusion, the result of the present experimental study has revealed that a temporary delay in physiological bone remodeling using single dose of ZA increases bone mineral content and makes the microarchitecture of the mandibular condyle more compact. These effects may be regarded as base data and considered in numerous clinical situations including TMJ.

Conflict of Interests

The authors declare that they have no conflict of interests regarding the publication of this paper.

Acknowledgment

This study was presented in the 7th International Oral and Maxillofacial Surgery Society Congress on May 29–June 2, 2013, in Antalya, Turkey, and won the Best Third Oral Presentation Award.

References

[1] L. C. Dijkgraaf, L. G. M. De Bont, G. Boering, and R. S. B. Liem, "The structure, biochemistry, and metabolism of osteoarthritic cartilage: a review of the literature," *Journal of Oral and Maxillofacial Surgery*, vol. 53, no. 10, pp. 1182–1192, 1995.

[2] R. de Leeuw, "Internal derangements of the temporomandibular joint," *Oral and Maxillofacial Surgery Clinics of North America*, vol. 20, no. 2, pp. 159–168, 2008.

[3] N. B. Helm, S. Padala, F. M. Beck, A. M. D'Atri, and S. S. Huja, "Short-term zoledronic acid reduces trabecular bone remodeling in dogs," *European Journal of Oral Sciences*, vol. 118, no. 5, pp. 460–465, 2010.

[4] K. Jiao, J. Dai, M.-Q. Wang, L.-N. Niu, S.-B. Yu, and X.-D. Liu, "Age- and sex-related changes of mandibular condylar cartilage and subchondral bone: a histomorphometric and micro-CT study in rats," *Archives of Oral Biology*, vol. 55, no. 2, pp. 155–163, 2010.

[5] K. S. Davison, K. Siminoski, J. D. Adachi et al., "The effects of antifracture therapies on the components of bone strength: assessment of fracture risk today and in the future," *Seminars in Arthritis and Rheumatism*, vol. 36, no. 1, pp. 10–21, 2006.

[6] M. T. Drake, B. L. Clarke, and S. Khosla, "Bisphosphonates: mechanism of action and role in clinical practice," *Mayo Clinic Proceedings*, vol. 83, no. 9, pp. 1032–1045, 2008.

[7] J. P. Walsh, L. C. Ward, G. O. Stewart et al., "A randomized clinical trial comparing oral alendronate and intravenous pamidronate for the treatment of Paget's disease of bone," *Bone*, vol. 34, no. 4, pp. 747–754, 2004.

[8] K. Wellington and K. L. Goa, "Zoledronic acid: a review of its use in the management of bone metastases and hypercalcaemia of malignancy," *Drugs*, vol. 63, no. 4, pp. 417–437, 2003.

[9] H. G. Bone, D. Hosking, J.-P. Devogelaer et al., "Ten years' experience with alendronate for osteoporosis in postmenopausal women," *The New England Journal of Medicine*, vol. 350, no. 12, pp. 1189–1199, 2004.

[10] J. M. Lane, S. N. Khan, W. J. O'Connor et al., "Bisphosphonate therapy in fibrous dysplasia," *Clinical Orthopaedics and Related Research*, no. 382, pp. 6–12, 2001.

[11] J.-P. Devogelaer, "New uses of bisphosphonates: osteogenesis imperfecta," *Current Opinion in Pharmacology*, vol. 2, no. 6, pp. 748–753, 2002.

[12] H. J. Lehmann, U. Mouritzen, S. Christgau, P. A. C. Cloos, and C. Christiansen, "Effect of bisphosphonates on cartilage turnover assessed with a newly developed assay for collagen type II degradation products," *Annals of the Rheumatic Diseases*, vol. 61, no. 6, pp. 530–533, 2002.

[13] W. P. Maksymowych, "Bisphosphonates for arthritis—a confusing rationale," *Journal of Rheumatology*, vol. 30, no. 3, pp. 430–434, 2003.

[14] A. A. Pampu, D. Dolanmaz, H. H. Tüz, and A. Karabacakoglu, "Experimental evaluation of the effects of zoledronic acid on regenerate bone formation and osteoporosis in mandibular distraction osteogenesis," *Journal of Oral and Maxillofacial Surgery*, vol. 64, no. 8, pp. 1232–1236, 2006.

[15] A. Yildiz, E. Esen, M. Kürkçü, I. Damlar, K. Dağlioğlu, and T. Akova, "Effect of zoledronic acid on osseointegration of titanium implants: an experimental study in an ovariectomized rabbit model," *Journal of Oral and Maxillofacial Surgery*, vol. 68, no. 3, pp. 515–523, 2010.

[16] U. Tatli, Y. Üstün, M. Kürkçü et al., "Effects of zoledronic acid on healing of mandibular fractures: an experimental study in rabbits," *Journal of Oral and Maxillofacial Surgery*, vol. 69, no. 6, pp. 1726–1735, 2011.

[17] E. B. W. Giesen, M. Ding, M. Dalstra, and T. M. G. J. Van Eijden, "Mechanical properties of cancellous bone in the human mandibular condyle are anisotropic," *Journal of Biomechanics*, vol. 34, no. 6, pp. 799–803, 2001.

[18] D. W. Dempster, J. E. Compston, M. K. Drezner et al., "Standardized nomenclature, symbols, and units for bone histomorphometry: a 2012 update of the report of the ASBMR Histomorphometry Nomenclature Committee," *Journal of Bone and Mineral Research*, vol. 28, no. 1, pp. 2–17, 2013.

[19] F. von Knoch, C. Jaquiery, M. Kowalsky et al., "Effects of bisphosphonates on proliferation and osteoblast differentiation of human bone marrow stromal cells," *Biomaterials*, vol. 26, no. 34, pp. 6941–6949, 2005.

[20] M. R. Allen, D. J. Kubek, and D. B. Burr, "Cancer treatment dosing regimens of zoledronic acid result in near-complete suppression of mandible intracortical bone remodeling in beagle dogs," *Journal of Bone and Mineral Research*, vol. 25, no. 1, pp. 98–105, 2010.

[21] F. P. Coxon, K. Thompson, and M. J. Rogers, "Recent advances in understanding the mechanism of action of bisphosphonates," *Current Opinion in Pharmacology*, vol. 6, no. 3, pp. 307–312, 2006.

[22] A. Naidu, P. C. Dechow, R. Spears, J. M. Wright, H. P. Kessler, and L. A. Opperman, "The effects of bisphosphonates on osteoblasts in vitro," *Oral Surgery, Oral Medicine, Oral Pathology, Oral Radiology and Endodontology*, vol. 106, no. 1, pp. 829–837, 2008.

Effects of Zoledronic Acid on Physiologic Bone Remodeling of Condylar Part of TMJ: A Radiologic and
Histomorphometric Examination in Rabbits

119

[23] N. Amanat, M. McDonald, C. Godfrey, L. Bilston, and D. Little, "Optimal timing of a single dose of zoledronic acid to increase strength in rat fracture repair," *Journal of Bone and Mineral Research*, vol. 22, no. 6, pp. 867–876, 2007.

[24] K. D. Kristensen, E.-M. Hauge, M. Dalstra et al., "Association between condylar morphology and changes in bony microstructure and sub-synovial inflammation in experimental temporomandibular joint arthritis," *Journal of Oral Pathology & Medicine*, vol. 40, no. 1, pp. 111–120, 2011.

[25] A. Laib, D. C. Newitt, Y. Lu, and S. Majumdar, "New model-independent measures of trabecular bone structure applied to in vivo high-resolution MR images," *Osteoporosis International*, vol. 13, no. 2, pp. 130–136, 2002.

[26] R. Müller, H. Van Campenhout, B. Van Damme et al., "Morphometric analysis of human bone biopsies: a quantitative structural comparison of histological sections and micro-computed tomography," *Bone*, vol. 23, no. 1, pp. 59–66, 1998.

[27] L. Dalle Carbonare, M. T. Valenti, F. Bertoldo et al., "Bone microarchitecture evaluated by histomorphometry," *Micron*, vol. 36, no. 7-8, pp. 609–616, 2005.

[28] P. Tengvall, B. Skoglund, A. Askendal, and P. Aspenberg, "Surface immobilized bisphosphonate improves stainless-steel screw fixation in rats," *Biomaterials*, vol. 25, no. 11, pp. 2133–2138, 2004.

[29] R. E. Marx, Y. Sawatari, M. Fortin, and V. Broumand, "Bisphosphonate-induced exposed bone (osteonecrosis/osteopetrosis) of the jaws: risk factors, recognition, prevention, and treatment," *Journal of Oral and Maxillofacial Surgery*, vol. 63, no. 11, pp. 1567–1575, 2005.

[30] J.-J. Body, I. Diel, and R. Bell, "Profiling the safety and tolerability of bisphosphonates," *Seminars in Oncology*, vol. 31, no. 10, pp. 73–78, 2004.

[31] T. Chen, J. Berenson, R. Vescio et al., "Pharmacokinetics and pharmacodynamics of zoledronic acid in cancer patients with bone metastases," *Journal of Clinical Pharmacology*, vol. 42, no. 11, pp. 1228–1236, 2002.

15

Extensive Introgression among Ancestral mtDNA Lineages: Phylogenetic Relationships of the Utaka within the Lake Malawi Cichlid Flock

Dieter Anseeuw,[1,2] Bruno Nevado,[3,4] Paul Busselen,[1] Jos Snoeks,[5,6] and Erik Verheyen[3,4]

[1] *Interdisciplinary Research Centre, K. U. Leuven Campus Kortrijk, Etienne Sabbelaan 53, 8500 Kortrijk, Belgium*
[2] *KATHO, Wilgenstraat 32, 8800 Roeselare, Belgium*
[3] *Vertebrate Department, Royal Belgian Institute of Natural Sciences, Vautierstraat 29, 1000 Brussels, Belgium*
[4] *Evolutionary Ecology Group, University of Antwerp, Middelheimcampus G.V. 332, Groenenborgerlaan 171, 2020 Antwerp, Belgium*
[5] *Zoology Department, Royal Museum for Central Africa, Leuvensesteenweg 13, 3080 Tervuren, Belgium*
[6] *Laboratory of Biodiversity and Evolutionary Genomics, K. U. Leuven, Charles Deberiotstraat 32, 3000 Leuven, Belgium*

Correspondence should be addressed to Dieter Anseeuw, dieter.anseeuw@katho.be

Academic Editor: Stephan Koblmüller

We present a comprehensive phylogenetic analysis of the Utaka, an informal taxonomic group of cichlid species from Lake Malawi. We analyse both nuclear and mtDNA data from five Utaka species representing two (*Copadichromis* and *Mchenga*) of the three genera within Utaka. Within three of the five analysed species we find two very divergent mtDNA lineages. These lineages are widespread and occur sympatrically in conspecific individuals in different areas throughout the lake. In a broader taxonomic context including representatives of the main groups within the Lake Malawi cichlid fauna, we find that one of these lineages clusters within the non-Mbuna mtDNA clade, while the other forms a separate clade stemming from the base of the Malawian cichlid radiation. This second mtDNA lineage was only found in Utaka individuals, mostly within *Copadichromis* sp. "virginalis kajose" specimens. The nuclear genes analysed, on the other hand, did not show traces of divergence within each species. We suggest that the discrepancy between the mtDNA and the nuclear DNA signatures is best explained by a past hybridisation event by which the mtDNA of another species introgressed into the ancestral *Copadichromis* sp. "virginalis kajose" gene pool.

1. Introduction

The Lake Malawi cichlid fauna comprises over 800 species [1] offering a spectacular example of adaptive radiation with virtually all niches in the lake being filled by members of this family [2, 3]. With a few exceptions, all Lake Malawi cichlids form a monophyletic group as supported by mitochondrial [4–6] and nuclear ([7–9] but see [10]) markers as well as allozymes [11, 12].

The phylogenetic reconstruction of Lake Malawi cichlid fauna has recovered six main mitochondrial DNA (mtDNA) lineages [5, 6, 10, 13]. Two of these lineages correspond to the *Rhamphochromis* and *Diplotaxodon* genera. A third lineage contains the nonendemic riverine *Astatotilapia calliptera*. The remaining cichlid fauna has been traditionally divided into two groups: one containing predominantly the rock-dwelling species commonly called Mbuna and the second containing the remaining Lake Malawi cichlids. However, phylogenetic reconstructions have shown that both groups are artificial [5, 10, 13, 14]. Several *Lethrinops, Aulonocara,* and *Alticorpus* species (ecologically and morphologically typically assigned to the non-Mbuna) cluster within the Mbuna clade. Furthermore, the non-Mbuna genus *Copadichromis* has been shown to have representatives belonging to both the non-Mbuna clade, as well as to a separate lineage. The genus *Copadichromis*, together with the genus *Nyassachromis* and the newly erected genus *Mchenga*, constitute the Utaka, a species assemblage of midwater-feeding zooplanktivorous cichlid species. The phylogenetic position of this group remains unclear with Moran et al. [5] and Turner et al. [13] not recovering mtDNA monophyly within this assemblage: *M. eucinostomus* and *C. borleyi* were placed within the non-Mbuna clade, while *C. mloto* (reidentified as *Copadichromis* sp. "virginalis kajose", J. Snoeks pers. obs.) and some other individuals of the *Copadichromis virginalis* complex seemed to represent a different, well-diverged lineage.

However only few Utaka specimens have been included in phylogenetic analyses so far. Therefore, currently available mtDNA phylogenies are inconclusive as to whether the Utaka are genetically associated with the non-Mbuna clade, whether they constitute an originally separate ancestral lineage, or whether only one or a few species or specimens cluster in a separate lineage. If specimens of a species cluster in genetically distant lineages, this may be a result of the retention of ancestral polymorphism, the existence of a cryptic species, or traces of a past hybridisation/introgression event. Support for these alternative hypotheses may be gained by using a multilocus approach (e.g., [10, 14–16]). We therefore combined mtDNA gene sequences with data from nuclear microsatellite loci. If the nuclear genetic signature is concordant with the mtDNA in subdividing a species into genetically separated units, this may point towards a cryptic species. On the other hand, if a mtDNA split within a species is not supported by the nuclear genetic data, this may be an indication of introgression of genetic material from another species, or of shared ancestral polymorphism.

Whereas the resolution of the specific interrelationships within the major clades remains problematic, the six main mtDNA clades of the Malawi cichlid flock are clearly delineated [5, 6, 13]. Shared polymorphism within taxa might result from incomplete lineage sorting, taxonomic inaccuracies, and/or hybridisation. While the other possibilities cannot be completely ruled out, there is a growing number of studies acknowledging the important role of hybridisation in the evolutionary history of adaptive radiations (e.g., [17–19]). In this study we present the most comprehensive mtDNA phylogeny of the Utaka assemblage so far. We aim at elucidating the phylogenetic position of the Utaka within the Malawian cichlid radiation and shed light on the causes for its taxonomic and molecular assignment inconsistency.

2. Material and Methods

2.1. Taxonomic Sampling. We examined individuals of five Utaka species (*Copadichromis* sp. "virginalis kajose", *C. quadrimaculatus*, *M. eucinostomus*, *C. chrysonotus*, and *C. borleyi*) from twelve localities throughout Lake Malawi and one locality in Lake Malombe (Figure 1). Pelvic fin clips were preserved in 100% ethanol and stored at room temperature. Voucher specimens were fixed in 10% formalin and are curated at the Royal Museum for Central Africa in Tervuren, Belgium. We included additional *Copadichromis* species that were sampled during the SADC/GEF project [1] and previously published mtDNA control region (complete D-loop) sequences of Lake Malawi cichlids, which we obtained from GenBank.

2.2. DNA Extraction, mtDNA Amplification, and Sequencing. Whole genomic DNA was extracted from ethanol-preserved fin clips using proteinase K digestion and salt precipitation,

according to Aljanabi and Martinez [20]. DNA extracts were resuspended in $100\,\mu L$ of autoclaved Milli Q water. The first fragment of the mtDNA control region was sequenced for 412 Utaka specimens (179 *Copadichromis* sp. "virginalis kajose", Genbank Accession EF211832-EF211945 and EF647210-EF647271; 55 *C. quadrimaculatus*, Genbank Accession EF647341-EF647438 and EF647578-EF647579; 67 *M. eucinostomus*, Genbank Accession EF647356-EF647390, EF647439-EF647460, EF647498-EF647505 and EF647581-EF647582; 70 *C. chrysonotus*, Genbank Accession EF647273-EF647340, and EF647571-EF647572; 41 *C. borleyi*, Genbank Accession EF647470-EF647497, EF647520-EF647531, and EF647548), using published primers by Meyer et al. [4]. We additionally sequenced the second fragment of the control region using the primers by Salzburger et al. [21] and Lee et al. [22] for 14 individuals, selected on the basis of the results of the phylogenetic reconstruction for the first fragment of the control region. Polymerase chain reactions (PCRs) were carried out in $25\,\mu L$ buffered reaction mixtures, containing $5\,\mu L$ template DNA, $5\,\mu L$ of each primer ($2\,\mu M$), $200\,\mu M$ of each dNTP, $2.5\,\mu L$ of 10x buffer (1 mM MgCl2), and 0.65 units of Red Taq Polymerase (Sigma Aldrich). PCRs were performed under the following conditions: $94°C$ for $120\,s$, followed by 35 cycles of $94°C$ for $60\,s$, $52°C$ for $60\,s$, $72°C$ for $120\,s$, followed by $72°C$ for 10 min. PCR products were purified following the TMQiaquick PCR purification Kit protocol and sequenced on an ABI 3130 automatic sequencer (Applied Biosystems) using standard protocols.

2.3. Microsatellite Variation. A total of 179 *C.* sp. "virginalis kajose", 230 *C. chrysonotus*, 252 *C. quadrimaculatus,* and 344 *M. eucinostomus* individuals were screened for genetic variation at nine microsatellite markers: Pzeb1, Pzeb3, Pzeb4, Pzeb5 [23], UNH002 [24], TmoM5, TmoM11, TmoM27 [25], and UME003 [26]. PCRs were performed under the following conditions: $94°C$ for $120\,s$, followed by 5 cycles of $94°C$ for $45\,s$; $55°C$ for $45\,s$; $72°C$ for $45\,s$, followed by 30 cycles of $90°C$ for $30\,s$; $55°C$ for $30\,s$; $72°C$ for $30\,s$, followed by $72°C$ for 10 min. $10\,\mu L$ reaction mixes included $1\,\mu L$ template DNA, $0.5\,\mu M$ of each primer, $200\,\mu M$ of each dNTP, 0.26 units Taq polymerase (Sigma Aldrich, Germany), $1\,\mu L$ $10\times$ reaction buffer (Sigma Aldrich). PCR amplification products were run on 6% denaturing polyacrylamide gels using an ALF Express DNA Sequencer (Amersham Pharmacia Biotech). Fragment sizes were scored with ALFWin Fragment Analyser v1.0 (Amersham Pharmacia Biotech), using M13mp8 DNA standards as external references, following van Oppen et al. [23].

2.4. Phylogenetic Reconstructions. For the reconstruction of the phylogenetic relationships of the Utaka, two datasets were analysed. Both were aligned using CLUSTALW [27] and visually checked afterwards using the program SEAVIEW [28]. The first dataset contained 412 *Copadichromis* spp. and *Mchenga* sp. sequences of the first fragment of the control region (328 bp). The program COLLAPSE v1.2 [29] was used to reduce this dataset to one individual sequence

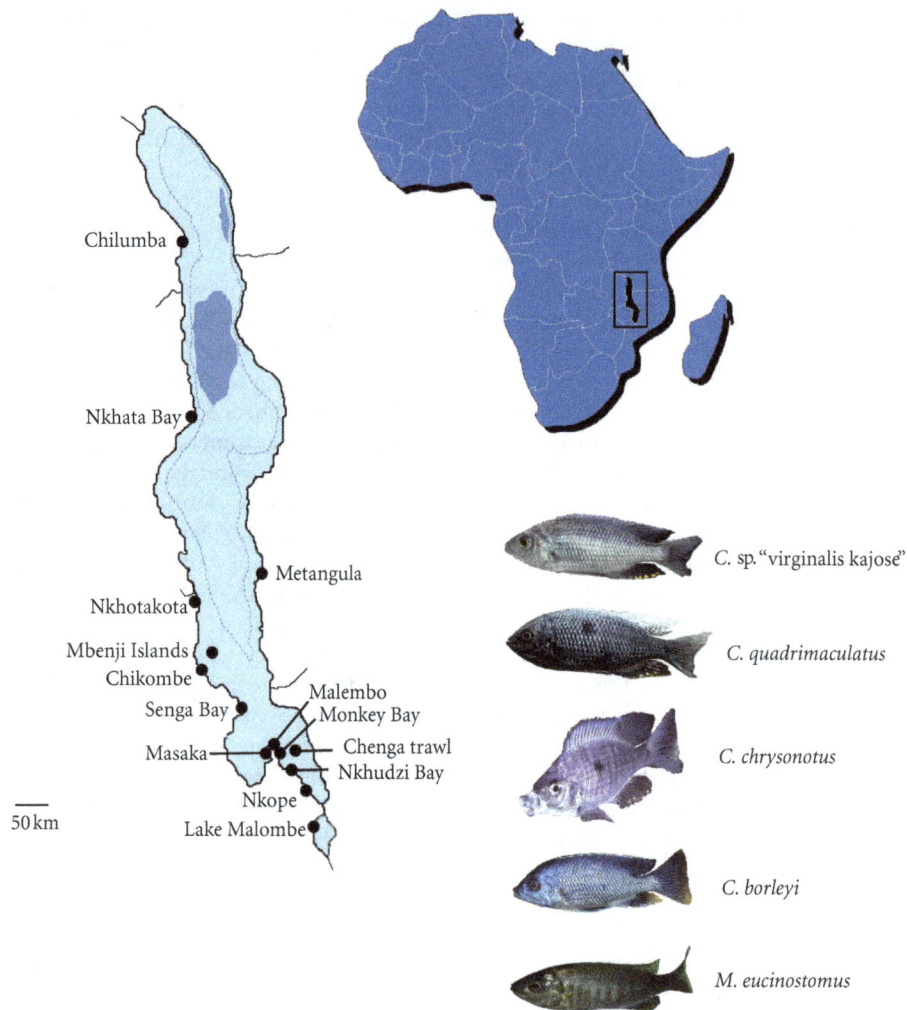

FIGURE 1: Map of Lake Malawi showing the localities sampled. A detailed listing on the origin of each specimen is presented in the Supplementary Table of the Supplementary Material available online at doi:10.1155/2012/865603.

per haplotype for further analyses. GTR+G+I model was the best-fitted model of sequence evolution inferred by MODELTEST v3.7 [30] according to the Akaike Information Criterion (AIC). A maximum likelihood (ML) heuristic search was performed with PHYML [31] starting from a neighbour-joining (NJ) tree. Parameters of the tree and of the substitution model were optimised sequentially until no increase in likelihood was found. The program TCS [32] was used to generate a haplotype network using statistical parsimony. Based on the results of the short control region phylogenetic reconstruction, we performed a second, more computationally intensive phylogenetic analysis with a smaller dataset containing 47 representatives of the different main lineages in the Lake Malawi cichlid flock (both new sequences and sequences extracted from GenBank) to test the interrelationships between these main lineages. Sites that could not be unambiguously aligned were removed prior to analysis. The final dataset, 837 bp long, was first run through MODELTEST, which selected the TrN+I+G model (AIC criterion).

Phylogenetic inferences were carried out using maximum-parsimony (MP, 100 replicates starting from random stepwise addition trees; TBR branch swapping) with different transition-transversion weights (1:1, 2:1 and 3:1) in PAUP* v4.0 [27]. ML reconstructions (100 replicates starting from random stepwise addition trees; TBR branch swapping) were run in PAUP*. Sequential searches were performed by reestimating the substitution model parameters upon the best tree found and then running a new search with these parameters. This was done until no change in the likelihood of the tree or in the estimated parameters was found. Support for the internal branches in the ML tree was assessed by analysing 100 bootstrapped replicates in the program PHYML. For Bayesian inference (BI) analyses, the GTR+G+I model was used since the TrN+G+I is not implemented in MRBAYES v.3.1 [33]. Markov Chain Monte Carlo samplings were run for 25 million generations. Two runs with four chains for each run were sampled every thousandth generation until the average standard deviation of split frequencies between runs reached ~0.003. Inspection

Extensive Introgression among Ancestral mtDNA Lineages: Phylogenetic Relationships of the Utaka within the Lake Malawi Cichlid Flock

123

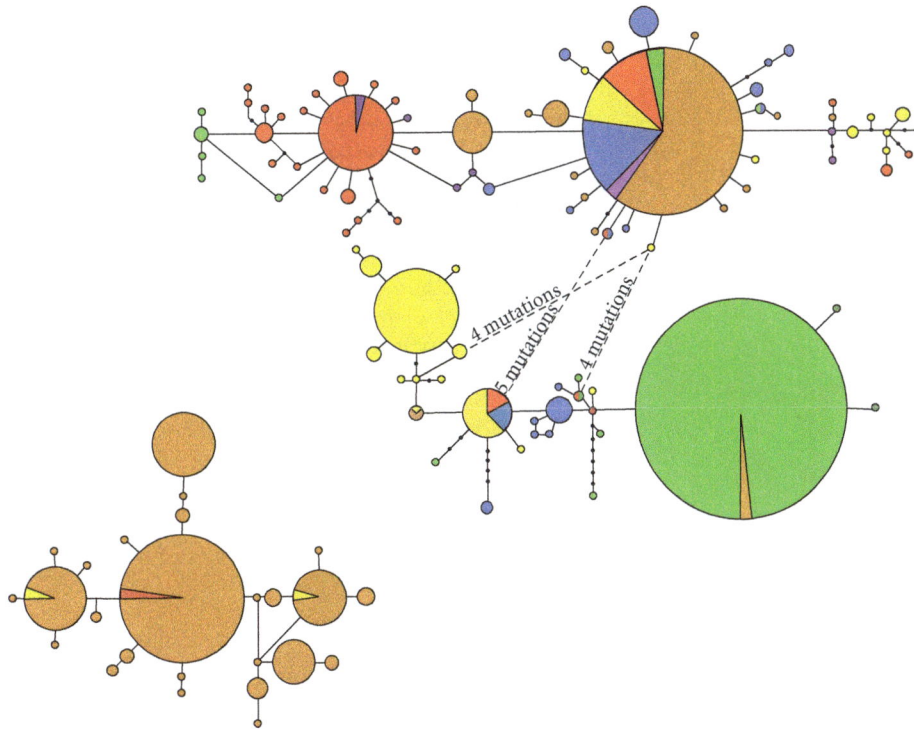

FIGURE 2: Haplotype networks of the Utaka obtained in this study. The upper network contains the majority of the specimens analysed and is separated from the lower network (named virginalis clade in text) by more than seven mutations. Each circle represents a haplotype and is coloured according to the respective species: blue: *C. borleyi*; orange: *C.* sp. "virginalis kajose"; red: *C. quadrimaculatus*; green: *C. chrysonotus* yellow: *M. eucinostomus*; purple: *C. mloto* and *C.* sp. "meta". Size of the circles is proportional to the frequency of each haplotype as indicated in the scaled circles. Small black circles in branches represent missing haplotypes. Dashed lines represent alternative connections with the number of missing haplotypes written above the lines.

of plot of likelihood versus generation revealed that the runs had reached stability and so did the analysis of the Potential Scale Reduction Factors.

Using the Shimodaira-Hasegawa test [34], as implemented in PAUP*, we tested the relative fit of two alternative tree topologies: the forced monophyly of all Utaka specimens was compared to the best, unconstrained tree. Significance of the difference in log-likelihood between the two trees was assessed by means of the Resample Estimated Log-Likelihood test (RELL).

2.5. Microsatellite Data Analysis. Linkage disequilibrium between loci was tested using exact tests as implemented by GENEPOP 3.3 [35]. We estimated the number of populations present in our microsatellite dataset using the program STRUCTURE [36, 37]. We calculated the posterior probability for different numbers of putative populations (K from 1 to 18 populations) using a model-based assignment. Burn-in was set at 100,000 steps followed by 300,000 MCMC iterations at each K. Simulations were run five times for each K to check for convergence of the MCMC. We performed clustering both under the admixture model without prior population information and with correlated allele frequencies between populations. To determine the most likely number of clusters, the rate of change in the log probability of data and in the statistic ΔK [38] between

successive K values was estimated using StructureHarvester [39].

3. Results

3.1. Phylogenetic Reconstructions. The purpose of our phylogenetic analyses was twofold. First, we assessed the phylogenetic relationships among as many specimens as available from the five Utaka species that we collected throughout the lake. For this extensive dataset, we sequenced the short (328 bp) but most variable part of the mtDNA control region. By this analysis we aimed to detect specific or geographical patterns among the Utaka species studied. Second, we attempted to resolve the phylogenetic position of the Utaka species within the Lake Malawi cichlid flock. For this purpose we sequenced the complete mitochondrial control region for representative specimens (n = 14) of the previous dataset and included published sequences from species representing the main lineages in the Malawian cichlid flock. A total of 115 haplotypes were found amongst the 412 Utaka short mtDNA control region sequences (Figure 2). The ML tree presented two divergent clades within the Utaka: a large clade containing circa 70% of all sequences, and a smaller group. The latter almost exclusively contained *C.* sp. "virginalis kajose" individuals (125 *C.* sp. "virginalis kajose" individuals out of 179 sequenced clustered within this clade),

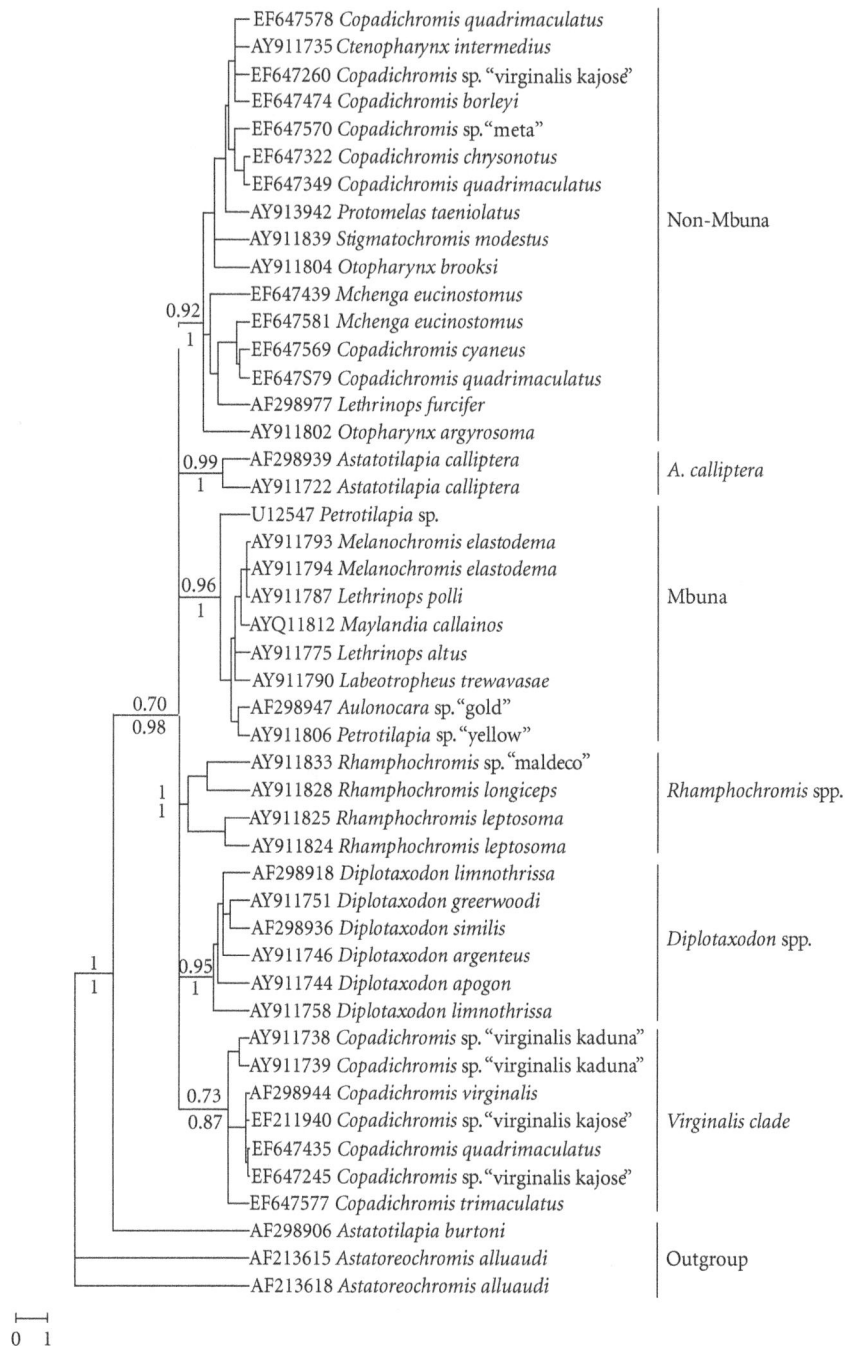

FIGURE 3: Maximum-likelihood reconstruction (PhyML) of the Lake Malawi cichlid flock using the complete mtDNA control region. Numbers next to the branches show bootstrap percentage support (upper) and bayesian posterior probabilities (below) of the main clades. Clade names follow denomination given in text. *Astatoreochromis alluaudi* and *Astatotilapia burtoni* represent the outgroups. Scale bar indicates substitutions per nucleotide site.

together with two (out of 67) *M. eucinostomus* and one (out of 55) *C. quadrimaculatus* individuals. Both mtDNA clades were present lake-wide in nearly all localities sampled, and within each lineage distinct geographic structuring was absent.

The complete mtDNA control region phylogenetic reconstructions using MP (with the different weighting schemes), ML, and BI consistently recovered the 6 main

clades among the Malawian cichlids (Figure 3): (i) a lineage containing most non-Mbuna and Utaka species (non-Mbuna clade hereafter); (ii) a clade containing only *Copadichromis* individuals (virginalis clade hereafter); (iii) a Mbuna clade containing all Mbuna species plus some deepwater *Lethrinops* species and an *Aulonocara* specimen; (iv) a *Diplotaxodon* clade; (v) *Rhamphochromis* clade; (vi) a clade containing *A. calliptera*. For these clades, bootstrap values

Extensive Introgression among Ancestral mtDNA Lineages: Phylogenetic Relationships of the Utaka within the Lake Malawi Cichlid Flock

125

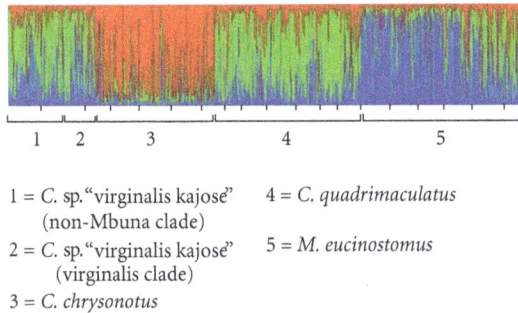

1 = C. sp. "virginalis kajose"   4 = C. quadrimaculatus
    (non-Mbuna clade)
                                 5 = M. eucinostomus
2 = C. sp. "virginalis kajose"
    (virginalis clade)

3 = C. chrysonotus

FIGURE 4: Bar plot result of the STRUCTURE assignment test for $K = 3$ under the nonadmixture model. The two mtDNA groups within C. sp. "virginalis kajose" (non-Mbuna and virginalis clade) cannot be distinguished from each other based on the nuclear markers, whereas C. chrysonotus and M. eucinostomus clearly differentiate although they share the same mtDNA haplotype lineage.

and posterior probabilities displayed high node support values, except for the virginalis clade, which had a bootstrap support of 73 and a posterior probability of 0.87 (Figure 3). The phylogenetic relationships between the clades remained, however, unresolved: their branching order was variable, depending on the reconstruction methods used and was even resolved as a polytomy in the Bayesian analysis. The Shimodaira-Hasegawa test indicated a significant difference in likelihood score between the two topologies examined ($P = 0.03$), giving preference to the unconstrained topology (where Utaka are paraphyletic) over the best tree obeying to the monophyly of all Utaka specimens analysed.

3.2. Microsatellites. Linkage disequilibrium tests across loci and populations revealed no significant allelic associations. The model-based clustering approach implemented in the program STRUCTURE yielded estimated Ln probabilities for $1 \leq K \leq 18$ ranging from −37678 to −34160 with the highest posterior probability and ΔK (=20.605) for $K = 3$. In the most likely scenario, STRUCTURE assigned M. eucinostomus and C. chrysonotus to two different groups, while C. quadrimaculatus and C. sp. "virginalis kajose" formed a third cluster (Figure 4).

4. Discussion

The phylogenetic inferences of the Utaka assemblage performed herein showed that it contains two genetically distant and geographically widespread mtDNA lineages. The two lineages have been observed before [5, 10, 13, 14] but this is the first study to reveal the paraphyly not only of the genus *Copadichromis* in individuals from throughout Lake Malawi, but also of three (of the five analysed) Utaka species (based on the short mtDNA sequences). In a wider taxonomic context involving the other Malawian cichlid lineages, the most abundant of the two lineages in the Utaka clustered within the non-Mbuna mtDNA clade, while the other formed a separate clade containing exclusively Utaka specimens, mostly C. sp. "virginalis kajose" individuals. The

paraphyly of the Utaka does not represent an artefact in our analyses, as corroborated by the long and well-supported branches that connect the non-Mbuna and the virginalis clades, as well as by the significant result of the Shimodaira-Hasegawa test.

One possible explanation is that the Utaka share ancestral polymorphic alleles and/or represent a truly paraphyletic group containing multiple lineages that have undergone convergent evolution. Importantly, the occurrence of two divergent mtDNA lineages within the Utaka is related neither to taxonomic clustering, nor to geographical structuring. If the two haplogroups observed within the Utaka indeed correspond to two ancestral lineages that are genetically isolated for such a long time that their mtDNA genotypes have become so deeply diverged, we would expect this to be also reflected in the nuclear genome of the species. However, we did not find any subdivision of nuclear gene pools that corresponds to the deep mtDNA divergence, neither across the Utaka species, nor within C. sp. "virginalis kajose" which yields the majority of the individuals in the divergent virginalis clade as well as a large number of individuals in the non-Mbuna clade. It thus seems unlikely that the presence of a cryptic species is the cause of the mtDNA divergence within C. sp. "virginalis kajose." Recently published phylogenetic reconstructions using AFLP loci [10, 14] also showed a discordance between the nuclear and mitochondrial placement of *Copadichromis virginalis* within the Malawi cichlid radiation, supporting our finding that the observed paraphyly of the Utaka and of C. sp. "virginalis kajose" is unlikely to be the result of incomplete ancestral lineage sorting or true paraphyly.

Alternatively, a disparate pattern of divergence between mitochondrial and nuclear DNA among conspecific individuals may be the result of a past hybridisation and introgression event, a process which has been documented in Malawian cichlids before (e.g., [40–42]) and for which evidence is accumulating (e.g., [10, 14, 16, 43]). Under this hypothesis we advance two possibilities regarding the original position of the Utaka within the Malawi cichlid phylogeny. A first scenario assumes that all Utaka species formerly constituted a separate ancestral clade within the Malawi cichlid flock, corresponding with the current virginalis clade. Subsequent unidirectional introgression of mtDNA from non-Mbuna into the Utaka could then explain the observed clustering of Utaka specimens within the non-Mbuna lineage. This scenario would involve that either all, or the ancestors of the current Utaka species, would have been extensively hybridised with a non-Mbuna species, resulting in the almost complete replacement of the original mtDNA of the Utaka. A second scenario assumes that all Utaka species initially belonged to the non-Mbuna lineage and a species from a distant mtDNA lineage hybridised with *Copadichromis* species. The mtDNA detected in the virginalis clade may then represent the introgressed mtDNA.

Interestingly and despite our extensive taxonomic sampling, the maternal species involved in the putative hybridisation event remains unidentified as the virginalis clade only contained representatives of the Utaka assemblage. It would seem that the species with which *Copadichromis* spp.

hybridised either has thus far not been subjected to molecular studies or may no longer be present in the lake. Empirical evidence for or against the above scenarios can be gained by examining mtDNA of supplementary Utaka species to validate whether the majority of the taxa cluster is within the non-Mbuna clade or within the virginalis clade. The more Utaka species cluster within the non-Mbuna clade, the less probable becomes the first scenario. Regardless of which of the two mtDNA lineages is the original or the introgressing one, and irrespective of the maternal species involved in the hybridisation event, our results show that the two mtDNA lineages have persisted within the gene pool of *Copadichromis* sp. "virginalis kajose" for a rather long period, as suggested by the diversity displayed by either of these two lineages (Figure 2). It thus suggests that either the population size of this species has remained very high since the hybridisation event (such that genetic drift would represent a lesser issue) or that some other mechanism is maintaining the two lineages within the same species (e.g., balancing or frequency-dependent selection).

Interspecific gene flow is increasingly recognized as an important factor in shaping speciation (e.g., [17–19, 44, 45]). Progressively more examples for hybridization are known from African cichlid fish: among Lake Tanganyika's cichlids evidence is found for ancient introgression (e.g., [15, 46–48]) and a complete replacement [49] of mtDNA in multiple tribes of the cichlid assemblage. From Lake Malawi, evidence for deep introgression leaving a long-term signal in its haplochromine radiation [10, 14, 43], as well as evidence for more recent natural hybridisation [16, 50, 51] among Malawi cichlids, has been provided. In the Lake Victoria cichlid flock recent or ongoing hybridisation [52–54] presumably affects large parts of the species' genomes by homogenization [54, 55], hampering the reconstruction of its young evolutionary history [54, 56, 57], yet potentially seeding the process of speciation [58] but see [55]. In Cameroonian crater lakes the hybridisation of two ancient lineages resulted in the formation of a new and ecologically highly distinct species [59]. Also for *Steatocranus* cichlids from the Congo basin it was recently shown that ancient as well as recent introgression of genes and hybridisation produced a genomic network that potentially promoted divergence and speciation [60]. Our results chime well with previous studies reporting hybridisation in the early stages of a cichlid radiation. Our findings reconcile with the recently reported evidence for ancient introgression between Mbuna and deep-benthic cichlids at the base of the Malawi radiation [14]. Remarkably, in our study we could not separate *Copadichromis* sp. "virginalis kajose" and *C. quadrimaculatus*, two phenotypically distinct taxa, by our microsatellite markers. However, it has already been reported that the performance of a clustering method may become poor for F_{st}'s below 0.05 ([61], J. Pritchard, *pers. comm.*). The estimates of population differentiation were low in both species ($\theta = 0.006$ in *C.* sp. "virginalis kajose" and $\theta = 0.007$ in *C. quadrimaculatus*, reported in [62]), and slightly higher among the two species ($\theta = 0.01$). Whether this observation might yield a demonstration of the relative ease of hybridisation among phenotypically well-differentiated

taxa [14, 43] or be the result of an insufficient resolution of the markers used, deserves further research.

Acknowledgments

The field work for this project was supported by the Interdisciplinary Research Centre of the KU Leuven Campus Kortrijk, a grant from the King Leopold III Fund for Nature Exploration and Conservation to D. Anseeuw, and grants from the Fund for Scientific Research (FWO) and the Stichting tot bevordering van het wetenschappelijk onderzoek in Afrika to J. Snoeks. B. Nevado was supported by a Grant no. SFRH/BD/17704/2004 from Fundacão para a Ciência e Tecnologia. The authors wish to thank the Malawi Fisheries Department, and especially Sam Mapila and Orton Kachinjika, for permission to conduct fieldwork in Malawi, the Department for International Development (DFID) of the British High Commission in Malawi and especially George Turner (University of Hull, presently at Bangor University), for putting a 4×4 vehicle at our disposal for the sampling trip. They thank the late Davis Mandere for his valuable assistance in the field, the Fisheries Research Unit in Monkey Bay and Nkhata Bay for their assistance and the late Stuart Grant for his hospitality.

References

[1] J. Snoeks, "Material and methods," in *The Cichlid Diversity of Lake Malawi/Nyasa/Niassa: Identification, Distribution and Taxonomy*, J. Snoeks, Ed., pp. 12–19, Cichlid Press, El Paso, Tex, USA, 2004.

[2] G. Fryer and T. D. Iles, *The Cichlid Fishes of the Great Lakes of Africa: Their Biology and Evolution*, TFH Publications, UK, Edinburgh, 1972.

[3] D. H. Eccles and E. Trewavas, *Malawian Cichlid Fishes: The Classification of Some Haplochromine Genera*, Lake Fish Movies, Herten, Germany, 1989.

[4] A. Meyer, T. D. Kocher, P. Basasibwaki, and A. C. Wilson, "Monophyletic origin of Lake Victoria cichlid fishes suggested by mitochondrial DNA sequences," *Nature*, vol. 347, no. 6293, pp. 550–553, 1990.

[5] P. Moran, I. Kornfield, and P. N. Reinthal, "Molecular systematics and radiation of the haplochromine cichlids (Teleostei: Perciformes) of Lake Malawi," *Copeia*, no. 2, pp. 274–288, 1994.

[6] P. W. Shaw, G. F. Turner, M. R. Idid, R. L. Robinson, and G. R. Carvalho, "Genetic population structure indicates sympatric speciation of Lake Malawi pelagic cichlids," *Proceedings of the Royal Society B*, vol. 267, no. 1459, pp. 2273–2280, 2000.

[7] R. C. Albertson, J. A. Markert, P. D. Danley, and T. D. Kocher, "Phylogeny of a rapidly evolving clade: the cichlid fishes of Lake Malawi, East Africa," *Proceedings of the National Academy of Sciences of the United States of America*, vol. 96, no. 9, pp. 5107–5110, 1999.

[8] C. J. Allender, O. Seehausen, M. E. Knight, G. F. Turner, and N. Maclean, "Divergent selection during speciation of Lake Malawi cichlid fishes inferred from parallel radiations in nuptial coloration," *Proceedings of the National Academy of Sciences of the United States of America*, vol. 100, no. 2, pp. 14074–14079, 2003.

Extensive Introgression among Ancestral mtDNA Lineages: Phylogenetic Relationships of the Utaka within the Lake
Malawi Cichlid Flock

127

[9] M. R. Kidd, C. E. Kidd, and T. D. Kocher, "Axes of differentiation in the bower-building cichlids of Lake Malawi," *Molecular Ecology*, vol. 15, no. 2, pp. 459–478, 2006.

[10] D. A. Joyce, D. H. Lunt, M. J. Genner, G. F. Turner, R. Bills, and O. Seehausen, "Repeated colonization and hybridization in Lake Malawi cichlids," *Current Biology*, vol. 21, no. 3, pp. R108–R109, 2011.

[11] I. L. Kornfield, "Evidence for rapid speciation in African cichlid fishes," *Experientia*, vol. 34, no. 3, pp. 335–336, 1978.

[12] I. L. Kornfield, K. R. McKaye, and T. D. Kocher, "Evidence for the immigration hypothesis in the endemic cichlid fauna of Lake Tanganyika," *Isozyme Bulletin*, vol. 18, p. 76, 1985.

[13] G. F. Turner, R. L. Robinson, P. W. Shaw, and G. R. Carvalho, "Identification and biology of *Diplotaxodon, Rhamphochromis* and *Pallidochromis*," in *The Cichlid Diversity of Lake Malawi/Nyasa/Niassa: Identification, Distribution and Taxonomy*, J. Snoeks, Ed., Cichlid Press, El Paso, Tex, USA, 2004.

[14] M. J. Genner and G. F. Turner, "Ancient hybridization and phenotypic novelty within Lake Malawi's cichlid fish radiation," *Molecular Biology and Evolution*, vol. 29, pp. 195–206, 2012.

[15] L. Rüber, A. Meyer, C. Sturmbauer, and E. Verheyen, "Population structure in two sympatric species of the Lake Tanganyika cichlid tribe *Eretmodini*: evidence for introgression," *Molecular Ecology*, vol. 10, no. 5, pp. 1207–1225, 2001.

[16] M. C. Mims, C. D. Hulsey, B. M. Fitzpatrick, and J. T. Streelman, "Geography disentangles introgression from ancestral polymorphism in Lake Malawi cichlids," *Molecular Ecology*, vol. 19, no. 5, pp. 940–951, 2010.

[17] O. Seehausen, "Hybridization and adaptive radiation," *Trends in Ecology and Evolution*, vol. 19, no. 4, pp. 198–207, 2004.

[18] L. H. Rieseberg, M. A. Archer, and R. K. Wayne, "Transgressive segregation, adaptation and speciation," *Heredity*, vol. 83, no. 4, pp. 363–372, 1999.

[19] M. Barrier, B. G. Baldwin, R. H. Robichaux, and M. D. Purugganan, "Interspecific hybrid ancestry of a plant adaptive radiation: allopolyploidy of the Hawaiian silversword alliance (Asteraceae) inferred from floral homeotic gene duplications," *Molecular Biology and Evolution*, vol. 16, no. 8, pp. 1105–1113, 1999.

[20] S. M. Aljanabi and I. Martinez, "Universal and rapid salt-extraction of high quality genomic DNA for PCR-based techniques," *Nucleic Acids Research*, vol. 25, no. 22, pp. 4692–4693, 1997.

[21] W. Salzburger, A. Meyer, S. Baric, E. Verheyen, and C. Sturmbauer, "Phylogeny of the Lake Tanganyika cichlid species flock and its relationship to the Central and East African haplochromine cichlid fish faunas," *Systematic Biology*, vol. 51, no. 1, pp. 113–135, 2002.

[22] W. J. Lee, J. Conroy, W. H. Howell, and T. D. Kocher, "Structure and evolution of teleost mitochondrial control regions," *Journal of Molecular Evolution*, vol. 41, no. 1, pp. 54–66, 1995.

[23] M. J. H. Van Oppen, C. Rico, J. C. Deutsch, G. F. Turner, and G. M. Hewitt, "Isolation and characterization of microsatellite loci in the cichlid fish *Pseudotropheus zebra*," *Molecular Ecology*, vol. 6, no. 4, pp. 185–186, 1997.

[24] K. A. Kellogg, J. A. Markert, J. R. Stauffer, and T. D. Kocher, "Microsatellite variation demonstrates multiple paternity in lekking cichlid fishes from Lake Malawi, Africa," *Proceedings of the Royal Society B*, vol. 260, no. 1357, pp. 79–84, 1995.

[25] R. Zardoya, D. M. Vollmer, C. Craddock, J. T. Streelman, S. Karl, and A. Meyer, "Evolutionary conservation of microsatellite flanking regions and their use in resolving the phylogeny of cichlid fishes (Pisces: Perciformes)," *Proceedings of the Royal Society B*, vol. 263, no. 1376, pp. 1589–1598, 1996.

[26] A. Parker and I. Kornfield, "Polygynandry in *Pseudotropheus zebra*, a cichlid fish from Lake Malawi," *Environmental Biology of Fishes*, vol. 47, no. 4, pp. 345–352, 1996.

[27] J. D. Thompson, D. G. Higgins, and T. J. Gibson, "CLUSTAL W: improving the sensitivity of progressive multiple sequence alignment through sequence weighting, position-specific gap penalties and weight matrix choice," *Nucleic Acids Research*, vol. 22, no. 22, pp. 4673–4680, 1994.

[28] N. Galtier, M. Gouy, and C. Gautier, "SEA VIEW and PHYLO_WIN: Two graphic tools for sequence alignment and molecular phytogeny," *Computer Applications in the Biosciences*, vol. 12, no. 6, pp. 543–548, 1996.

[29] D. Posada, "Collapse: Describing haplotypes from sequence alignments," 2004, http://darwin.uvigo.es/software/collapse.html.

[30] D. Posada and K. A. Crandall, "MODELTEST: Testing the model of DNA substitution," *Bioinformatics*, vol. 14, no. 9, pp. 817–818, 1998.

[31] S. Guindon and O. Gascuel, "A simple, fast and accurate algorithm to estimate large phylogenies by maximum likelihood," *Systematic Biology*, vol. 52, no. 5, pp. 696–704, 2003.

[32] M. Clement, D. Posada, and K. A. Crandall, "TCS: A computer program to estimate gene genealogies," *Molecular Ecology*, vol. 9, no. 10, pp. 1657–1659, 2000.

[33] F. Ronquist and J. P. Huelsenbeck, "MrBayes 3: Bayesian phylogenetic inference under mixed models," *Bioinformatics*, vol. 19, no. 12, pp. 1572–1574, 2003.

[34] H. Shimodaira and M. Hasegawa, "Multiple comparisons of log-likelihoods with applications to phylogenetic inference," *Molecular Biology and Evolution*, vol. 16, no. 8, pp. 1114–1116, 1999.

[35] M. Raymond and F. Rousset, "GENEPOP Version 1.2: Populations genetics software for exact tests and ecumenicism," *Journal of Heredity*, vol. 86, pp. 248–249, 1995.

[36] J. K. Pritchard, M. Stephens, and P. Donnelly, "Inference of population structure using multilocus genotype data," *Genetics*, vol. 155, no. 2, pp. 945–959, 2000.

[37] D. Falush, M. Stephens, and J. K. Pritchard, "Inference of population structure using multilocus genotype data: Dominant markers and null alleles," *Molecular Ecology Notes*, vol. 7, no. 4, pp. 574–578, 2007.

[38] G. Evanno, S. Regnaut, and J. Goudet, "Detecting the number of clusters of individuals using the software STRUCTURE: A simulation study," *Molecular Ecology*, vol. 14, no. 8, pp. 2611–2620, 2005.

[39] D. A. Earl and B. M. von Holdt, "STRUCTURE HARVESTER: a website and program for visualizing STRUCTURE output and implementing the Evanno method," *Conservation Genetics Resources*. In press.

[40] P. F. Smith, A. Konings, and I. Kornfield, "Hybrid origin of a cichlid population in Lake Malawi: implications for genetic variation and species diversity," *Molecular Ecology*, vol. 12, no. 9, pp. 2497–2504, 2003.

[41] J. T. Streelman, S. L. Gmyrek, M. R. Kidd et al., "Hybridization and contemporary evolution in an introduced cichlid fish from Lake Malawi National Park," *Molecular Ecology*, vol. 13, no. 8, pp. 2471–2479, 2004.

[42] R. C. Albertson and T. D. Kocher, "Genetic architecture sets limits on transgressive segregation in hybrid cichlid fishes," *Evolution*, vol. 59, no. 3, pp. 686–690, 2005.

[43] Y. H. E. Loh, L. S. Katz, M. C. Mims, T. D. Kocher, S. V. Yi, and J. T. Streelman, "Comparative analysis reveals signatures of differentiation amid genomic polymorphism in Lake Malawi cichlids," *Genome Biology*, vol. 9, no. 7, article R113, 2008.

[44] D. Garant, S. E. Forde, and A. P. Hendry, "The multifarious effects of dispersal and gene flow on contemporary adaptation," *Functional Ecology*, vol. 21, no. 3, pp. 434–443, 2007.

[45] P. A. Larsen, M. R. Marchán-Rivadeneira, and R. J. Baker, "Natural hybridization generates mammalian lineage with species characteristics," *Proceedings of the National Academy of Sciences of the United States of America*, vol. 107, no. 25, pp. 11447–11452, 2010.

[46] W. Salzburger, S. Baric, and C. Sturmbauer, "Speciation via introgressive hybridization in East African cichlids?" *Molecular Ecology*, vol. 11, no. 3, pp. 619–625, 2002.

[47] S. Koblmüller, N. Duftner, K. M. Sefc et al., "Reticulate phylogeny of gastropod-shell-breeding cichlids from Lake Tanganyika—the result of repeated introgressive hybridization," *BMC Evolutionary Biology*, vol. 7, article 7, 2007.

[48] S. Koblmüller, B. Egger, C. Sturmbauer, and K. M. Sefc, "Rapid radiation, ancient incomplete lineage sorting and ancient hybridization in the endemic Lake Tanganyika cichlid tribe Tropheini," *Molecular Phylogenetics and Evolution*, vol. 55, no. 1, pp. 318–334, 2010.

[49] B. Nevado, S. Koblmüller, C. Sturmbauer, J. Snoeks, J. Usano-Alemany, and E. Verheyen, "Complete mitochondrial DNA replacement in a Lake Tanganyika cichlid fish," *Molecular Ecology*, vol. 18, no. 20, pp. 4240–4255, 2009.

[50] J. R. Stauffer, N. J. Bowers, T. D. Kocher, and K. R. McKaye, "Evidence of hybridisation between *Cyanotilapia afra* and *Pseudotropheus zebra* (Teleostei: Cichlidae) following an intralacustrine translocation in Lake Malawi," *Copeia*, vol. 1996, no. 1, pp. 203–208, 1996.

[51] P. F. Smith, A. Konings, and I. Kornfield, "Hybrid origin of a cichlid population in Lake Malawi: Implications for genetic variation and species diversity," *Molecular Ecology*, vol. 12, no. 9, pp. 2497–2504, 2003.

[52] I. Van Der Sluijs, J. J. M. Van Alphen, and O. Seehausen, "Preference polymorphism for coloration but no speciation in a population of Lake Victoria cichlids," *Behavioral Ecology*, vol. 19, no. 1, pp. 177–183, 2008.

[53] M. E. Maan, O. Seehausen, and J. J. M. Van Alphen, "Female mating preferences and male coloration covary with water transparency in a Lake Victoria cichlid fish," *Biological Journal of the Linnean Society*, vol. 99, no. 2, pp. 398–406, 2010.

[54] I. E. Samonte, Y. Satta, A. Sato, H. Tichy, N. Takahata, and J. Klein, "Gene flow between species of Lake Victoria haplochromine fishes," *Molecular Biology and Evolution*, vol. 24, no. 9, pp. 2069–2080, 2007.

[55] O. Seehausen, G. Takimoto, D. Roy, and J. Jokela, "Speciation reversal and biodiversity dynamics with hybridization in changing environments," *Molecular Ecology*, vol. 17, no. 1, pp. 30–44, 2008.

[56] S. Nagl, H. Tichy, W. E. Mayer, N. Takezaki, N. Takahata, and J. Klein, "The origin and age of haplochromine fihes in Lake Victoria, East Africa," *Proceedings of the Royal Society of London B*, vol. 267, pp. 1049–1061, 2000.

[57] A. Sato, N. Takezaki, H. Tichy, F. Figueroa, W. E. Mayer, and J. Klein, "Origin and speciation of haplochromine fishes in East African crater lakes investigated by the analysis of their mtDNA, Mhc genes, and SINEs," *Molecular Biology and Evolution*, vol. 20, no. 9, pp. 1448–1462, 2003.

[58] O. Seehausen, "Hybridization and adaptive radiation," *Trends in Ecology and Evolution*, vol. 19, no. 4, pp. 198–207, 2004.

[59] U. K. Schliewen and B. Klee, "Reticulate sympatric speciation in Cameroonian crater lake cichlids," *Frontiers in Zoology*, vol. 1, article 5, 2004.

[60] J. Schwarzer, B. Misof, and U. K. Schliewen, "Speciation within genomic networks: a case study based on *Steatocranus* cichlids of the lower Congo rapids," *Journal of Evolutionary Biology*, vol. 25, pp. 138–148, 2012.

[61] D. E. Pearse and K. A. Crandall, "Beyond FST: Analysis of population genetic data for conservation," *Conservation Genetics*, vol. 5, no. 5, pp. 585–602, 2004.

[62] D. Anseeuw, G. E. Maes, P. Busselen, D. Knapen, J. Snoeks, and E. Verheyen, "Subtle population structure and male-biased dispersal in two *Copadichromis* species (Teleostei, Cichlidae) from Lake Malawi, East Africa," *Hydrobiologia*, vol. 615, no. 1, pp. 69–79, 2008.

16

Microsatellites Cross-Species Amplification across Some African Cichlids

Etienne Bezault,[1, 2, 3] Xavier Rognon,[2, 4] Karim Gharbi,[2, 5] Jean-Francois Baroiller,[1] and Bernard Chevassus[2]

[1] UMR 110, Cirad-Ifremer INTREPID, 34398 Montpellier, France
[2] INRA, UMR 1313 Génétique Animale et Biologie Intégrative, 78352 Jouy-en-Josas, France
[3] Department of Biology, Reed College, Portland, OR 97202, USA
[4] AgroParisTech, UMR 1313, Génétique Animale et Biologie Intégrative, 75231 Paris, France
[5] Institute of Evolutionary Biology, School of Biological Sciences, University of Edinburgh, Edinburgh EH9 3JT, UK

Correspondence should be addressed to Etienne Bezault, ebezault@yahoo.fr

Academic Editor: Kristina M. Sefc

The transfer of the genomic resources developed in the Nile tilapia, *Oreochromis niloticus*, to other Tilapiines *sensu lato* and African cichlid would provide new possibilities to study this amazing group from genetics, ecology, evolution, aquaculture, and conservation point of view. We tested the cross-species amplification of 32 *O. niloticus* microsatellite markers in a panel of 15 species from 5 different African cichlid tribes: Oreochromines (*Oreochromis, Sarotherodon*), Boreotilapiines (*Tilapia*), Chromidotilapines, Hemichromines, and Haplochromines. Amplification was successfully observed for 29 markers (91%), with a frequency of polymorphic (P_{95}) loci per species around 70%. The mean number of alleles per locus and species was 3.2 but varied from 3.7 within *Oreochromis* species to 1.6 within the nontilapia species. The high level of cross-species amplification and polymorphism of the microsatellite markers tested in this study provides powerful tools for a wide range of molecular genetic studies within tilapia species as well as for other African cichlids.

1. Introduction

African cichlid fish are of extreme interest for both evolutionary biology and applied genetics purposes, including amazing models for speciation, adaptation, behaviour and neurosciences [1–5] as well as groups of major importance for aquaculture and fisheries (strain selection and improvement, stock assessment, etc.) [6–10]. A wide range of structural and functional genomic resources have been developed for cichlids in the past 15 years, predominantly in the Nile tilapia, *Oreochromis niloticus* [11–14]. While genome sequencing projects are in progress for several African cichlids, the transfer of genomic resources from *O. niloticus* across the entire group of tilapias *sensu lato* as well as other African cichlid tribes would provide powerful tools to support a wide range of evolutionary biology studies, including comparative phylogenetics, genome mapping, evolution of gene family sequence and expression, candidate gene analyses for adaptation, and population genetics.

Microsatellite markers are one of the most interesting resources to transfer across lineages, as they can provide numerous locus-specific molecular markers and putatively homologous sequences across taxa. In addition to their high level of polymorphism, the evolutionary conservation of the flanking region of microsatellite loci allows large-scale heterospecific amplification [15, 16], as previously shown in various animal groups, particularly fish [17–19]. However, the rate of cross-species amplification varies widely among taxonomic groups and loci [18, 20]. In addition to their application in population genetics, conserved microsatellite markers are particularly useful for population, species or hybrid identification (especially at early developmental stages) and candidate-marker analysis, comparative genetic mapping, and QTL analysis. Furthermore, compared to anonymous multilocus genomic markers (RFLP, AFLP, ISSR) and SNPs, microsatellites present the important advantages of (i) being highly reproducible and very easily transferable between laboratory (with limited equipment and computational requirement), (ii) providing a high polymorphism information contain (PIC) per locus, and (iii) being highly cost efficient when only a small number of loci are needed. For these reasons, microsatellites markers are likely to remain popular for a wide range of ecology and evolutionary studies (e.g., relatedness and parentage analysis, population diversity and demography assessment, noninvasive genetic analysis, and conservation).

Since the first publication of microsatellite markers cloned in *O. niloticus* [13], thousands have been published and more than 500 have been positioned onto the genetic map of *O. niloticus* and the closely related *O. aureus* [14, 21]. These microsatellites have been used to map traits of interest, such as sex determination factors [22, 23], and have also been found to influence the expression of genes associated to physiological adaptation [24].

Outside the tilapias, microsatellite markers have been developed in a few different Haplochromines species: *Copadichromis cyclicos* [25], *Tropheus moorii* [19], *Pseudotropheus zebra* [26], *Astatoreochromis alluaudi* [27], *Pundamilia pundamilia* [28], *Metriaclima zebra* [29], *Pseudocrenilabrus multicolor* [30], *Paralabidochromis chilotes* [31], and *Astatotilapia burtoni* [32]. However these studies reported a smaller number of markers than that in Nile tilapia. The use of microsatellite markers in Haplochromines has been almost strictly restricted to descriptive population genetics and parentage/relatedness analysis, which represent only a subset of the possibilities offered by having a large set of genome-anchored microsatellite markers, as available for *O. niloticus*.

Additionally, microsatellites developed outside tilapias were derived exclusively from the most species-rich group of African cichlids and there are very limited genomic resources in all the other "under-studied" African cichlid tribes [33–35].

Considering the central position occupied by the Tilapiines *sensu lato* in the African cichlid phylogeny [38], their large diversity within at least 3 monophyletic clades [39–41], and the important number of species involved in population transfers, hybridisation, and/or invasion [8, 42], we decided to investigate the cross-species amplification efficiency of Nile tilapia microsatellites among the different groups of the Tilapiines *sensu lato* as well as three other African cichlid tribes, to extend the use/availability of this resource across a wide range of African cichlid species, including "under-studies" groups. The panel of species investigated then spans a large section of the African cichlid radiation, with an estimated overall divergence time of 33.4–63.7 Myrs [41, 43].

2. Material and Methods

Tests of cross-species amplification were conducted in a panel of 15 African cichlid species, representing all three major genera of Tilapiines *sensu lato*: 7 *Oreochromis*, 2 *Sarotherodon*, both genera belonging to the Oreochromines, and 3 *Tilapia* (*Coptodon*), belonging to the Boreotilapiines; as well as representatives of 3 other African cichlid tribes, including the derived Haplochromines, and two more basal tribes, the Chromidotilapiines and the Hemichromines (see details in Table 1). Analyses were conducted using 3 to 9 individuals per species (Table 1). Genomic DNA was extracted from fin clips stored in ethanol using a standard phenol-chloroform protocol [44].

The panel of 32 microsatellites was selected from the markers isolated in *O. niloticus* [13]. Genotyping was obtained by PCR amplification with radioactive (P[33]) labeled primers [44, 45]. Allele variants were separated on 6% acrylamide gel electrophoresis. For each marker, the annealing temperature and $MgCl_2$ concentration were adjusted to optimise the efficiency of PCR amplification based on *O. niloticus* and two others species: one closely related among *Oreochromis* (*O. mossambicus*) and one distantly related among the Oreochromines (*S. melanotheron*). Cross-species amplifications were carried out using these conditions in the 15 studied species (Table 2). For each microsatellite marker, the amplification success has been estimated qualitatively on a 4-level scale based on the quality of the electrophoresis pattern across the test individuals (i.e., "++" for strong and sharp amplification pattern, "+" for good quality pattern with some stutters, echo-alleles or low intensity, "−" for high variance of amplification quality across individuals, very high level of stutter, and/or high frequency of null alleles, and "−−" very poor quality pattern, nonspecific or lack of, amplification). For each locus by species combination ($n = 480$), we assessed the amplification success and counted the number of different alleles among individuals. The presence of putative null alleles (i.e., nonamplified alleles) was inferred when a few individuals consistently showed an absence of allele amplification while other individuals from the same species showed high-quality amplification pattern or in the complete absence of heterozygous individuals. Echo-alleles (i.e., supplementary allele coamplifying across individuals producing amplification pattern consistently representing 2 or 4 alleles per individuals, with the longest allele separated from the shortest "cosegregating" allele by an identical length across individuals/alleles) were also identified. Furthermore, the rate of amplification success, the frequency of polymorphic loci (P_{95}), and the mean number of allele per locus were calculated per species, genus, and tribe across all studied microsatellites markers.

3. Results and Discussion

Very high rates of microsatellite amplification and polymorphism were observed (both 97%), in the Nile tilapia, with a mean number of alleles per locus of 4.3. Across the 14 other test species, 29 loci gave good quality amplifications (91%-Tables 2 and 3), while 3 markers (9%) showed a high discrepancy of amplification efficiency and/or unclear amplification pattern (Table 2; *see details in supplementary material* which

TABLE 1: Species studied for cross-species amplification tests, with geographic origin, and number of samples analysed per species.

Lineages Genus Species	Geographic origin	n
Oreochromines		
Oreochromis		
O. (Oreochromis) niloticus	Bouake (Cote d'Ivoire)*	9
O. (Oreochromis) aureus	Lake Manzala (Egypt)	5
O. (Oreochromis) mossambicus	Mozambique	5
O. (Oreochromis) shiranus	Lake Malawi	5
O. (Nyasalapia) macrochir	Bouake (Cote d'Ivoire)**	5
O. (Nyasalapia) saka	Lake Malawi	5
O. (Nyasalapia) squamipinnis	Lake Malawi	5
Sarotherodon		
S. (Sarotherodon) galilaeus	Bamako (Niger)	3
S. (Sarotherodon) melanotheron	Ébrié Lagoon (Ivory Cost)	5
Boreotilapiines		
Tilapia		
T. (Coptodon) dageti	Bamako (Niger)	5
T. (Coptodon) guineensis	Ivory Cost/Senegal	4
T. (Coptodon) zillii	Lake Manzala (Egypt)	5
Haplochromines		
Haplochromis		
Haplochromis sp. "rock kribensis"	Lake Victoria	3
Chromidotilapines		
Chromidotilapia		
Chromidotilapia guntheri	Bamako (Niger)	3
Hemichromines		
Hemichromis		
Hemichromis bimaculatus	Bandama (Ivory Cost)	5

Introduced stocks: *with mixed origin (Volta and Nile) [36]; **from wild population (RDC) [37].

is available online at doi:10.1155/2012/870935: Table S1). Excluding the Nile tilapia, the average intraspecific rate of successful amplification and polymorphism across the panel of 32 markers was more than 70% (Table 3).

The expected relationship between the success of cross-species amplifications and evolutionary distance from marker cloning species [15, 20] was observed, reflecting the phylogenetic relationships between the different groups of African cichlids [39–41] (Table 3; *see details in supplementary material*: Table S2). Within the Tilapiines *sensu lato*, species from both mouth-brooder genera (i.e., *Oreochromis* and *Sarotherodon*), constitutive of the monophyletic clade of the Oreochromines diverged 12.8–21.4 Myrs ago, showed very high and similar amplification (88% and 86%, resp.) and polymorphism (76% and 85%, resp.) rates, whereas species from the genus *Tilapia*, belonging to the Boreotilapiines with a divergence time from Oreochromines of 30.6–39.6 Myrs, showed lower rates of amplification (67%) and polymorphism (59%). The three other African cichlid tribes exhibited lower values for amplification and polymorphism rates: 38% and 50%, respectively, in the more derived lineage, Haplochromines, whereas a more heterogeneous pattern was found for the two more basal lineages, Chromidotilapiines

(i.e., 47% and 20%, resp.) and Hemichromines (i.e., 19% and 50%, resp.). Allelic diversity varied with the same trends with a mean number of alleles per locus and species ranging from 3.7 and 3.3, respectively, for *Oreochromis* spp. and *Sarotherodon* spp. to 2.4 for *Tilapia* spp. and 1.6 in average (from 1.4 to 2.3) for the non-Tilapiines groups. The frequency of loci with putative null alleles also appeared to increase in the more distant species (*supplementary material*: Table S2). Rather than strictly reflecting reductions in polymorphism and/or the loss of the marker loci with increasing phylogenetic distance from the species in which the marker was cloned, these relationships are caused by mutations in the flanking regions complementary to the PCR primers. The conservation of microsatellites loci in the genomes has been shown to be potentially very long, and anyway much longer than the divergence time allowing successful cross-species amplification based on a given pair of primers, generally designed based on the only knowledge of the locus sequence in the species of cloning. The global success of cross-species amplification of a given microsatellite marker and/or the recovery of its different allelic variant (i.e., elimination of null allele) could then be enhanced in target species by either a specific optimisation

Table 2: Microsatellite loci tested for cross-species amplification with indications of repeat structure observed in *O. niloticus* (according to Lee and Kocher, [13]), allele size range of the amplified fragment across all tested species, PCR and electrophoresis conditions (labeled primer, annealing temperature/magnesium concentration (mM)/electrophoresis Volt-hour), and amplification quality obtained after PCR optimisation tests (from very good ++ to poor −−; see detail of the categories in main text); loci presenting a wide cross-species amplification efficiency are in bold.

Loci	GenBank access No.	Structure	Range (bp)	PCR and electrophoresis conditions	Amplification efficiency
UNH-008	G31346	Perfect	196–236	R* 56/1.2/6000	++
UNH-102	G12255	Perfect	132–185	R* 50/1.2/4500	++
UNH-103	G12256	Perfect	171–260	R* 48/1.2/6000	+
UNH-106	G12259	Compound	115–189	R* 50/1.2/3500	+
UNH-115	G12268	Compound	100–146	F* 50/1.5/3500	++
UNH-117	G12270	Interrupted	108–146	R* 5411.2/4500	++
UNH-120	G12273	Compound	—	R* 48/2/—	−−
UNH-123	G12276	Perfect	142–232	F* 48/1.2/4500	++
UNH-124	G12277	Perfect	295–324	F* 54/1.2/7500	++
UNH-125	G12278	Compound	134–198	R* 48/1.5/4500	+
UNH-129	G12282	Interrupted	180–253	R* 48/1.2/4500	+
UNH-130	G12283	Perfect	174–242	R* 50/1.2/4500	+
UNH-131	G12284	Perfect	283–303	F* 48/2/6000	−
UNH-132	G12285	Perfect	100–134	R* 52/1.2/3500	+
UNH-135	G12287	Interrupted	124–284	R* 50/1.5/4500	+
UNH-138	G12290	Perfect	144–250	R* 48/1.5/4500	+
UNH-142	G12294	Interrupted	142–192	F* 48/1.2/4500	++
UNH-146	G12298	Interrupted	111–149	F* 60/1/3500	++
UNH-149	G12301	Perfect	143–225	R* 48/1.5/4500	+
UNH-154	G12306	Perfect	98–176	R* 50/1.2/3500	++
UNH-159	G12311	Perfect	205–267	R* 55/1.2/6000	++
UNH-162	G12314	Perfect	125–252	R* 48/1.5/6000	++
UNH-169	G12321	Interrupted	124–240	R* 54/1.2/3500	++
UNH-173	G12325	Perfect	124–188	F* 55/1.2/4500	+
UNH-174	G12326	Perfect	146–187	F* 48/1.5/4500	++
UNH-189	G12341	Perfect	135–208	R* 52/1.2/4500	+
UNH-190	G12342	Compound	133–202	R* 60/1/4500	+
UNH-193	G12386	Perfect	—	R* 48/2/3500	−−
UNH-197	G12348	Interrupted	154–228	R* 50/1.2/4500	+
UNH-207	G12358	Interrupted	90–198	R* 60/1.2/3500	++
UNH-211	G12362	Perfect	82–194	R* 48/1.5/3500	++
UNH-216	G12367	Perfect	126–212	R* 52/1.2/3500	++

of the amplification conditions or the modification of the sequence of the primers. This is especially appropriate when target species are distantly related to the cloning species of the markers and initial cross-species tests reveal low level of polymorphism with potentially high frequency of null allele (which would heavily bias any allele frequency-based estimates).

To represent the multi-locus pattern of genetic diversity across the 15 study species, we performed a population-based correspondence analysis using the software Genetix [46]. This multivariate analysis conducted on the genotype matrix allows to represent the clustering pattern among the different species groups, as well as among individuals within each of them in a factorial space (F1, F2, F3). This analysis allowed to clearly resolve the different species, except for

O. saka and *O. squamipinnis* which are highly overlapping in the factorial space (Figure 1). Three separate groups of species were defined: the Oreochromines species, with all *Oreochromis* and *Sarotherodon* species, the Tilapia species, and all non-Tilapiines species. This clustering pattern reflects the phylogenetic relationships between the two tribes of Tilapiines *sensu lato*, that is, Oreochromines and Boreotilapiines. However the clustering of the three other tribes, which represent the most distant taxa from the source species, reveals the influence of the overall reduced polymorphism in highly distant taxa. This points out the limits of microsatellite size polymorphisms to estimate genetic divergence and/or phylogenetic relationship between too distantly related taxa, due to allele size homoplasy and/or increase of null allele frequency [19].

TABLE 3: Results of cross-species amplification performed over the 32 tested microsatellite loci on the 15 African cichlid species studied, including amplification rate, polymorphism rate, and mean number of alleles per locus, estimated per genus and tribe.

Groups	N species	Amplification rate	Polymorphism (P_{95})	Mean allele number per locus Per group	Per species	% shared alleles per
O. niloticus		97%	97%	—	4.3	—
Oreochromis spp.*	6	88%	76%	17.8	3.7	37%
Sarotherodon spp.	2	86%	85%	6.4	3.3	9.2%
Tilapia spp.	3	67%	59%	6	2.4	19.7%
Tilapiines*	11	82%	74%	24.3	3.7	20.5%
non-Tilapiines	3	34%	36%	3.4	1.6	2.3%
Haplochromines	1	38%	50%	—	1.6	—
Chromidotilapines	1	47%	20%	—	1.4	—
Hemichromines	1	19%	50%	—	2.3	—
Total*		72%	70%	25.7	3.2	5.3%

*Excluding *O. niloticus*.

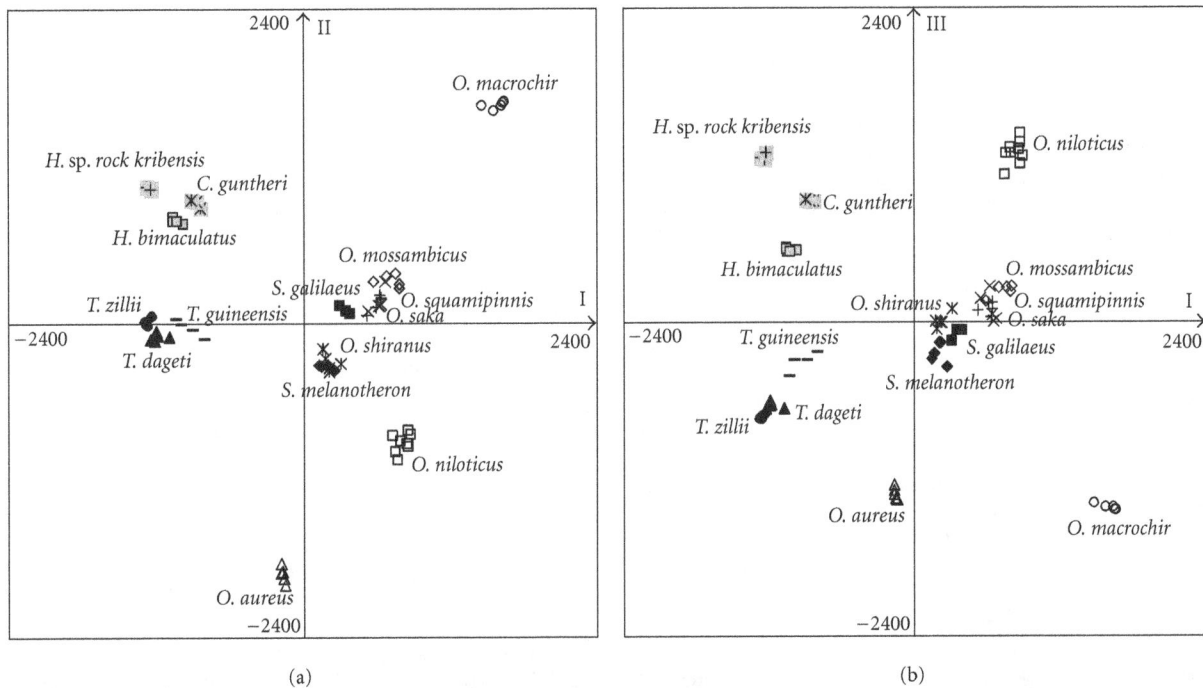

FIGURE 1: Clustering of the 15 study species based on multilocus diversity: correspondence analysis based on the individual genotypes over the 29 microsatellites loci successfully amplified and performed on the barycentre of the species: (a) factorial plane F1-F2 and (b) factorial planes F1–F3.

4. Conclusion

This study provides a quantitative estimate of the transferability of *O. niloticus*-derived microsatellites markers across 5 divergent African cichlid tribes, from the highly studied Haplochromines group to less studied tribes as Oreochromines, Boreotilapiines, Chromidotilapiines, and Hemichromines. The high rate of cross-species amplification and polymorphism highlights the usefulness of microsatellites markers for comparative genetic studies within Oreochromines and other African cichlids tribes, including stock/species identification, comparative genome mapping, candidate genes, or hybridisation surveys. Despite the fast growing opportunities to produce large-scale genomic data in nonmodel organisms, we believe that highly polymorphic, locus-specific markers such as microsatellites will continue to be useful for a wide range of genetic analyses in African cichlids.

Acknowledgments

The authors would like to thank Stéphane Mauger and René Guyomard for laboratory assistance; Frederic Clota, Martial Derivaz, Philippe Morisens, and Jérome Lazard for their support in the scientific program of inter-generic hybridisation of tilapias, as well as the anonymous referee and the editor, Kristina M. Sefc, for their valuable comments on the manuscript. This study was supported by grants from the CIRAD and INRA (France).

References

[1] T. D. Kocher, "Adaptive evolution and explosive speciation: the cichlid fish model," *Nature Reviews Genetics*, vol. 5, no. 4, pp. 288–298, 2004.

[2] I. Kornfield and P. F. Smith, "African cichlid fishes: model systems for evolutionary biology," *Annual Review of Ecology and Systematics*, vol. 31, pp. 163–196, 2000.

[3] O. Seehausen, "African cichlid fish: a model system in adaptive radiation research," *Proceedings of the Royal Society B*, vol. 273, no. 1597, pp. 1987–1998, 2006.

[4] G. W. Barlow, *Cichlid Fishes: Nature's Grand Experiment in Evolution*, Perseus Books, Cambridge, UK, 2000.

[5] H. A. Hofmann, "Functional genomics of neural and behavioral plasticity," *Journal of Neurobiology*, vol. 54, no. 1, pp. 272–282, 2003.

[6] A. E. Eknath and G. Hulata, "Use and exchange of genetic resources of Nile tilapia (*Oreochromis niloticus*)," *Reviews in Aquaculture*, vol. 1, no. 3-4, pp. 197–213, 2009.

[7] H. Josupeit, *World Market of Tilapia. Globefish Research Programme*, FAO, Rome, Italy, 2005.

[8] R. S. V. Pullin, "Cichlids in aquaculture," in *Cichlid Fishes: Behaviour, Ecology and Evolution*, M. Keenleyside, Ed., pp. 280–300, Chapman & Hall, London, UK, 1991.

[9] O. L. F. Weyl, A. J. Ribbink, and D. Tweddlel, "Lake Malawi: fishes, fisheries, biodiversity, health and habitat," *Aquatic Ecosystem Health and Management*, vol. 13, no. 3, pp. 241–254, 2010.

[10] F. Witte and W. L. T. van Densen, *Fish Stocks and Fisheries; A Handbook for Field Observations*, Samara House, Cardigan, UK, 1995.

[11] T. D. Kocher, "Status of genomic resources in tilapia (*Oreochromis spp.*)," in *Proceedings of the Plant & Animal Genome Conference*, San Diego, Calif, USA, January 2001.

[12] T. D. Kocher, "Genome sequence of a cichlid fish: the Nile Tilapia (*Oreochromis niloticus*)," Proposal to the JGI Community Sequencing Program by he Cichlid Genome Consortium, 2005.

[13] W. J. Lee and T. D. Kocher, "Microsatellite DNA markers for genetic mapping in *Oreochromis niloticus*," *Journal of Fish Biology*, vol. 49, no. 1, pp. 169–171, 1996.

[14] T. D. Kocher, W. J. Lee, H. Sobolewska, D. Penman, and B. McAndrew, "A genetic linkage map of a cichlid fish, the tilapia (*Oreochromis niloticus*)," *Genetics*, vol. 148, no. 3, pp. 1225–1232, 1998.

[15] A. Estoup and B. Angers, "Microsatellites and minisatellites for molecular ecology: theoretical and empirical considerations," in *Advances in Molecular Ecology*, G. R. Carvalho, Ed., pp. 55–79, IOS Press & Ohmsha, 1998.

[16] P. Jarne and P. J. L. Lagoda, "Microsatellites, from molecules to populations and back," *Trends in Ecology and Evolution*, vol. 11, no. 10, pp. 424–429, 1996.

[17] J. A. DeWoody and J. C. Avise, "Microsatellite variation in marine, freshwater and anadromous fishes compared with other animals," *Journal of Fish Biology*, vol. 56, no. 3, pp. 461–473, 2000.

[18] C. Rico, I. Rico, and G. Hewitt, "470 million years of conservation of microsatellite loci among fish species," *Proceedings of the Royal Society B*, vol. 263, no. 1370, pp. 549–557, 1996.

[19] R. Zardoya, D. M. Vollmer, C. Craddock, J. T. Streelman, S. Karl, and A. Meyer, "Evolutionary conservation of microsatellite flanking regions and their use in resolving the phylogeny of cichlid fishes (Pisces: Perciformes)," *Proceedings of the Royal Society B*, vol. 263, no. 1376, pp. 1589–1598, 1996.

[20] C. R. Primmer, A. P. Moller, and H. Ellegren, "A wide-range survey of cross-species microsatellite amplification in birds," *Molecular Ecology*, vol. 5, no. 3, pp. 365–378, 1996.

[21] B. Y. Lee, W. J. Lee, J. T. Streelman et al., "A second-generation genetic linkage map of tilapia (*Oreochromis spp.*)," *Genetics*, vol. 170, no. 1, pp. 237–244, 2005.

[22] B. Y. Lee, G. Hulata, and T. D. Kocher, "Two unlinked loci controlling the sex of blue tilapia (*Oreochromis aureus*)," *Heredity*, vol. 92, no. 6, pp. 543–549, 2004.

[23] B. Y. Lee, D. J. Penman, and T. D. Kocher, "Identification of a sex-determining region in Nile tilapia (*Oreochromis niloticus*) using bulked segregant analysis," *Animal Genetics*, vol. 34, no. 5, pp. 379–383, 2003.

[24] J. T. Streelman and T. D. Kocher, "Microsatellite variation associated with prolactin expression and growth of salt-challenged tilapia," *Physiological Genomics*, vol. 2002, no. 9, pp. 1–4, 2002.

[25] K. A. Kellogg, J. A. Markert, J. R. Stauffer, and T. D. Kocher, "Microsatellite variation demonstrates multiple paternity in lekking cichlid fishes from Lake Malawi, Africa," *Proceedings of the Royal Society B*, vol. 260, no. 1357, pp. 79–84, 1995.

[26] M. J. H. Van Oppen, C. Rico, J. C. Deutsch, G. F. Turner, and G. M. Hewitt, "Isolation and characterization of microsatellite loci in the cichlid fish *Pseudotropheus zebra*," *Molecular Ecology*, vol. 6, no. 4, pp. 387–388, 1997.

[27] L. Wu, L. Kaufman, and P. A. Fuerst, "Isolation of microsatellite markers in *Astatoreochromis alluaudi* and their cross-species amplifications in other African cichlids," *Molecular Ecology*, vol. 8, no. 5, pp. 895–897, 1999.

[28] M. I. Taylor, F. Meardon, G. Turner, O. Seehausen, H. D. J. Mrosso, and C. Rico, "Characterization of tetranucleotide microsatellite loci in a Lake Victorian, haplochromine cichlid fish: a *Pundamilia pundamilia x Pundamilia nyererei* hybrid," *Molecular Ecology Notes*, vol. 2, no. 4, pp. 443–445, 2002.

[29] R. C. Albertson, J. T. Streelman, and T. D. Kocher, "Directional selection has shaped the oral jaws of Lake Malawi cichlid fishes," *Proceedings of the National Academy of Sciences of the United States of America*, vol. 100, no. 9, pp. 5252–5257, 2003.

[30] E. Crispo, C. Hagen, T. Glenn, G. Geneau, and L. J. Chapman, "Isolation and characterization of tetranucleotide microsatellite markers in a mouth-brooding haplochromine cichlid fish (Pseudocrenilabrus multicolor victoriae) from Uganda," *Molecular Ecology Notes*, vol. 7, no. 6, pp. 1293–1295, 2007.

[31] K. Maeda, H. Takeshima, S. Mizoiri, N. Okada, M. Nishida, and H. Tachida, "Isolation and characterization of microsatellite loci in the cichlid fish in Lake Victoria, *Haplochromis chilotes*," *Molecular Ecology Resources*, vol. 8, no. 2, pp. 428–430, 2008.

[32] M. Sanetra, F. Henning, S. Fukamachi, and A. Meyer, "A microsatellite-based genetic linkage map of the cichlid fish, *Astatotilapia burtoni* (Teleostei): a comparison of genomic architectures among rapidly speciating cichlids," *Genetics*, vol. 182, no. 1, pp. 387–397, 2009.

[33] J. A. Markert, R. C. Schelly, and M. L. Stiassny, "Genetic isolation and morphological divergence mediated by high-energy rapids in two cichlid genera from the lower Congo rapids," *BMC Evolutionary Biology*, vol. 10, no. 1, article 149, 2010.

[34] D. Neumann, M. L. J. Stiassny, and U. K. Schliewen, "Two new sympatric *Sarotherodon* species (Pisces: Cichlidae)

endemic to Lake Ejagham, Cameroon, west-central Africa, with comments on the *Sarotherodon galilaeus* species complex," *Zootaxa*, no. 2765, pp. 1–20, 2011.

[35] M. L. J. Stiassny and U. K. Schliewen, "Congochromis, a new cichlid genus (Teleostei: Cichlidae) from Central Africa, with the description of a new species from the upper Congo River, democratic Republic of Congo," *American Museum Novitates*, no. 3576, pp. 1–14, 2007.

[36] X. Rognon and R. Guyomard, "Large extent of mitochondrial DNA transfer from *Oreochromis aureus* to *O. niloticus* in West Africa," *Molecular Ecology*, vol. 12, no. 2, pp. 435–445, 2003.

[37] R. S. V. Pullin, *Ressources Génétiques en Tilapias pour L'aquaculture*, ICLARM, Manila, Philippines, 1988.

[38] E. Trewavas, *Tilapiine Fishes of the Genera* Sarotherodon, Oreochromis and Danakilia, British Museum Natural History, London, UK, 1983.

[39] S. Nagl, H. Tichy, W. E. Mayer, I. E. Samonte, B. J. McAndrew, and J. Klein, "Classification and phylogenetic relationships of African tilapiine fishes inferred from mitochondrial DNA sequences," *Molecular Phylogenetics and Evolution*, vol. 20, no. 3, pp. 361–374, 2001.

[40] V. Klett and A. Meyer, "What, if anything, is a Tilapia? Mitochondrial ND2 phylogeny of tilapiines and the evolution of parental care systems in the African cichlid fishes," *Molecular Biology and Evolution*, vol. 19, no. 6, pp. 865–883, 2002.

[41] J. Schwarzer, B. Misof, D. Tautz, and U. K. Schliewen, "The root of the East African cichlid radiations," *BMC Evolutionary Biology*, vol. 9, no. 1, article 186, 2009.

[42] G. W. Wohlfarth, "The unexploited potential of tilapia hybrids in aquaculture," *Aquaculture & Fisheries Management*, vol. 25, no. 8, pp. 781–788, 1994.

[43] M. J. Genner, O. Seehausen, D. H. Lunt et al., "Age of cichlids: new dates for ancient lake fish radiations," *Molecular Biology and Evolution*, vol. 24, no. 5, pp. 1269–1282, 2007.

[44] "Marqueurs microsatellites: isolement à l'aide de sondes non-radioactives, caractérisation et mise au point," http://www.agroparistech.fr/svs/genere/microsat/microsat.htm.

[45] A. Estoup, K. Gharbi, M. SanCristobal, C. Chevalet, P. Haffray, and R. Guyomard, "Parentage assignment using microsatellites in turbot (*Scophthalmus maximus*) and rainbow trout (*Oncorhynchus mykiss*) hatchery populations," *Canadian Journal of Fisheries and Aquatic Sciences*, vol. 55, no. 3, pp. 715–725, 1998.

[46] K. Belkhir, P. Borsa, L. Chikhi, N. Raufaste, and F. Bonhomme, *GENETIX 4.02, Logiciel Sous Windows pour la Génétique des Populations*, Laboratoire Génome, Populations, Interactions, CNRS UMR 5000, Université de Montpellier II, Montpellier, France, 2001.

17

Vehicles, Replicators, and Intercellular Movement of Genetic Information: Evolutionary Dissection of a Bacterial Cell

Matti Jalasvuori[1, 2]

[1] Department of Biological and Environmental Science, Center of Excellence in Biological Interactions, University of Jyväskylä, 40014 Jyväskylä, Finland
[2] Division of Evolution, Ecology and Genetics, Research School of Biology, Australian National University, Canberra, ACT 0200, Australia

Correspondence should be addressed to Matti Jalasvuori, matti.jalasvuori@jyu.fi

Academic Editor: Hiromi Nishida

Prokaryotic biosphere is vastly diverse in many respects. Any given bacterial cell may harbor in different combinations viruses, plasmids, transposons, and other genetic elements along with their chromosome(s). These agents interact in complex environments in various ways causing multitude of phenotypic effects on their hosting cells. In this discussion I perform a dissection for a bacterial cell in order to simplify the diversity into components that may help approach the ocean of details in evolving microbial worlds. The cell itself is separated from all the genetic replicators that use the cell vehicle for preservation and propagation. I introduce a classification that groups different replicators according to their horizontal movement potential between cells and according to their effects on the fitness of their present host cells. The classification is used to discuss and improve the means by which we approach general evolutionary tendencies in microbial communities. Moreover, the classification is utilized as a tool to help formulating evolutionary hypotheses and to discuss emerging bacterial pathogens as well as to promote understanding on the average phenotypes of different replicators in general. It is also discussed that any given biosphere comprising prokaryotic cell vehicles and genetic replicators may naturally evolve to have horizontally moving replicators of various types.

1. Introduction

Viruses that infect prokaryotic cells are known to be enormously diverse in terms of genetic information [1, 2]. Most novel viral isolates are likely to have at least some genes that have no homologues among any of the previously known genes, including those in the genomes of related viruses [3]. Yet, there has been a dispute whether or not new genes may actually emerge in viruses [3]. Viruses are dependent on cellular resources such as nucleotides, amino acids, and lipids for producing more viruses; therefore it seems justified to ask whether they also use cellular genes for their genetic information. Yet, when viral genes are compared to other genes in databases, it often appears that they have no cellular counterparts [2]. Where then do these viral genes come from? Have they been acquired from a cellular host that we simply have not sequenced before? Or alternatively, are the cellular genes perhaps just evolving rapidly in viral genomes

so that their common ancestry with the host genes can no longer be derived? Or perhaps, is it indeed possible that new genes actually emerge in viruses themselves?

Forterre and Prangishvili from Pasteur Institute argued that the core of the dispute appears to be in the notion that viruses are often considered to be just their protein-encapsulated extracellular forms [4] that are only stealing cellular resources (including genes) for their own purposes [3, 5, 6]. Take any textbook on viruses and majority of the pictures representing viruses are of the various types of viral shells composed of proteins (and sometimes lipids) that enclose the viral genome. But these infectious virus particles, or virions, are inert in all respects unless they encounter a susceptible host cell [7]. And due to this inertness of virions it is difficult to understand how a virus could ever come up with completely new genes.

The answer is, naturally, that viruses cannot produce new genes during their extracellular state, and thus any potential event for the emergence of a new viral gene must still occur within a cell during the replication cycle of a virus [5]. But if the gene emerges in the genome of a virus, then would it rather be the virus, and not the cell, that was the originator of that gene? Or, to put it differently, was it not the virus that benefited from the emergence of new genetic information? The actual process that causes the genetic information to acquire the status of a gene would still be due to similar processes as the origin of genes within chromosomes (these being different types of genetic changes, such as point mutations, insertions, deletions, gene duplications, etc.), but these changes would be selected due to their improvements on the fitness of the virus. This reasoning has made Forterre to propose a model where viruses are seen essentially as a cellular life form that can also have an extracellular state [7, 8]. Virus is not strictly equivalent to the protein-enclosed viral genome. Rather, the extracellular form of a virus should be denoted as a virion, and this virion should not be mistaken for a virus. Viruses, in a complete sense, are organisms that live within cells (i.e., ribosome-encoding organisms) and can transform other cells into virus-cell organisms by producing more virions. In other words, viruses can utilize an extracellular encapsulated form to transfer its genetic information from one cell to another. Forterre coined a term *virocell*, which refers to the stage of viral life during which the virus is within a cell [7]. The virocell organism is indeed both a (capsid encoding) virus and a (ribosome encoding) chromosome, and the actual phenotype of the virocell is encoded by both of these genetic entities. The virocells are entirely capable of coming up with novel genetic information just as cells are, and thus approaching viruses from this perspective should clear any controversies about the emergence of new genetic information in viruses.

Forterre's line of reasoning along with my own studies on various different genetic elements (including character- ization of temperate and virulent viruses [9, 10]; deter- mination of common ancestor between plasmids, viruses and chromosomal elements [11]; conduction of evolution experiments with bacteria, viruses, and plasmids [12, 13]; as well as more theoretical work on horizontal movement of genetic information [14, 15]) has served as an inspiration for this paper. Indeed, it could be argued in more general terms what it means that prokaryotic cells can be (and often are) chimeras of various types of genetically reproducing elements. Virocell concept clears effectively many of the con- fusions between viruses and virions and their relationship with cells. Nonetheless, virocell is only a special case among all the possible types of prokaryotic organisms. Bacterial and archaeal cells can also contain conjugative plasmids, various types of transposons, defective prophages, and many other independent replicators that are distinct from the ribosome encoding prokaryotic chromosome. Together these replica- tors can produce organisms in all possible combinations. In order for the arguments about virocells to be consistent with the other potential chimeras of genetic replicators, the cell itself must be considered as a separate entity from all the genetic replicators (including chromosomes) that exploit the cell structure for replication. In the following chapters I will perform an evolutionary dissection to a bacterial cell. This will lead into the separation of cell vehicles and replicators from each other and thus provide one potential way to approach the evolution of bacterial organisms.

2. Vehicles and Replicators

"*A vehicle is any unit, discrete enough to seem worth naming, which houses a collection of replicators and which works as a unit for the preservation and propagation of those replicators*", Richard Dawkins wrote in *Extended Phenotype*. Dawkins utilized the concepts of replicators and vehicles in an argument which stated that evolution ultimately operated on the level of genetic information and not on the level of populations of organisms, species, or even cells. Replicators refer to packages of genetic information that are responsible for any effective phenotype of the vehicle. Vehicle itself can be a cell, a multicellular organism, or, for example, the host organism of a parasite. "*A vehicle is not a replicator*", argued Dawkins in an attempt to underline that it is the replicator (like the chromosome of a parasite) and not the vehicle (like the parasitized cell) that evolves. This difference, however, may sometimes be seemingly trivial, which is why it has caused some dissonance among evolutionary biologists.

Nevertheless, Dawkins' work focused mostly on explain- ing evolutionary issues of eukaryotic organisms, but the replicator-centered evolution naturally operates also within and between prokaryotic cells. Indeed, there is a vast diversity of different forms of genetic replicators that use prokaryotic cell vehicles for their preservation and propagation. Any particular prokaryote that lives in this biosphere, being that a bacterium on your forehead or an archaeon in the bottom of Pacific Ocean, harbors a chromosome but may also host a collection of other replicators, including plasmids, transposons, and viruses. Some of the replicators, like conjugative plasmids and viruses, are able to actively move between available vehicles in its environment, thus making these replicators less dependent on the survival of any particular lineage of cell vehicles. Therefore they are not an inherent part of any particular bacterium and may thus be considered as distinct forms of genetically replicating entities that utilize cells for their propagation and survival (similarly with the viruses in Forterre's virocell concept).

The continuous struggle for existence within and between prokaryotic vehicles modifies the phenotypes of the replicators. A lot of theoretical and experimental work has been done in order to clarify the functions and the evolutionary trajectories of viruses, bacterial cells, and plasmids in different ecological contexts and under various selection pressures. However, in this discussion I take a step away from any particular type of a replicator or an organism and explore from a general perspective whether the lateral movement potential (or lack of it) of the replicators could help illuminate some evolutionary aspects of the prokaryotic biosphere. This discussion attempts to provide an intuitive view on the selfish genes and various types of replicators in bacterial and archaeal cells. It is my intention to keep the text simple and readable regardless of the reader's expertise on bacteria, viruses, plasmids, or, for that matter, evolutionary theory. Moreover, given the vast amount of details in microbial world, I hope that the readers realize that certain corners had to be cut in various places in order to keep the text within realistic length.

Furthermore, in an attempt to maintain the simplicity, the following nomenclature and definitions are used throughout this paper. A *cell vehicle* denotes a prokaryotic cell with membranes, resources, and everything else but excludes any genetic material. *Cell-vehicle lineage* indicates a single vehicle and its direct descendant that emerge by cell division. A *replicator* is any discrete enough collection of genetic material (that seems worth naming), which utilizes the cell vehicle for its preservation and propagation. Replicators are replicated as distinct units forming a coherent collection of genetic material that can be separated with reasonable effort from other replicators. Replicators may be replicated as a part of the replication of other replicators, as integrative viruses are replicated along with host-chromosome multiplication, but essentially these two replicators can be denoted as two distinct entities given that the integrative virus can replicate its genetic information also separately from the replication of the chromosome. The mean by which the genetic information of a replicator is replicated is not relevant. However, I prefer to not make a too strict definition for a replicator as it is likely to lead to unproductive hair-splitting arguments. Yet, it must be noted that replicators do not include ribosomes or other nucleic acids containing molecules that essentially have an enzymatic function but that are not used as template for their own replication. *Vertical relationship* or vertical inheritance of a replicator indicates that this genetic replicator preserves itself within a dividing lineage of cell vehicles. *Horizontal movement potential* denotes that the replicator is able to introduce itself into a cell-vehicle lineage where the replicator was previously absent. Any feature that is encoded or induced by a replicator is denoted as a *phenotype*. Figure 1 links these terms with their biological counterparts.

3. Laterally Moving Replicators

Prokaryotic world contains a number of different types of replicators that have potential for lateral movement between cell-vehicle lineages. Here I briefly introduce the basic types of laterally moving replicators.

3.1. Conjugative Plasmids. Conjugative plasmids are extrachromosomal assemblies of genetic material that replicate independently within their host vehicles [16, 17]. Conjugative plasmids may encode complex toxin-antitoxin systems and other effectors that ensure that the dividing cell vehicles harbor copies of the plasmid [18]. Conjugative plasmids also encode proteins that facilitate the transfer of the conjugative plasmid from one cell vehicle to another [19]. Conjugative plasmids can spread between distantly related cell vehicles, but one copy of the plasmid is always maintained within the donating cell. Conjugative plasmids have no extracellular stage and are thus dependent on the host cell at all times.

3.2. Integrative and Conjugative Elements (ICEs). Similarly with conjugative plasmids, ICEs can force the host cell vehicle to form a cell-to-cell contact with other cells in the present environment and use this contact for transporting the genetic element from one cell to another [16, 20, 21]. ICEs can spread between distantly related cell vehicles and replicate therein. ICEs integrate into the chromosome during their life cycle and differ from conjugative plasmids in this respect. This integration may often lead to the transfer of some chromosomal genes from one host to another.

Conjugative plasmids and ICEs are known for their antibiotic resistance genes [22]. Arguably the lateral movement of conjugative plasmids and ICEs is responsible for majority of novel drug-resistant bacterial phenotypes in hospitals and other clinically important environments [16]. Conjugative plasmids and ICEs contain variety of different types of genes including those encoding for virulence factors. However, detailed analysis of this genetic variability and their exact functions and/or roles in certain ecological contexts are beyond the scope of this paper.

3.3. Temperate Viruses. Viruses are replicators that enclose their genetic material within a protective protein capsid [3]. This capsid can leave the host cell and introduce the genetic material into a new cell vehicle far away from the initial host. Thus viruses (unlike plasmids) can be transiently independent from the survival of any particular host cell vehicle. The extracellular state of viruses is known as virion, and it should not be mistaken for a virus [7, 8]. Differences between virions and viruses were discussed in the introduction.

The assembly of viral particles often leads to the destruction of the cell vehicle. However, temperate viruses are able to exist peacefully within their host cell as a so-called provirus [23]. During the provirus state no viral particles are produced. Yet, this lysogenic cycle can be interrupted, which then leads into reigniting the virus particle production. Viruses can become integrated into the host chromosome or exist as extrachromosomal genetic elements during the provirus state [24–26]. A lysogenized cell vehicle is (usually) resistant to infections of other related viruses.

3.4. Virulent Viruses. Virulent viruses are incapable of lysogenic life cycle as they do not maintain regulation machinery that would allow them to retain from virus particle production. Virulent viruses destroy the infected cell vehicle at the end of their replication cycle. However, some virulent viruses can sometimes halt their replication cycle when the host cell is going into dormant state [27].

3.5. Passive Movement of Other Replicators. Prokaryotic cell vehicles can harbor other replicators that can occasionally move horizontally between cell-vehicle lineages, but they do not actively encode functions that would facilitate horizontal movement. These replicators include genetic elements like nonconjugative plasmids, and transposons. Plasmids, transposons, and even complete chromosomes can become transferred from one vehicle to another through the same

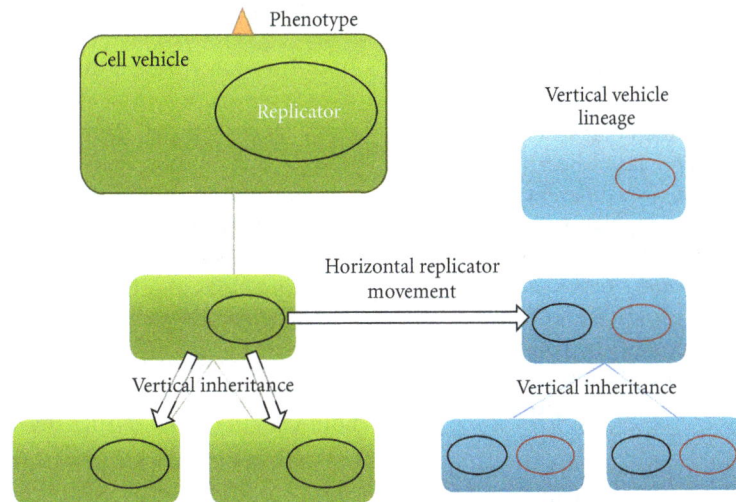

FIGURE 1: The basic terminology used throughout the paper and their biological counterparts.

conjugation channels that conjugative plasmids and ICEs use. Some plasmids or plasmid-like elements can spread from one cell to another within virus capsid [28] or cell-to-cell connecting nanotubes [29, 30]. Moreover, the natural competence of certain cell vehicles allows the uptake of foreign genetic material from the environment [31, 32], which can lead into horizontal movement of replicators between unrelated vehicle lineages. I will not perform thorough analysis of these various types of ways by which genetic information may become transferred between cell vehicles, but it is important to note that such events occur in natural systems.

4. Replicator Dependency on Vertical Survival of Cell Vehicles

Vertical lineage of a cell vehicle indicates a single prokaryotic cell vehicle and all of its direct descendants that emerge via cell division. If a replicator is exclusively dependent on the survival of a certain cell-vehicle lineage, the replicator would inevitably die along with the lineage. Chromosomes exemplify such a replicator. Yet, certain genes of chromosomes may become horizontally transferred even if the cell-vehicle lineage in general would go extinct (e.g., by transposon-induced transfer and recombination). However, for the clarity of this paper, all such (relatively) random potentials that are not general (enough) features of the replicators are being ignored.

Virulent viruses represent a class of replicators that are not bounded by the vertical survival of the cell vehicle. They even cause the demise of the particular cell vehicle as a part of their replication cycle. Yet, we must realize the limits of such definitions as these are just depictions of the average behaviors of biological entities within reasonable time frames. Naturally even virulent viruses are dependent on the survival of the particular vehicle they are infecting until new virus particles are completely assembled. They are

also dependent on the existence of susceptible vehicles in the environment. Nevertheless, it can be argued that, due to their survival strategy, virulent viruses are not dependent on any particular vehicle.

All other replicator types, like plasmids and temperate viruses, are intermediates between virulent viruses and chromosomes in respect to their dependencies on the vertical survival of their current vehicles. This relationship between replicators and vehicles is, naturally, reciprocal as the cell vehicle is not able to survive in absence of the chromosome whereas it fails to survive in presence of a virulent virus. Interestingly, however, this seemingly trivial notion allows us to position the different replicators on a scale where their dependency on cell vehicle appears to (negatively) correlate with their vehicle survival affecting phenotypes (see Section 10). In other words, it is possible that the average phenotype of any replicator matches its position on a chart where lateral movement is on one axis and the vehicle-benefitting phenotype is on the other. Of course, this is just a rough approximation and only an artificial depiction of the result of natural selection repeatedly acting on the replicators. Yet, it can provide a tool to describe the average behavior of prokaryotic replicators. Before addressing this aspect in greater detail, we need to analyze and classify the replicators in a more definitive manner.

5. Classification of Replicators

Most (if not all) of the different types of replicators that utilize prokaryotic cell vehicles for preservation and propagation can be classified according to their horizontal movement potential between individual cell-vehicle lineages and according to their vertical dependencies on cell vehicles. I will attempt to argue that certain phenotypic traits usually associate with replicators of the same class. Subsequently I will discuss the reasons behind this by analyzing few hypothetical scenarios where natural selection might favor the association of these phenotypic traits with horizontally moving replicators rather than with strictly vertically evolving chromosomes. The classification is presented in Table 1.

My attempt was to retain the classification as simple as possible while maintaining the essential insights that may be derivable from it. Yet, it must be noted that strict boundaries cannot be drawn between different classes because this classification seeks to group together highly different and usually unrelated biological entities. Indeed, there are numerous cases where replicators have changed their present classes and have done this rapidly in evolutionary terms. For example, many chromosome-integrating proviruses (*Class IV*) are known to have become defective viruses by conjoining with a *Class I* replicator (chromosome) and thus becoming only a vertically inherited element [23]. Conjugative plasmids (*Class III*) are known to have become conjugation-defective plasmids (turning into *Class II* replicators) [13], and it has been noted that homologous genetic elements can belong to multiple different classes [10, 11]. In other words, the classes do not represent any permanent characteristics of the replicators. Therefore it seems appropriate to ask whether assignment of replicators into any of the classes is able to catch any practical attributes of an evolving biosphere (and thus justify its formulation). In the remainder of the paper I will attempt to address this question from few different perspectives. For example, I will argue (with some examples) that by changing a *class* the replicator starts to evolve towards other replicators within that group. This suggests that repeated rounds of selection on the replicator can have a general trend in shaping the replicator into a typical member of its *class*. Nevertheless, the complexity of the actual natural systems and the shortcomings of such classifications in relation to this complexity are discussed to some extent.

Finally, I want to emphasize that this classification only attempts to provide a tool to improve our means for understanding and discussing the evolution of prokaryotes and their genetic elements. Some might find the classification trivial or obvious, but I believe that it can help some of us simplify the vastly diverse prokaryotic world into evolutionarily useful components. In any case, there are various types of genetic elements in this biosphere that express different phenotypes and vary in their potential for horizontal movement between cells. It seems very likely that the phenotype is associated at least for some parts with the movement potential. Assuming the opposite (that the horizontal movement potential is not related with the phenotype) seems impossible as you may consider the vehicle-terminating virulent viruses as an example of an expressed phenotype (the only mean by which a virulent virus may survive is due to its horizontal movement between vehicles). Therefore, whether we find it practical or not, it is possible to group these features to some extent (regardless of the usefulness of the presented classification). Naturally some extensions to the presented classification (like, e.g., inclusion of the notion of plasmid incompatibility with each other and with some chromosomes [33]) can be introduced, if found necessary. However, I tried to avoid any unnecessary complexity in order to keep the classification intuitively comprehensive.

6. Phenotypic Traits of Replicators

In this section I will go through the usual phenotypic traits of each replicator *classes*. However, it must be noted there are many replicators that have minor or major exceptions to the general traits within each *class*. In other words, replicators in general form a highly diverse group of genetic entities that utilize cell vehicles for replication and preservation in various environments and in various ecological contexts. Yet, general approximations may be done to some extent.

6.1. Class I: Prokaryotic Chromosomes. Chromosomes are the main genetic replicator in cell vehicles. It segregates into both daughter cells during division. It is often considered that any prokaryotic cell is "equal" to its chromosome. Indeed, when studies attempt to identify the genus of a bacterium, the ribosomal genes or some other highly essential chromosomal genes are selected for sequencing. By determining the divergence of sequence of that gene in comparison to other homologous genes in other cell vehicles, it is possible to assign the taxonomic position of the bacterium. This indicates that many chromosomal genes are absolutely essential for the survival of the cell vehicle, and therefore they can be reliably used to determine the evolutionary histories of both the chromosomes and their corresponding cell vehicles (even if I here treat chromosomes and vehicles as distinct and separate components of a cell organism).

The survival of the chromosome replicator is tightly interlocked with the survival of its current cell vehicle. Natural selection favors any phenotypic change in the chromosome that improves the reproductive success and survival of the cell vehicle. In other words, a favorable mutation (or other genetic change) in a chromosome should not decrease the fitness (or increase the reproductive cost) of the cell vehicle. However, evolutionary process within actual populations of prokaryotes is very complex process (even if other replicator types are not involved), and selection may operate on levels above individual cell vehicles. Yet, for the purposes of this discussion, the correlation of the fitness of the chromosomal replicator with the fitness of the cell vehicle is satisfying enough.

6.2. Class II: Plasmids and Transposons. Plasmids are circular or linear DNA molecules that replicate independently to chromosomes within cell vehicles. However, plasmids always require certain genetic products of chromosomes (being those ribosomes, DNA polymerases, or something else). The sizes of their genome vary from a few kilobases to hundreds of thousands of bases.

Plasmids rely on few different strategies to ensure their survival within the dividing host vehicles. They can encode molecular mechanisms that separate the plasmids along with the chromosomes. Some plasmids contain genes for a toxin-antitoxin system. Plasmid encodes both a stable toxin and unstable antitoxin. The stable toxin will destroy host vehicle, if the vehicle does not contain a copy of the antitoxin-producing plasmid. The plasmids that have

TABLE 1: Classification of replicators.

Class	Example replicators	Vertical dependency	Horizontal movement potential	Description of average phenotypes
I	Prokaryotic chromosomes	*Completely dependent*	*No potential*	Encodes the main functional units of all cell vehicles. Required for the binary fission of the cell vehicle.
II	Plasmids, transposons	*Highly dependent*	*Passive*	Low reproductive cost to host cell vehicle. Can encode opportunistically useful phenotypic traits.
III	Conjugative plasmids, integrative and conjugative elements (ICEs)	*Moderately dependent* (always requires a cell vehicle)	*Active* without an extracellular stage	Moderate or low reproductive cost to host cell vehicle. Usually encode opportunistically useful phenotypic traits.
IV	Temperate viruses	*Somewhat dependent* (can survive even if the cell-vehicle terminates)	*Active* with an extracellular stage	Moderate or low reproductive cost to host cell vehicle. Sometimes encode opportunistically useful phenotypic traits.
V	Virulent viruses	*Not dependent*	*Active* with an extracellular stage	Insurmountable reproductive cost that terminates the host cell vehicle. Does not encode cell-vehicle benefitting traits.

either segregation or toxin-antitoxin system (or both) usually control the copy number of plasmids within cell vehicles [34]. These plasmids are large in their size, and thus each copy of the plasmid is a burden to the general reproductive rate of the cell vehicle. Similarly with temperate viruses, plasmids are able to prevent other vehicles harboring similar plasmids to conjugate with their present vehicle [35].

Smaller plasmids may not encode sophisticated segregation mechanisms, but instead they can exist in high numbers within cell vehicles (tens to hundreds of copies) and are stably maintained due to the high probability that the dividing cell will contain a copy of the plasmid in both daughter cells.

Several studies have shown that the presence of plasmids in cell vehicles increases the reproductive cost of the cell. In other words, when cells without and with plasmids are grown in similar conditions, cells without plasmids are able to reproduce more rapidly. Moreover, cell vehicles themselves are generally not dependent on their plasmids. From this perspective it is obvious that the plasmid has to ensure its survival within the vehicle. Should the plasmids decrease the cost of reproduction of the cell vehicle, then selection would favor plasmid-containing cells over plasmid-free cells even without any encoded survival mechanisms.

However and despite the general burden of plasmid, they can sometimes greatly improve the reproductive success of the cell vehicle. Antibiotic resistance genes are often part of plasmid replicators [16, 36]. Other plasmids have genes that help the cell vehicle utilize rare resources when nutrients are scarce. Plasmids can also encode toxins that help the cell vehicle destroy surrounding cells, like human tissues, and thus utilize the resources from these cells for their own benefit [37]. The reasons behind the existence of these genes in *Class II* (and *III*) replicators are discussed later.

6.3. Class III: Conjugative Plasmids and Other Conjugative Elements.
Conjugative plasmids are extrachromosomal genetic elements similar to *Class II* plasmids. However, their existence within a cell vehicle changes the vehicle phenotype by such that the cell can form conjugation channel between its current vehicle and another vehicle in the surrounding environment. Through this channel *Class III* plasmid transfers itself into another cell vehicle. Conjugations put a reproductive cost on the hosting cell vehicle, and thus plasmids can regulate its repression as well as inhibit superconjugation with vehicles that already contain a copy of *Class III* plasmid [35]. Conjugative elements can respond to the stress of the host vehicle, like the presence of antibiotics in the environment, and ignite transfer of the element to other vehicles [38].

Conjugative plasmids use similar and homologous mechanisms for their stable maintenance within vertical cell-vehicle lineages with nonconjugative (*Class II*) counterparts. *Class III* replicators often contain antibiotic resistance genes, and studies suggest that *Class III* replicators are the main cause behind the emergence of clinically relevant bacteria resistant to antibiotics [16].

6.4. Class IV: Temperate Viruses.
Temperate viruses can produce virions, that is, the infectious virus particles, and therefore exist in a "dormant" state in the extracellular environment. However, they can also vertically coexist within cell-vehicle lineages along with *Class I, II,* and *III* replicators. Temperate viruses may integrate into the host chromosome and replicate as a part of *Class I* replicator during cell division. This integration, however, does not abolish the ability to move horizontally between vehicle lineages.

The genomes of temperate viruses may contain genes that are beneficial to the reproduction of their host vehicles (under certain conditions). Presence of a provirus can transform an avirulent bacterium into a virulent one by providing genes for different types of toxins [39]. These toxins can, for example, allow the bacterium to destroy host tissues. Proviruses may also change the host-vehicle phenotype so that it cannot be recognized by eukaryotic immune systems [40].

Class IV viruses are able to detect the malfunction, damage or stress of their host cell vehicles. Proviruses react to these signals by igniting the production of virus particles

[41]. In other words, temperate viruses can predict the upcoming interruption of the vertical cell-vehicle lineage and readily progress into expressing their horizontally moving phenotype. As temperate phages are not dependent on the survival of the host vehicle, they often destroy the doomed vehicle themselves as a part of their lytic life cycle.

6.5. Class V: Virulent Viruses. Virulent viruses also produce virions and thus spend part of their life cycle in the extracellular environment as inert particles. Virulent viruses exclusively destroy the host vehicle as part of their life cycle. Virulent viruses generally do not contain genes that would benefit the vertical survival of the host cell-vehicle. The genetic content of *Class V* replicators appears to aim to effectively utilize the resources of cell vehicles in order to produce multiple horizontally moving virus particles. This, however, does not mean that virulent viruses are simple. Many lytic viruses, like T4, can independently encode essential functions such as some transfer RNA genes, and, indeed, T4 is one of the most complex bacteriophages described to date [42].

7. How Replicators Benefit from the Horizontal Movement between Vehicle Lineages?

Why should a replicator change or move to another vehicle lineage? It is not always obvious why the horizontal movement can be beneficial for a replicator. Indeed, without acknowledging the horizontal movement potential, it appears difficult to understand why bacterial cells or independent replicators have certain types of genes or phenotypes. By realizing that bacterial cells themselves are not always the actual units that are targeted by natural selection can help adopting a truthful image of the microbial world. In this section I consider few simple hypothetical cases that exemplify the effects of horizontal movement on the evolution of replicators and on bacterial organisms.

However, it must be pointed out that this section does not aim to provide any general models or prove any concepts, but instead it is an attempt to intuitively promote the way by which we see the replicators as dynamic components of cell-vehicle populations. The following scenarios are artificial, but their simplicity may help grasping the essence behind the evolution of horizontal movement potential.

7.1. Benefit of Being a Plasmid (Class II and Class III). Imagine a world consisting of hundred independent cell-vehicle lineages. Each of these lineages contains only a single cell that reproduces as fast as it dies, keeping the number of each cell-vehicle type effectively at one. All the lineages replicate and die at identical rates in ultimate resources, and thus the proportions of each cell-vehicle type remains the same. In practice, there is no evolution in this system. By definition this means that the genetic composition of the population is not changing in respect to time.

However, assuming that one of the hundred lineages contains its genetic information in two independent replicators: a chromosome and a reproductively costless plasmid, given

that the plasmid has a potential for horizontal movement between vehicle lineages, then the separation of these replicators into two distinct entities already brings evolution to the system.

In the beginning the plasmid is present only in one percent of the cells in the world. Yet, sometimes after cell death the plasmid is released into the environment. From the environment it has a tiny chance to become introduced into a new cell-vehicle lineage. Each new transformed lineage increases the proportion of the plasmid by one percent and further contributes to the plasmid spread rate. Eventually the plasmid would be present in all of the cell vehicles, and therefore, in comparison to the initial chromosomal partner of the plasmid, the plasmid will be hundred times more successful in terms of prevalence among vehicle lineages. The simple existence of a replicator in an extrachromosomal form with tiny chance for horizontal movement has given it the potential to become by far the most abundant replicator in the system. This simple mind exercise can provide us with a glimpse of the underlying forces of natural selection that operates in actual biological systems. But why should natural selection favor the maintenance of the extrachromosomal form of the plasmid? Why not integrate with the chromosome after entering the cell? If some of the plasmids had permanently integrated to the chromosomes, they would have ceased transforming new cell vehicles into plasmid-containing lineages after the death of the bacterial organism. Thus, as long as there are plasmid-free vehicles available in the system, some of the plasmids may retain their extrachromosomal status as it facilitates the spread (as depicted in Figure 2).

Now consider how the introduction of reproductive cost on the plasmid replication would change the system. Or what if the plasmid somehow evolved a more effectively spreading phenotype and sometimes the plasmid could be lost due to segregation infidelity? Or if the plasmid contained genes that can sometimes increase the reproduction rate of the hosting cell vehicle while they put a general fitness cost on the host? Some of these questions are discussed below. Yet, such complexity is the reality of the ecological dynamics of plasmids in natural environments, and thus these mind games can only provide a platform from which to dive into the real world.

Nevertheless, it appears reasonable to assume that under certain conditions evolution may favor extrachromosomal genetic elements, such as plasmids, that can occasionally join previously plasmid-free cell vehicles. Plasmids also benefit from being as little reproductive cost to their host cells as possible. However, we immediately notice that the faster the plasmid can spread among plasmid-free cells, the faster it takes over the cell-vehicle populations. If there were hundred million cell-vehicle lineages instead of a hundred, even tiniest changes in the rate of spread would hugely affect the reproductive success of the plasmid (given some restricted time window for observing the success). Many studies have tackled the details of the interplay between the spread rates and reproductive costs of plasmids [19, 35, 43]. Theoretical work suggests that certain parameter values generally allow the stable maintenance of plasmids in a (sub) population of

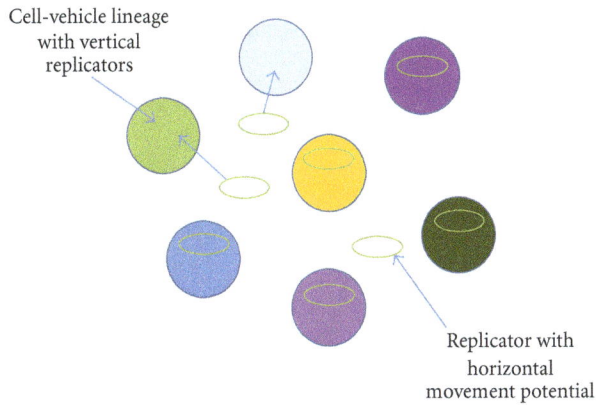

FIGURE 2: Replicators with horizontal movement potential can become common in various cell-vehicle lineages and therefore free of the survival of any particular lineage.

cell vehicles [44, 45]. Nevertheless, the rapid spread leads us to conjugative plasmids, which can actively force their host cell-vehicles to conjunct with plasmid-free cell vehicles in an attempt to transfer the plasmid.

Conjugative plasmids (generally) spread faster than nonconjugative plasmids, and thus, if the two plasmid types were equal in other respects, conjugative plasmids would apparently be evolutionarily more favorable plasmid type. However, the formation of conjugation channels between cell vehicles does not come without a reproductive cost. Indeed, evolutionary research of bacteria often focuses on studying such tradeoffs where one phenotype (e.g., conjugative) is favorable in certain conditions and the other phenotype (e.g., nonconjugative) in alternative conditions. In principle, the conjugative phenotype is practically useless if all cells in the population already harbor a copy of the conjugative plasmid and similarly highly useful when there are plenty of plasmid-free cell vehicles around [43]. Conjugative plasmids always require a cell-to-cell contact for plasmid transfer, indicating that only one (or few) cell(s) at the time can receive the plasmid.

However, as a mind exercise, consider a high copy number nonconjugative plasmid, which can release several plasmid replicators to the environment upon the death of the host vehicle. In principle, each of these replicators has a potential to become introduced into a new cell vehicle, and under ideal conditions high copy number plasmids could spread very fast in a plasmid-free population of cell vehicles. Yet, the naked DNA molecule is fragile in an extracellular environment and the uptake of the molecule requires a competence for plasmid intake from the cell vehicle. In other words, the plasmid will not survive long in the environment and it cannot force the cells to internalize the DNA molecule. Therefore it must be favorable from the perspective of the chromosome or other in-vehicle replicator (as they would encode the competent phenotype of the cell-vehicle) to introduce the new DNA molecule into the cell vehicle. Genes for antibiotic resistances and other beneficial functions can, under certain conditions, significantly increase the fitness of any cell-vehicle lineages. For this reason, opportunistic

genes do not need to only improve the survival of their present vehicles but may sometimes also indirectly improve the probability by which the plasmid can spread horizontally to a new cell vehicle lineage and survive within that lineage thereafter. Natural competence, or the uptake of genetic material into the cell-vehicle from its vicinity, is as the name indicates a natural trait of many bacteria [46]. However, there are also many reasons why natural competence can backfire, and, supposedly, for this reason it is not prevalent trait among bacteria.

Nevertheless, plasmids may evolve mechanisms that allow them to hitchhike through conjugation channels build by other plasmids. This allows them to utilize the horizontal transfer potential without the burden of maintaining genetic machinery for it. Plasmids may also favor evolution towards higher copy numbers within a single cell vehicle in order for the highest copy-number plasmid to have the highest chance for getting transferred into new host vehicle. Yet, the increased cost of maintaining most copies can become compensated on population level by the lower reproductive cost that the lower copy-number plasmid put on individual vehicles [47]. As these different aspects hopefully demonstrate, the actual evolution of the phenotypes of plasmids is a complex subject in which several aspects must be considered. It is not immediately obvious which traits are favorable, and thus I want to retain here the more distant perspective on plasmids and other genetic elements.

7.2. Benefit of Being a Virus (Class IV and V Replicators). In previous section it was considered how the release of high-copy-number plasmids into the environment could provide these replicators a high spread rate among vehicles, if the vehicles in the same environment are willing to take in these replicators. However, viruses are able to overcome this barrier of willing uptake by having the extracellular phenotype that forces the intrusion of the replicator into a suitable host vehicle.

Viral life strategy is dependent on the existence of suitable vehicles in the environment. However, given a susceptible population of cell vehicles, viral strategy is the fastest way by which the replicator can spread in the population. For this reason, all cellular organisms are under constant pressure to avoid viral infections. This, in turn, has led to the everlasting evolutionary arms race between viruses and their hosts [48, 49]. Viruses can obviously effectively maintain their life strategy despite the cost that they put on their current host. However, the ubiquity of viruses cannot be understood without taking the cell vehicles and the vehicle phenotypes into account. Indeed, virions, the extracellular forms of viruses, are the most abundant biological entities on our planet [6]. Yet, as Forterre has argued, virions themselves cannot be considered as living organisms in the same respect as cells can. Ultimately, viruses survive because their hosts survive [50].

7.3. Benefit of Being a Chromosome (Class I Replicator). The existence of chromosomes in any cellular organism is so profound to our concept of cells that we might not even come to think of them as one of the replicators that utilize the cell vehicle for its propagation and preservation. However, in order to distinct the vehicles and replicators from each other under natural selection, we must also address the benefit (and cost) of being a strictly vertically inherited replicator (e.g., a chromosome). To emphasize the reality behind the distinction of replicators from vehicles, it was recently shown that the genome of one bacterial cell vehicle can be replaced by a (closely related) chromosome from another cell vehicle or by an artificially synthesized chromosome [51, 52]. This indicates that the concept of bacterial cell vehicles and their chromosomes is compatible with experiments and therefore their separation is not just a theoretical notion. I discuss here one possible way to approach the evolution of replicators towards a strictly vertical phenotype.

As stated before, all replicators are dependent on cell vehicles for their propagation. The actual living systems have limited resources, and thus the number of cell vehicles rapidly advances to its maximum as the system can support only limited number of cells. This forces the population of cell vehicles to compete for resources. The vertical survival of the vehicle lineage depends on the competitive success of the vehicle. This indicates that for a replicator inhabiting the most successful vehicle *at the beginning of the competition* provides you with most descendants at the end of the experiment—unless, of course, the replicator can horizontally be transferred to other vehicles (as was argued above).

Now, for the sake of argument, let us play with this idea and consider a situation where all the genes within a cell vehicle are separate replicators (these being like very simple *Class II* plasmids). Each gene has a potential for being horizontally transferred between vehicles after cell destruction, but it also has a chance to become lost during cell-vehicle division (depicted in Figure 3). The reproductive success of the vehicle corresponds to the current combination of genes and other genetic information therein as they are responsible for the phenotype of the vehicle. Certain combinations are more successful than others, and therefore they have more descendants within certain timeframe. Some genes are essential for the survival and division of the vehicle, and thus loss of these replicators would terminate the vehicle lineage. Selection should focus on ensuring that the most essential genes are vertically stably maintained as any resources spent on an attempt to divide are wasted unless the essential genes are present in the new cell vehicle. Yet, maintenance of the faithful distribution of thousand individual molecules during a single cell division appears difficult to evolve or heavily costly (given that each of these molecules should have, e.g., an individual type of a segregation system or have regions for chromosome-like segregation), and selection should therefore intuitively progress towards the fusion of these genetic replicators into a single or as few molecules as possible (since this should help the robustness of the segregation during cell division). These replicators would be *Class I* replicators in the presented classification. This is very superficial analysis, yet, it might help grasp the idea that certain genetic functions need to be present within all vehicles at all times, and therefore they

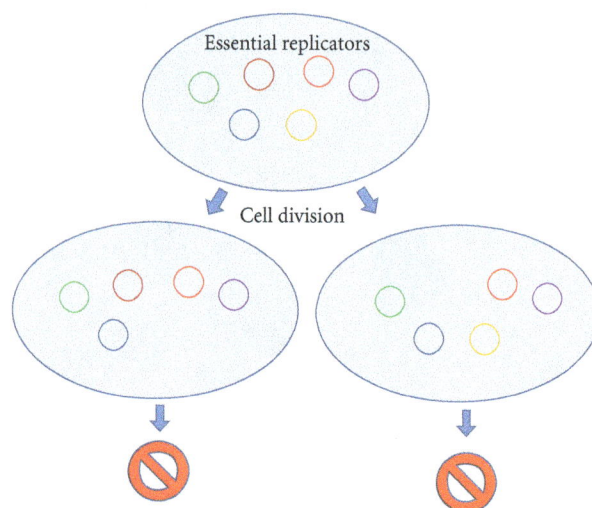

FIGURE 3: A cell vehicle, which contains its essential genetic information in multiple independent replicators, may be prone to lose some replicators during cell division and thus produce incompetent cells.

would be vertically inherited to all functional cell vehicles during vehicle division.

8. Replicators Evolving from One Class to Another

In Section 5 I briefly described few examples of replicators evolving into replicators of different *classes*. Now I will go through some examples where the ecological context favors the replicator to adopt the life strategy of replicators belonging to another *class*. Moreover, I will argue that the subsequent evolution of the replicator starts to favor phenotypes that resemble other replicators within its new *class*.

The general point for discussing this evolution is to illustrate that the classification can provide a framework for approaching complex evolutionary settings. Scientific classifications, however, may be harmful for profound understanding of systems, if we are unable to see beyond the classes themselves. Yet, I believe that a proper classification can give a simplifying touch on some of the acting forces of nature. It must be noted that the different classes of the presented classification do not have strict boundaries and replicators can readily change their classes. Still, the possibility to situate the replicators into these *classes* may reflect general evolutionary tendencies of complex microbial systems and thus prove practical in understanding microbial world.

8.1. Temperate Viruses Evolve into Virulent Viruses. Many bacteriophages are known to acquire mutations, which makes them unable to repress their lytic pathway [9, 24, 41]. These *Class IV* replicators lose their potential to exist vertically within a lineage of cell vehicles, and thus they transform into *Class V* replicators in the classification.

These virulent mutants (or so-called clear plaque mutants) enter bacterial cells, replicate their genomes, express their structural proteins, assemble new virions, and lyse the cell. This evolution of *Class IV* replicators into *Class V* replicators is commonly used in bacteriophage research as the "new" *Class V* replicators are devoid of vertical survival within lineages, and therefore their fitness correlates only with their potential for replicating in other vehicle lineages. This, in turn, often increases the production rate of virions [24], which therefore helps conducting experiments that require virus particles. In other words and from the viewpoint of the classification, *Class IV* replicators started to approach the typical phenotypes of *Class V* replicators due to their incapability for vertical existence within a vehicle lineage.

8.2. Conjugative Plasmids Evolve into Nonconjugative Plasmids. Dahlberg and Chao, 2003, cultivated bacterial cell vehicles containing certain conjugative plasmids for 1100 generations (about half a year) [53]. The system did not contain plasmid-free vehicles, and therefore there was essentially no selection for maintaining the horizontal transfer potential of the conjugative plasmid. Indeed, it was observed that some of the *Class III* replicators had lost their potential for conjugation or the rate of conjugation had decreased during the 1100 generations of their host vehicles. Moreover, the reproductive cost of the plasmid had decreased significantly, indicating that selection efficiently focused on improving the vertical survival of the element within its current vehicle lineage.

After invading the whole population of cell vehicles, horizontal movement had no benefits for *Class III* replicator whereas the vertical survival improved its reproductive success. Therefore, the phenotype of *Class III* replicator in this study started approaching that of *Class II* and *Class I* replicators.

8.3. Temperate Viruses Evolve into Chromosomal Elements. Defective bacteriophages are abundant in many bacterial chromosomes [23]. What good does permanent colonization of a certain vertical lineage of cell vehicles do for *Class IV* replicator? Why not maintain the potential for forming the extracellular viral particle and thus the horizontal transfer potential? Indeed, it has been shown that bacterial genomes harboring functional prophages can have advantage over relatives that lack the phage [54].

Given the modern genomics, natural selection operating repeatedly on microbial communities appears to sometimes favor bacterial chromosomes that have defective bacteriophages integrated into them [23]. Naturally, there must be some reason why it is more favorable for the chromosome to maintain a defective provirus rather than a functional one. One possible (and obvious) explanation considers the differences between functional and defective proviruses. A functional provirus can occasionally induce its lytic activity and thus destroy the host cell vehicle (and the chromosome). Those cells that maintain a prophage are immune to infections by other similar viruses as these defective viruses can encode mechanisms that prevent superinfection, that is,

multiple infections, of a single cell. However, given that the key elements for producing virions become in some way dysfunctional, then the defective virus becomes unable to destroy the host cell vehicle. In a population of cell vehicles where all chromosomes host a same provirus, then the ones hosting a defective provirus may have an advantage over the others [54].

Moreover, defective proviruses appear to start evolve a strictly vertical life strategy. Studies have demonstrated that the cost for carrying a provirus abates the longer the cells are grown in presence of the virus. Some of the proviral genes belonging to defective proviruses are still expressed within cells, suggesting that the provirus phenotype is benefitting only its present cell vehicle [23, 55]. This illustrates how replicators change their classes and utilize its previous genetic information in support to its new life style.

9. Why Antibiotic Resistance Genes Are Often Associated with *Class II* and *III* Replicators?

Why do bacteria help other, sometimes very distantly related, bacteria in their environment by sharing their antibiotic resistance genes with them? If you think that bacteria are generally competing with other bacteria for available resources, then it appears controversial to realize that the same bacteria are helping their rivals against antibiotic-producing organisms. Should it not be evolutionarily favorable for bacteria to let other bacteria die to antibiotics and thus allow them become the sole survivors of the system? This, however, is not the case when we observe bacteria in environments that are abundant with antibiotics. Have the bacteria allied against us just for the heck of it?

In order to realize why bacteria appear to be cooperating against our attempts to utilize antibiotics as an antimicrobial therapy, we must note that antibiotic resistance genes are often part of independent replicators which are not dependent on any particular bacterial cell [13, 16, 20, 21]. This scenario illustrates how and why the presented dissection of bacterial cell can be useful in comprehending bacterial evolution in environments where their evolution might be the matter of life and death.

It is known that majority of antibiotic resistance genes among clinical isolates of bacteria are actually part of conjugative or nonconjugative plasmids or transposons rather than being an inherent feature of any particular chromosome [16]. The spread of plasmids is considered the most common mean by which bacterial strains transform into drug-resistant phenotypes not only in clinical environments but also within other natural environments [20, 21, 38, 56, 57]. Indeed, antibiotic resistance provides a good example of natural selection where certain genes may become a part of horizontally moving replicators rather than vertical ones.

Once again, I will present a hypothetical scenario (adapted from [58]) that may illuminate how natural selection results in rapidly spreading antibiotic-resistance genes within communities of competing bacteria (depicted in Figure 4). Imagine a system containing ten different bacterial species occupying their individual niches. Each of

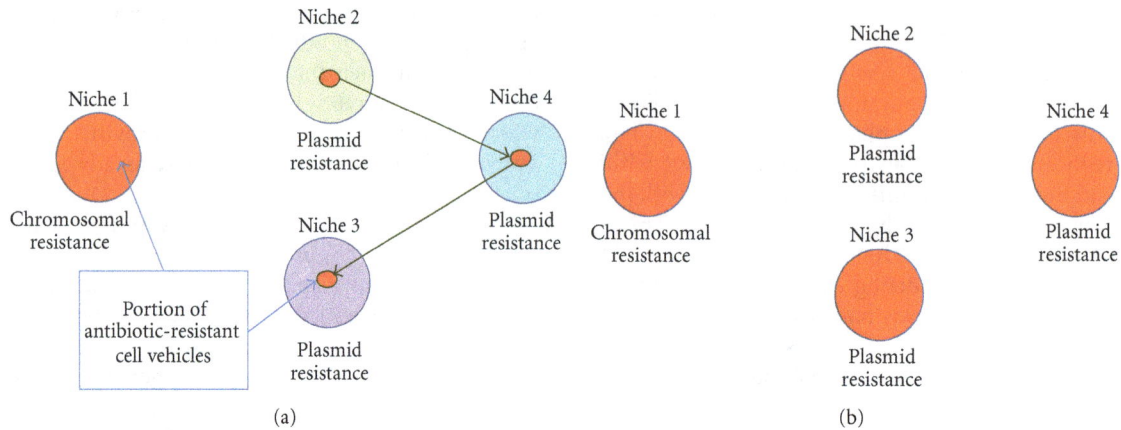

FIGURE 4: When plasmid- and chromosome-borne antibiotic resistances are compared, the plasmid-borne resistance can become more abundant after exposure to antibiotics.

the bacterial lineages is well adapted to their own niche, and none of the other nine lineages are able to invade these niches. One of the nine vehicle lineages contains an antibiotic-resistance gene in its chromosome whereas one of the lineages contains a conjugative plasmid which carries the same antibiotic resistance gene. The conjugative plasmid poses a reproductive cost to its host cell vehicle, but it moves seldom to other lineages. The plasmid does not become prevalent in any single lineage due to the cost, but a portion of cell-vehicles in each of the lineages ends up harboring the plasmid at all times.

Now, an antibiotics-producing organism enters the environment and subjects all bacterial cell vehicles in all of the ten ecological niches to antibiotic selection. The bacteria will either die or suffer a significant reproductive cost due to the antibiotics that disrupt or terminate the functionality of the cell vehicles. Only those vehicles that happen to contain the antibiotic-resistance gene go unaffected by the antibiotics. The selection results in the death of majority of cell vehicles in the system, leaving room for the remaining cells to repopulate each niche.

Which cell vehicles are likely to occupy the free niches? In this scenario we can imagine two possibilities: either it is the cell vehicle that contains the chromosome with the antibiotic resistance gene or it is one of the cell vehicles that harbor the conjugative plasmid. The fitness of the cell vehicle in any of the niches is likely to correlate with its evolutionary history. In other words, cell-vehicles that previously occupied a certain niche are supposedly best adapted to that niche despite of the presence the plasmid in those vehicles. For this reason the vehicle population containing the chromosomal resistance gene might be unable to conquer any of the niches that suffered from the antibiotic selection despite the fact that the chromosomal resistance lineage itself was not affected by the selection. The result would be that nine of the ten niches became occupied by cell vehicles in which the conjugative plasmid is prevalent due to the opportunistic antibiotic resistance gene, and only one of the ten niches contained the resistance gene in the chromosome.

Horizontally spreading replicators, like plasmids and conjugative plasmids, might not be able to become abundant in cell-vehicle lineages due to their cost to the vehicle reproduction. They can, however, be present in multiple lineages as a minority. This minority of plasmid harboring vehicles with opportunistic genes can provide sudden boost to the vehicle fitness (as described above) and therefore become dominant in the population [58].

10. Do the Replicator and Vehicle Dependencies on Each Other Reflect General Evolutionary Tendencies?

As was argued in Section 4, replicators depend on the vertical survival of vehicles to various degrees. Similarly vehicles fail to survive in absence of certain replicators whereas they fail to survive in presence of other replicators. Chromosomes, for example, are fully dependent on their present lineages while virulent viruses are independent from any particular lineage of vehicles. This allows us to plot these dependencies on an approximate scale where on one axis there is the dependency of the replicators on vehicle lineages and on the other axis there is the effect of the replicator on the survival of its present vehicle (Figure 5). I will attempt to demonstrate that this plot may be useful visualization for approaching the evolution of prokaryotic replicators.

First, we observe that the more dependent a replicator is on a certain vehicle lineage, the more dependent a vehicle lineage is on the replicator. Second, we see that the more harmful a replicator is to a lineage, the less it depends on the survival of any particular vehicle. This correlation may appear to be a trivial tautology, but I suggest that, when we know the replicators' position on one axis, we also know its position on the other. I intend to state here that natural selection may be "aware" of this plot and therefore replicators generally evolve towards the corresponding position on the two-dimensional chart. In other words, if a vehicle cannot survive without some particular replicator, then selection

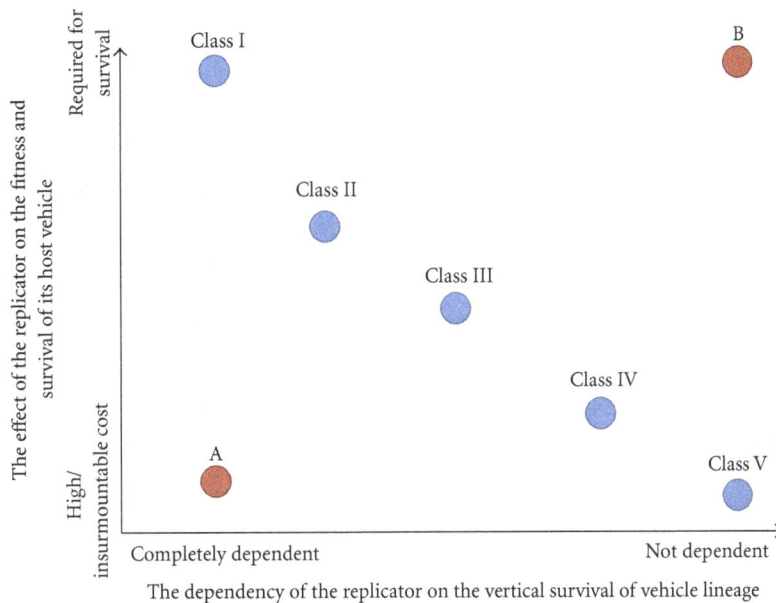

FIGURE 5: Positioning of the different *classes* of the classification into a two-dimensional plot where on one axis there is the horizontal movement potential of the class and on the other there is the effect of the replicator on its present cell vehicle.

favors changes that make it evolve towards a less horizontal form. Similarly, if a replicator is very costly in terms of reproductive success to its vehicle, then it survives best by being able to move horizontally between vehicle lineages or by evolving a less costly phenotype. While I will not attempt to prove this, I propose this as a hypothesis that may be used as a framework for predicting results of simulations or experiments and also for providing a general perspective on the evolution of prokaryotic biosphere.

For the sake of argument, I put letters A and B (representing imaginary replicators) on the plot (Figure 5) at positions that are free of *natural* replicators. Would it be possible that A and B actually existed in nature? I argue that the answer is negative. However, I want to emphasize that such replicators may, of course, exist transiently, but natural selection favors the change towards their correct positions on either of the axes or, alternatively, they will go extinct altogether. Therefore replicators A and B are not evolutionarily stable replicators with their present life strategies.

Replicator A decreases the reproductive fitness of its vehicle. Therefore any vehicle in the environment that lacks A is able to outreproduce vehicles containing A, leading into the extinction of A. However, A could also achieve potential to be transferred horizontally between vehicle lineages (by recombining with a conjugative plasmid, e.g.), which would make A less dependent on the survival of its current lineage. This means that A would be likely to move rightwards on the two-dimensional plot. The other possibility is that A could evolve into a less costly replicator, making it move upwards on the plot. You may consider a host-destroying virus that makes defective virions as an example of A. This virus should evolve either a phenotype that does not destroy the vehicle or it should form functional virions in order to survive.

The case of replicator B is a somewhat less obvious one. B is essential for its present vehicle, but it is not dependent on the vertical survival of any particular vehicle lineages. In other words, vehicles require B for survival, but B itself can freely move between vehicles. However, if we think of the situation, we realize that (by definition) all vehicles must contain a copy of B in order to survive and reproduce. Therefore B would be present in every single surviving vehicle lineage, and the horizontal movement potential would pose only an unnecessary reproductive cost for the current vehicle. From this perspective it appears logical that B will lose its horizontal movement potential as selection would favor the nonhorizontal and therefore less costly phenotype.

A and B depict two unnatural cases, but they provide an example how natural selection may be operating against these positions in the plot. However, the situation becomes increasingly more difficult when we consider any intermediates between A and B. In natural environments and ecological communities, the position of the replicator on y-axis is likely to constantly change depending on the current surrounding conditions of its present vehicle. In presence of antibiotic-producing organisms, the plasmid providing the resistance might be essential for the cell, but this essentiality ceases when the antibiotic producing organism disappears from the effective area of the vehicle. What is the position of such a replicator on y-axis? Similarly certain replicators might be relatively costly to their host vehicles, but they can sometimes give huge advantage to their vehicles due to seldom occurring conditions. However, the diversity and the complexity of natural environments is vast, and it is easy to get lost into the ocean of details. For this reason the plot should be seen as a tool, which may allow approaching complex phenotypes of multiple different types of replicators from a very general perspective.

11. An Example Case of the Emergence of a Relevant Bacterial Organism through Accumulation of Multiple Replicators into a Single Vehicle

Vehicle concept and replicators can provide a general way to approach and explain changing behaviors of bacterial organisms. Microbial world is often seen to consist of just bacteria (and archaea) and viruses. These microbes are living on this planet in any suitable habitat, that is being anything from a rectum of an animal to a hydrothermal vent in the mid-Atlantic ridge. This view is not wrong, and indeed it is the one that we observe with our microscopes. Similarly, general books about microbes generally describe a variety of different viruses and prokaryotes with their taxonomic families and evolutionary relationships. However, these books often credit other horizontally moving replicators to lesser extent despite the fact that they may play a significant role in biological systems and that they are arguably distinct entities in respect to any particular bacterium. Moreover, the general view fails to distinguish the different roles of temperate and lytic viruses. Indeed, the chimerical reality of multiple intercellular and extracellular replicators is a fundamental part of bacterial life, and thus acknowledging this diversity can help us realize why and how certain microbial organisms arise.

What kind of an organism was the enterohemorrhagic *Escherichia coli* (EHEC) that was responsible for the outbreak in Germany in 2011 and that tragically killed tens and caused a severe disease in thousands? From the perspective of this paper, it is interesting that replicators of various classes of the presented classification played a role in the outbreak [59].

Mainstream media described EHEC as a common human bacterium that happened to cause a dreadful disease. To people who are unaware of the details of microbial world, the overall image must have been that *Escherichia coli* can sometimes become extremely harmful. How do some of these naturally commensal bacteria happen to turn into hazardous or even lethal pathogens? Naturally, the evolution of virulence is a complicated matter with a variety of affecting factors. Yet, for a realistic approach, we must understand that it can be independent genetic elements that are responsible for forcing the commensal bacterial organisms to turn into the causative agents of epidemics. Indeed, sometimes news articles about EHEC mentioned that bacteria naturally swap genes with each other and this exchange is behind the emergence of this new lethal version of the bacterium. However, it may still appear unclear how bacteria know to transfer these nasty genes into other bacteria and why do they do that. As in case of antibiotic resistances, for profound understanding we must realize that it is not any actual bacterium transferring these genes, but instead that a bacterium is an organism that consists of a cell vehicle along with a chromosome and possibly some other genetic replicators. And these other replicators are the ones that induce the phenotype that is responsible for transferring horizontally genes into other bacteria. And they do it because it is beneficial for their own survival and reproduction.

EHEC behind the Germany outbreak contained temperate viruses (*Class IV*) that provided the Shiga toxin genes responsible for the pathogenic phenotype of the bacterium [60]. In other words, EHEC would not have caused the epidemic if there were no *Class IV* replicator that used the same cell vehicle with the chromosome for its propagation and preservation. Moreover, EHEC strain contained a large conjugative plasmid (*Class III*) that provided the vehicle with antibiotic resistances and some other useful phenotypes. However, EHEC infections are usually not treated with antibiotics anyway as antibiotics may increase Shiga toxin production of the bacterium. Nevertheless, the plasmid may have given the vehicle a potential to survive in environments where it would have naturally succumbed. Overall, by realizing that bacterial cells are combinations of various independent genetic entities, we may understand how new diseases and super bugs emerge from previously harmless organisms.

12. Examples of Using the Classification in Formulating Hypotheses for Evolutionary Experiments

I want to demonstrate that the presented classification could be used to provide a framework for formulating some practical scientific hypotheses (e.g., predicting outcomes of evolution experiments). I give few simple examples that essentially ask whether or not we may approach evolving bacterial populations from the viewpoint of various replicators with differing potentials for horizontal movement and with differing effects on the survival of the cell vehicles.

Opportunistic genes that only sometimes but significantly improve the survival or reproductive rate of a cell vehicle are likely to become associated with horizontally moving replicators rather than Class I vertical replicators in natural communities of bacteria. In principle, this hypothesis could be tested by cultivating a diverse bacterial population in an environment where there are multiple niches available. One of the bacteria would contain the opportunistic gene (like antibiotic resistance) in its chromosome replicator, and one of them would have the gene in a horizontally moving replicator (like conjugative plasmid). The system would be let to grow for some time before and after introducing the antibiotic selection to the system. The prevalence of the opportunistic gene in horizontal replicator instead of the chromosomal replicator could be measured.

If an opportunistic gene associated with a horizontally moving replicator becomes mandatory for the survival of the cell vehicles in the environment, then the replicator associated with the opportunistic gene evolves towards a (more) vertical phenotype or the gene becomes part of one of the vertical replicators. In principle, the hypothesis could be tested by introducing a conjugative plasmid (containing an opportunistic gene, like antibiotic resistance) to a population of bacteria. Then, lethal doses of antibiotic selection would be stably maintained in the system over several bacterial generations. After the selection, the cost of the plasmid to the cell-vehicle and its conjugation rate could be measured.

If selection focuses against a replicator on which the cell vehicle is not dependent, then the complete replicator can become eliminated. If selection focuses on an essential replicator, then the replicator is likely to only change its phenotype or the whole vehicle lineage becomes terminated. This was the actual hypothesis in a recently published experimental paper by me and my colleagues [13]. We tested what happens when plasmid-dependent phages were cultivated with bacteria harboring plasmids in presence and absence of the selection for the plasmid. In absence of the selection, the plasmid was shown to become lost. In presence of the selection, the plasmid (or the selected parts of it) survived but its phenotype changed.

13. Host Range and the Replicators in the Evolution of Biospheres

In this final section I will consider what the possibility to classify replicators according to their effects on the survival of host cell vehicle and their horizontal movement potential might implicate about the evolution of vehicle- and replicator-based biospheres. For those interested in pondering the development of hypothetical forms of life, this discussion can serve as a (testable) hypothesis about the general trends in the evolution of any living system in this universe.

Replicators that move between vehicles have varying host ranges. By the term host range is meant the portion of cell vehicles into which a replicator can transfer to and subsequently replicate in. A virulent virus can usually infect only a tiny fraction of closely related cells whereas conjugative plasmid can be transferred successfully to a much wider range of unrelated cells. The host range of virulent viruses is narrow whereas the host range of a plasmid is large. Naturally, this is not a coincidence.

Virulent viruses terminate the cell vehicles wherein they replicate. Therefore all the other replicators, especially chromosomes, become eliminated due to virus replication. Selection therefore favors those chromosomes among a population of cell vehicles that produce phenotypes which are unrecognizable by viruses. This has been confirmed in various studies that demonstrate the coevolutionary arms race between viruses and their hosts [61]. On the other hand, conjugative plasmids have been shown to be able to transfer and replicate in a variety of different types of cells. There is stronger selection pressure for chromosomal replicators to avoid viruses than to avoid plasmids. Sometimes avoidance of a plasmid can be lethal whereas avoidance of a virus is rarely harmful. In other words, evolutionary dynamics, in general, force the replicators with higher cost on the host cell vehicle to have narrower host range. Now, it can be asked whether this notion may provide any insights about evolving biosystems. There are already numerous papers about coevolutionary dynamics of viruses and cells [61], about virus-driven evolution [62], and about host ranges

[63]. My intent is not to repeat them but instead to try applying a more general perspective on the issue.

Our biosphere is abundant with all the types of replicators of the proposed classification, and therefore we may not consider it relevant to think whether or not this is mere coincidence or a direction towards which *any given biosphere* progresses. But what if we take another independently emerged and evolved (although hypothetical) living system which contains vehicles and replicators? If we go through the replicators in that system, are we able to use this same classification for them as we are for replicators on Earth? Do all the *classes* have at least some representatives in the foreign biosphere? Or are there systems where, for example, only chromosomes or just chromosomes and plasmids thrive?

We need to note that the considered biosphere must be large enough in order for this question to be relevant. When we take a small sample of microbes in our world, we may find that some of the replicators, like conjugative plasmids, cannot be found. Therefore tiny cellular communities may not be able to support the full variety of replicators. But what is the case when we take, let us say, a planet full of microbial life? Can we say with relative certainty that we are going to find plasmids, conjugative elements, and viruses just because that is how natural selection in general tends to shape evolving biospheres that are abundant with single-celled organisms?

In order to approach this question, we may consider biospheres where replicators of some of the classes are absent and evaluate whether or not it is possible that some other replicators will inevitably evolve to represent the missing *class*. In Section 10, I argued that replicators may be evolving towards the *correct* position on the two-dimensional plot presented in Figure 5. Now, if one of the classes depicted in Figure 5 had no representatives in a given biosphere, like there were no *Class V* replicators at all, would some of the other replicators be likely to evolve to fill this free *niche*? I will not go through all the possible cases or scenarios but instead address few general ideas.

If a foreign biosphere completely lacked viruses (that can directly cause the demise of their hosting cell vehicles), what would likely to be different in comparison to our biosphere? Naturally, one can think of a huge number of things. However, perhaps one of the most relevant for our considerations is the notion that there would be no evolutionary arms race between viruses and hosts. Cellular populations would not need to maintain variation against constantly evolving virosphere, and, therefore, in absence of viral-induced selection for variance there might be a huge number of cells that maintain, for example, highly conserved surface components. This could indicate that if a virus emerged, it would be likely to be able to reproduce within a huge population of hosts. In other words, any crudest form of a virus would be likely to have a very wide host range and thus be highly successful in producing copies of itself. Therefore, the naivety of the biosphere due to the lack of previous exposure to viruses might render it highly vulnerable to viral invasion. Given a large biosphere and long-enough timeframe, viral strategy might be bound to emerge sooner or later. Experiments have shown that bacterial populations unexposed to viral selection tend to be more homogenous in comparison to those with viral predators [62].

What if a system had viruses and chromosomes but was devoid of plasmids? Would plasmids be likely to emerge? To address this question, we may need to consider what the usual characteristics of plasmids in our biosphere are, and then ask whether these characteristics should also become associated with plasmid-like replicators (with higher horizontal movement potential than chromosomes) in any other biosphere. Indeed, plasmids often appear to harbor opportunistic genes, like those conferring antibiotic resistance. Such genes may also be likely to exist in foreign living systems, given that biospheres anywhere should inhabit environments where selection pressures are likely to change according to the current ecological and environmental conditions of the particular cell vehicles. If such opportunistic genes are present, then by reconsidering the mind exercise presented in Section 6 and Figure 4, we may find it logical that *Class II* or *Class III* like replicators may emerge due to local evolutionary dynamics. In other words, opportunistic genes may provide an evolutionarily favorable path for the appearance of smaller low-cost replicators that have increased potential for horizontal movement.

In more general terms, I suggest that it is possible that in large biospheres evolution may progress towards various types of replicators with varying potential for horizontal movement, perhaps even to fill all the slots in the presented classification. Naturally, this suggestion can and must be subjected to variety of different types of experimental tests. Nevertheless, in our biosphere all the different classes appear to be evolutionarily stable strategies as they are abundant and ancient. Therefore, given a sizable enough frame from which to observe evolving systems with cell vehicles and replicators, similar stability may be inevitable to emerge. However, it is still very much possible that these classes may be a feature solely of our type of microbial life. Either way, improved knowledge of the underlying issues would help us understand evolving systems nonetheless.

Finally, I want to emphasize that all of the replicator types we now observe in our biosphere may have emerged before the formation of the first consistently reproducing cell vehicle and chromosome. However, discussing the emergence of all the *classes* as a part of an evolving primordial community is far beyond the scope and length of this paper (although being previously discussed to some extent [14, 15, 64–66]). It may, nevertheless, be possible that the early evolutionary dynamics of emerging life anywhere in this universe may naturally generate replicators with varying potential for horizontal movement between cell vehicles. And as the life advances, the replicators remain as a permanent part of the system.

To conclude, horizontal movement and replicator phenotypes may be approached from a general perspective where we do not pay attention to exact details but rather observe the overall characteristics of replicators in an attempt to understand why and how evolving systems, such as prokaryotic biospheres, may appear to be constructed the way they are. At this time, however, it might be impossible to say whether or not this would be of any practical use or lead to meaningful insights.

Acknowledgments

This work was supported by the Academy of Finland. The author wants to thank the reviewer for helpful comments especially on the host ranges of replicators. WISE is thanked for interesting discussions and sessions (47UMa).

References

[1] E. Hambly and C. A. Suttle, "The viriosphere, diversity, and genetic exchange within phage communities," *Current Opinion in Microbiology*, vol. 8, no. 4, pp. 444–450, 2005.

[2] Y. Yin and D. Fischer, "Identification and investigation of ORFans in the viral world," *BMC Genomics*, vol. 9, article 24, 2008.

[3] P. Forterre and D. Prangishvili, "The origin of viruses," *Research in Microbiology*, vol. 160, no. 7, pp. 466–472, 2009.

[4] F. Jacob and E. L. Wollman, "Viruses and genes," *Scientific American*, vol. 204, pp. 93–107, 1961.

[5] J. M. Claverie, "Viruses take center stage in cellular evolution," *Genome Biology*, vol. 7, no. 6, article 110, 2006.

[6] P. Forterre and D. Prangishvili, "The great billion-year war between ribosome- and capsid-encoding organisms (cells and viruses) as the major source of evolutionary novelties," *Annals of the New York Academy of Sciences*, vol. 1178, pp. 65–77, 2009.

[7] P. Forterre, "Manipulation of cellular syntheses and the nature of viruses: the virocell concept," *Comptes Rendus Chimie*, vol. 14, pp. 392–399, 2011.

[8] P. Forterre, "Giant viruses: conflicts in revisiting the virus concept," *Intervirology*, vol. 53, no. 5, pp. 362–378, 2010.

[9] S. Sozhamannan, M. McKinstry, S. M. Lentz et al., "Molecular characterization of a variant of Bacillus anthracis-specific phage AP50 with improved bacteriolytic activity," *Applied and Environmental Microbiology*, vol. 74, no. 21, pp. 6792–6796, 2008.

[10] M. Jalasvuori, S. T. Jaatinen, S. Laurinavičius et al., "The closest relatives of icosahedral viruses of thermophilic bacteria are among viruses and plasmids of the halophilic archaea," *Journal of Virology*, vol. 83, no. 18, pp. 9388–9397, 2009.

[11] M. Jalasvuori, A. Pawlowski, and J. K. H. Bamford, "A unique group of virus-related, genome-integrating elements found solely in the bacterial family Thermaceae and the archaeal family Halobacteriaceae," *Journal of Bacteriology*, vol. 192, no. 12, pp. 3231–3234, 2010.

[12] V. P. Friman, T. Hiltunen, M. Jalasvuori et al., "High temperature and bacteriophages can indirectly select for bacterial pathogenicity in environmental reservoirs," *PLoS ONE*, vol. 6, no. 3, article e17651, 2011.

[13] M. Jalasvuori, V. P. Friman, A. Nieminen, J. K. H. Bamford, and A. Buckling, "Bacteriophage selection against a plasmid-encoded sex apparatus leads to the loss of antibioticresistance plasmids," *Biology Letters*, vol. 7, no. 6, pp. 902–905, 2011.

[14] M. Jalasvuori, M. P. Jalasvuori, and J. K. H. Bamford, "Dynamics of a laterally evolving community of ribozyme-like agents as studied with a rule-based computing system," *Origins of Life and Evolution of Biospheres*, vol. 40, no. 3, pp. 319–334, 2010.

[15] M. Jalasvuori and J. K. H. Bamford, "Structural co-evolution of viruses and cells in the primordial world," *Origins of Life and Evolution of Biospheres*, vol. 38, no. 2, pp. 165–181, 2008.

[16] P. M. Bennett, "Plasmid encoded antibiotic resistance: acquisition and transfer of antibiotic resistance genes in bacteria," *British Journal of Pharmacology*, vol. 153, no. 1, pp. S347–S357, 2008.

[17] H. Hradecka, D. Karasova, and I. Rychlik, "Characterization of *Salmonella enterica* serovar Typhimurium conjugative plasmids transferring resistance to antibiotics and their interaction with the virulence plasmid," *Journal of Antimicrobial Chemotherapy*, vol. 62, no. 5, pp. 938–941, 2008.

[18] M. Adamczyk and G. Jagura-Burdzy, "Spread and survival of promiscuous IncP-1 plasmids," *Acta Biochimica Polonica*, vol. 50, no. 2, pp. 425–453, 2003.

[19] F. Dionisio, I. Matic, M. Radman, O. R. Rodrigues, and F. Taddei, "Plasmids spread very fast in heterogeneous bacterial communities," *Genetics*, vol. 162, no. 4, pp. 1525–1532, 2002.

[20] V. Burrus, G. Pavlovic, B. Decaris, and G. Guédon, "Conjugative transposons: the tip of the iceberg," *Molecular Microbiology*, vol. 46, no. 3, pp. 601–610, 2002.

[21] M. I. Ansari, E. Grohmann, and A. Malik, "Conjugative plasmids in multi-resistant bacterial isolates from Indian soil," *Journal of Applied Microbiology*, vol. 104, no. 6, pp. 1774–1781, 2008.

[22] M. I. Ansari, E. Grohmann, and A. Malik, "Conjugative plasmids in multi-resistant bacterial isolates from Indian soil," *Journal of Applied Microbiology*, vol. 104, no. 6, pp. 1774–1781, 2008.

[23] S. Casjens, "Prophages and bacterial genomics: what have we learned so far?" *Molecular Microbiology*, vol. 49, no. 2, pp. 277–300, 2003.

[24] A. Gaidelyte, S. T. Jaatinen, R. Daugelavičius, J. K. H. Bamford, and D. H. Bamford, "The linear double-stranded DNA of phage Bam35 enters lysogenic host cells, but the late phage functions are suppressed," *Journal of Bacteriology*, vol. 187, no. 10, pp. 3521–3527, 2005.

[25] C. Canchaya, G. Fournous, S. Chibani-Chennoufi, M. L. Dillmann, and H. Brüssow, "Phage as agents of lateral gene transfer," *Current Opinion in Microbiology*, vol. 6, no. 4, pp. 417–424, 2003.

[26] H. Brüssow, C. Canchaya, and W. D. Hardt, "Phages and the evolution of bacterial pathogens: from genomic rearrangements to lysogenic conversion," *Microbiology and Molecular Biology Reviews*, vol. 68, no. 3, pp. 560–602, 2004.

[27] W. J. J. Meijer, V. Castilla-Llorente, L. Villar, H. Murray, J. Errington, and M. Salas, "Molecular basis for the exploitation of spore formation as survival mechanism by virulent phage φ29," *EMBO Journal*, vol. 24, no. 20, pp. 3647–3657, 2005.

[28] Y. Wang, Z. Duan, H. Zhu et al., "A novel Sulfolobus non-conjugative extrachromosomal genetic element capable of integration into the host genome and spreading in the presence of a fusellovirus," *Virology*, vol. 363, no. 1, pp. 124–133, 2007.

[29] T. A. Ficht, "Bacterial exchange via nanotubes: lessons learned from the history of molecular biology," *Frontiers in Microbiology*, vol. 2, article 179, 2011.

[30] G. P. Dubey and S. Ben-Yehuda, "Intercellular nanotubes mediate bacterial communication," *Cell*, vol. 144, no. 4, pp. 590–600, 2011.

[31] M. G. Lorenz and W. Wackernagel, "Bacterial gene transfer by natural genetic transformation in the environment," *Microbiological Reviews*, vol. 58, no. 3, pp. 563–602, 1994.

[32] J. Davison, "Genetic exchange between bacteria in the environment," *Plasmid*, vol. 42, no. 2, pp. 73–91, 1999.

[33] L. A. Marraffini and E. J. Sontheimer, "CRISPR interference limits horizontal gene transfer in staphylococci by targeting DNA," *Science*, vol. 322, no. 5909, pp. 1843–1845, 2008.

[34] J. Paulsson and M. Ehrenberg, "Molecular clocks reduce plasmid loss rates: the R1 case," *Journal of Molecular Biology*, vol. 297, no. 1, pp. 179–192, 2000.

[35] M. P. Garcillán-Barcia and F. de la Cruz, "Why is entry exclusion an essential feature of conjugative plasmids?" *Plasmid*, vol. 60, no. 1, pp. 1–18, 2008.

[36] A. Saeed, H. Khatoon, and F. A. Ansari, "Multidrug resistant gram-negative bacteria in clinical isolates from Karachi," *Pakistan Journal of Pharmaceutical Sciences*, vol. 22, no. 1, pp. 44–48, 2009.

[37] M. Sengupta and S. Austin, "Prevalence and significance of plasmid maintenance functions in the virulence plasmids of pathogenic bacteria," *Infection and Immunity*, vol. 79, no. 7, pp. 2502–2509, 2011.

[38] P. J. Hastings, S. M. Rosenberg, and A. Slack, "Antibiotic-induced lateral transfer of antibiotic resistance," *Trends in Microbiology*, vol. 12, no. 9, pp. 401–404, 2004.

[39] C. L. Gyles, "Shiga toxin-producing *Escherichia coli*: an overview," *Journal of animal science*, vol. 85, no. 13, pp. E45–62, 2007.

[40] M. R. King, R. P. Vimr, S. M. Steenbergen et al., "*Escherichia coli* K1-specific bacteriophage CUS-3 distribution and function in phase-variable capsular polysialic acid O acetylation," *Journal of Bacteriology*, vol. 189, no. 17, pp. 6447–6456, 2007.

[41] N. Fornelos, J. K. H. Bamford, and J. Mahillon, "Phage-borne factors and host LexA regulate the lytic switch in phage GIL01," *Journal of Bacteriology*, vol. 193, no. 21, pp. 6008–6019, 2011.

[42] V. M. Petrov, S. Ratnayaka, J. M. Nolan, E. S. Miller, and J. D. Karam, "Genomes of the T4-related bacteriophages as windows on microbial genome evolution," *Virology Journal*, vol. 7, article 292, 2010.

[43] J. smith, "Superinfection drives virulence evolution in experimental populations of bacteria and plasmids," *Evolution*, vol. 65, no. 3, pp. 831–841, 2011.

[44] B. R. Levin, M. Lipsitch, V. Perrot et al., "The population genetics of antibiotic resistance," *Clinical Infectious Diseases*, vol. 24, no. 1, supplement, pp. S9–S16, 1997.

[45] F. M. Stewart and B. R. Levin, "The population biology of bacterial plasmids: a priori conditions for the existence of conjugationally transmitted factors," *Genetics*, vol. 87, no. 2, pp. 209–228, 1977.

[46] D. Dubnau, "DNA uptake in bacteria," *Annual Review of Microbiology*, vol. 53, pp. 217–244, 1999.

[47] M. M. Watve, N. Dahanukar, and M. G. Watve, "Sociobiological control of plasmid copy number in bacteria," *PLoS ONE*, vol. 5, no. 2, Article ID e9328, 2010.

[48] J. S. Weitz, H. Hartman, and S. A. Levin, "Coevolutionary arms races between bacteria and bacteriophage," *Proceedings of the National Academy of Sciences of the United States of America*, vol. 102, no. 27, pp. 9535–9540, 2005.

[49] A. Buckling and P. B. Rainey, "Antagonistic coevolution between a bacterium and a bacteriophage," *Proceedings of the Royal Society B*, vol. 269, no. 1494, pp. 931–936, 2002.

[50] F. M. Stewart and B. R. Levin, "The population biology of bacterial viruses: why be temperate," *Theoretical Population Biology*, vol. 26, no. 1, pp. 93–117, 1984.

[51] D. G. Gibson, J. I. Glass, C. Lartigue et al., "Creation of a bacterial cell controlled by a chemically synthesized genome," *Science*, vol. 329, no. 5987, pp. 52–56, 2010.

[52] C. Laitigue, S. Vashee, M. A. Algire et al., "Creating bacterial strains from genomes that have been cloned and engineered in yeast," *Science*, vol. 325, no. 5948, pp. 1693–1696, 2009.

[53] C. Dahlberg and L. Chao, "Amelioration of the cost of conjugative plasmid carriage in *Eschericha coli* K12," *Genetics*, vol. 165, no. 4, pp. 1641–1649, 2003.

[54] L. Bossi, J. A. Fuentes, G. Mora, and N. Figueroa-Bossi, "Prophage contribution to bacterial population dynamics," *Journal of Bacteriology*, vol. 185, no. 21, pp. 6467–6471, 2003.

[55] A. Campbell, "Comparative molecular biology of lambdoid phages," *Annual Review of Microbiology*, vol. 48, pp. 193–222, 1994.

[56] A. Malik, E. K. Çelik, C. Bohn, U. Böckelmann, K. Knobel, and E. Grohmann, "Detection of conjugative plasmids and antibiotic resistance genes in anthropogenic soils from Germany and India," *FEMS Microbiology Letters*, vol. 279, no. 2, pp. 207–216, 2008.

[57] E. Kristiansson, J. Fick, A. Janzon et al., "Pyrosequencing of antibiotic-contaminated river sediments reveals high levels of resistance and gene transfer elements," *PLoS ONE*, vol. 6, no. 2, article e17038, 2011.

[58] C. T. Bergstrom, M. Lipsitch, and B. R. Levin, "Natural selection, infectious transfer and the existence conditions for bacterial plasmids," *Genetics*, vol. 155, no. 4, pp. 1505–1519, 2000.

[59] O. Bezuidt, R. Pierneef, K. Mncube, G. Lima-Mendez, and O. N. Reva, "Mainstreams of horizontal gene exchange in enterobacteria: consideration of the outbreak of enterohemorrhagic *E. coli* O104:H4 in Germany in 2011," *PLoS ONE*, vol. 6, no. 10, article e25702, 2011.

[60] D. A. Rasko, D. R. Webster, J. W. Sahl et al., "Origins of the E. coli strain causing an outbreak of hemolytic-uremic syndrome in Germany," *The New England Journal of Medicine*, vol. 365, no. 8, pp. 709–717, 2011.

[61] J. S. Weitz, H. Hatman, and S. A. Levin, "Coevolution arms races between bacteria and bacteriophage," *Proceedings of the National Academy of Sciences of the United States of America*, vol. 102, pp. 9535–9540, 2011.

[62] S. Paterson, T. Vogwill, A. Buckling et al., "Antagonistic coevolution accelerates molecular evolution," *Nature*, vol. 464, no. 7286, pp. 275–278, 2010.

[63] C. O. Flores, J. R. Meyer, S. Valverde, L. Farr, and J. S. Weitz, "Statistical structure of host-phage interactions," *Proceedings of the National Academy of Sciences of the United States of America*, vol. 108, no. 28, pp. E288–E297, 2011.

[64] M. Jalasvuori and J. K. H. Bamford, "Viruses and life: can there be one without the other?" *Journal of Cosmology*, vol. 10, pp. 3446–3454, 2010.

[65] M. Jalasvuori and J. K. H. Bamford, "Did the ancient crenarchaeal viruses from the dawn of life survive exceptionally well the eons of meteorite bombardment?" *Astrobiology*, vol. 9, no. 1, pp. 131–137, 2009.

[66] E. V. Koonin, T. G. Senkevich, and V. V. Dolja, "The ancient virus world and evolution of cells," *Biology Direct*, vol. 1, article 29, 2006.

18

Divergence in Defence against Herbivores between Males and Females of Dioecious Plant Species

Germán Avila-Sakar and Cora Anne Romanow

Department of Biology, The University of Winnipeg, Winnipeg, MB, Canada R3B 2G3

Correspondence should be addressed to Germán Avila-Sakar, gasakar@gmail.com

Academic Editor: Jeremy L. Marshall

Defensive traits may evolve differently between sexes in dioecious plant species. Our current understanding of this process hinges on a partial view of the evolution of resistance traits that may result in male-biased herbivory in dioecious populations. Here, we present a critical summary of the current state of the knowledge of herbivory in dioecious species and propose alternative evolutionary scenarios that have been neglected. These scenarios consider the potential evolutionary and functional determinants of sexual dimorphism in patterns of resource allocation to reproduction, growth, and defence. We review the evidence upon which two previous reviews of sex-biased herbivory have concluded that male-biased herbivory is a rule for dioecious species, and we caution readers about a series of shortcomings of many of these studies. Lastly, we propose a minimal standard protocol that should be followed in any studies that intend to elucidate the (co)evolution of interactions between dioecious plants and their herbivores.

1. Introduction

Sexual systems in angiosperms range from hermaphroditism (monomorphic populations of plants with bisexual flowers) to dioecy (dimorphic populations of male and female individuals) and include almost all imaginable combinations and gradations (Table 1; [1, 2] and references therein). Such remarkable diversity of sexual systems has perplexed naturalists and evolutionary biologists for a long time [3–6]. The evolution of dioecy from a hermaphroditic ancestor has been particularly difficult to understand because the invasion and maintenance of unisexual mutants in a population of hermaphrodites require that the loss of fitness resulting from the loss of one sexual function be compensated by increased fitness gains through the remaining sexual function of the unisexual mutant [7, 8]. This requirement seems very restrictive, and therefore considerable effort has been devoted towards understanding the conditions under which dioecy can evolve [5, 8–18]. In contrast, the evolution of sexually dimorphic traits following the evolution of dioecy (successful establishment of only two reciprocal unisexual morphs in a population) has received less attention. Consequently, our current understanding of the evolution of

sex-related traits ultimately leading to morphological or physiological differences between unisexual morphs (i.e., sexual dimorphism or secondary sexual traits) is still limited, despite recent advances and excellent syntheses on the topic [5, 6, 19–21]. This paper focuses on one set of traits subject to becoming sexually dimorphic upon the evolution of dioecy: those traits that provide plants with defence against herbivores.

1.1. Herbivory and the Evolution of Dioecy. Sex-biased herbivory may be one of the selective pressures conducive to the evolution of dioecy, and it can also be a consequence of sex-specific selection on patterns of resource allocation in dioecious species. Considering only the gynodioecy pathway of the evolution of dioecy, we can think of three possible scenarios regarding the role of herbivory in each of the two steps involved in this pathway (Figure 1). The first step in the gynodioecy pathway to dioecy is the successful establishment of females (male-sterile mutants) in a population of hermaphrodites, thus resulting in a gynodioecious population. As mentioned above, this step requires that females compensate for the fitness loss incurred with the

TABLE 1: Terminology for flowers and sexual systems.

Term	Description
Flowers	
Pistillate	Unisexual flower with functional pistils only (female flower; may have vestigial, sterile stamens (staminodia))
Staminate	Unisexual flower with functional stamens only (male flower; may have vestigial, sterile pistils (pistilodia))
Bisexual, perfect	Bisexual flower with both functional pistils and stamens
Sexual system	
Monomorphic	One kind of plant (floral morph) in the population
Hermaphrodite	Most commonly applied to plants with bisexual flowers, but all monomorphic populations consist of hermaphrodite individuals
Monoecious	Pistillate and staminate flowers on same plant
Gynomonecious	Both bisexual and pistillate flowers on same plant
Andromonoecious	Both bisexual and staminate flowers on same plant
Trimonecious	Bisexual, pistillate, and staminate flowers on same plant
Dimorphic	Two kinds of plants (floral morphs) in the population
Dioecious	One morph male (with staminate flowers only); the other female (with pistillate flowers only)
Gynodioecious	One morph female, the other hermaphrodite (with either bisexual flowers or both pistillate and staminate flowers)
Androdioecious	One morph male, the other hermaphrodite (as above)
Trimorphic	Three floral morphs in the population
Trioecious	Males, females, and hermaphrodites

Modified from Dellaporta and Calderon-Urrea 1993.

loss of the male function. The reallocation of resources freed from the male function towards defence may contribute towards fitness compensation if increased defence results in greater fitness for the females [9]. Increased defence may result in lower herbivore damage on females than on hermaphrodites (Figure 1, path B). However, this is not the only possibility. Defence may be achieved through resistance: traits that reduce the rate of herbivore attack such as low nutritional content of tissues (particularly, N content), secondary metabolites, trichomes, cutin, waxy cuticles, lignin, and volatiles that attract natural enemies of herbivores [22]; and also through tolerance: traits that mitigate the negative effects of damage on fitness, including higher or lower growth rates, mobilization of stored resources, and activation of apical meristems [23]. If females reallocate resources to tolerance traits, they could be the morph with greater herbivore damage (Figure 1, path C).

The second step in the gynodioecy pathway to dioecy is the successful establishment of male individuals (female-sterile mutants) in a gynodioecious population followed by the loss of the hermaphroditic morph, thus resulting in a dioecious population. Upon the evolution of two unisexual morphs, defensive traits may evolve differently in each sex and eventually become sex linked [84–86]. The particular way in which defensive traits diverge between sexes will depend on the costs and benefits derived from the specific pattern of resource allocation to growth, reproduction, and defence in each sex. Currently, it is thought that females generally evolve greater resistance than males (see the following; Figure 1, path b).

This paper focuses on the origin of sex-biased herbivory in dioecious species. Therefore, we will not delve into morph-biased herbivory in gynodioecious species, which would be the topic of a different essay. However, we do recognize that sex-biased herbivory—indicative of sexual dimorphism in resistance against herbivores—is likely related to morph-biased herbivory in the ancestral gynodioecious population from which it evolved (Figure 1, path B-b). Thus, male-biased herbivory may have its origin in a gynodioecious population where hermaphrodites (functionally, the male morph) bear greater levels of herbivory than females.

1.2. Male-Biased Herbivory. The above view for the origin of greater resistance against herbivores in females is based directly on the principle of allocation: resources freed from the male function are used for the female function, growth, and defence. In contrast to this view, the finding of male-biased herbivory in dioecious populations has been explained on the basis of sex-specific selection of resistance traits, where the main difference between sexes that drives the sex-specific selection is the cost of reproduction. In this alternative view, female individuals of dioecious species are expected to have lower herbivory levels than males because the higher cost of reproduction of females confers a selective advantage to females with traits that reduce herbivore attack [66]. The logic of this argument is as follows: since females invest more in reproduction than males, they are left with a smaller pool of resources for growth and therefore must grow more slowly than males [20, 87]. According to

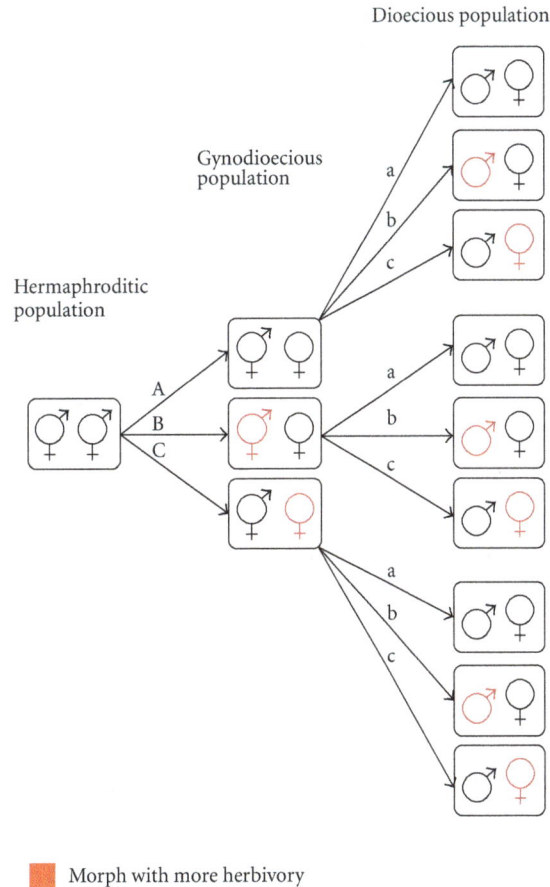

FIGURE 1: Possible scenarios of the inception of morph- or sex-biased herbivory in the evolution of dioecy via the gynodioecy pathway. Symbols represent hermaphrodite, male, or female morphs in a population (rectangle). Arrows represent evolutionary pathways between populations with different sexual systems. The first step in the pathway is the transition from a hermaphroditic to a gynodioecious population by the successful invasion of females (left-most set of arrows). The second step in the pathway is the transition from a gynodioecious to a dioecious population following the successful invasion of males and the disappearance of hermaphrodites (right-most arrows). Letters indicate different evolutionary paths.

the resource availability hypothesis, the fitness cost of losing tissue to herbivores is greater for plants that grow more slowly, thus favouring the evolution of increased defence against herbivores in slow-growing plants [88]. Since females tend to grow more slowly, they should be better defended against herbivores than males [88]. Consequently, dioecious species should experience male-biased herbivory. We must note that the argument for greater defence levels in females is usually understood in terms of resistance, but it could also be interpreted in terms of tolerance, in which case the predictions of sex-biased herbivory would be the opposite (Figure 1, path c), as developed below.

The first review of the empirical studies on the topic of sex-biased herbivory concluded that "males are more likely than females to be preferentially used by herbivores" and suggested that male-biased herbivory was widespread among dioecious species [72]. The authors, however, recognized that sex-biased herbivory was by no means a unanimous finding across all the dioecious species examined to that date, and that the relative susceptibility of each sex to

herbivory could be influenced, among other factors, by fluctuations in ecological tradeoffs between functions (rather than evolutionary changes in patterns of allocation), such as phenological changes in resource allocation to reproduction and growth [75, 89]. Therefore, the life stage at which damage measurements are taken can determine whether a study concludes that herbivory is sex biased or not.

In addition, Ågren et al. cautioned against publication bias, whereby studies that found differences between genders could be more likely to be published than those that did not; and taxonomic bias, an overabundance of studies from certain genera or families. In this instance, the taxonomic bias is correlated with an ecological bias for studies of temperate species, despite dioecy being more prevalent in tropical ecosystems [90]. Ågren et al. called for future studies that (1) examine the causes of differential palatability between males and females, (2) measure the fitness consequences of natural levels of herbivory in both sexes, and (3) determine whether herbivore pressure can actually cause adaptive changes in tissue palatability. In addition, they urged for broadening the taxonomical scope of the studies. In spite of such encouragement, there still is a paucity of studies that address these issues.

More recently, sex-biased herbivory in dioecious species was tested by means of a meta-analysis of 33 studies encompassing 30 species, 19 of which were previously included in Ågren et al.'s 1999 review [73]. The authors tested for publication bias and found it to be minimal. However, they did not emphasize other shortcomings of the dataset and concluded that male-biased herbivory in dioecious species is a rule.

Here we propose alternative evolutionary scenarios that could result in female-biased herbivory or lack of intersexual differences in herbivory levels. We invite the reader to reconsider the evidence for male-biased herbivory in dioecious plants and recommend a standard protocol for evolutionary-ecological studies of sex-biased herbivory in dioecious species that addresses the shortcomings listed below. We contend that taking for granted the generality of male-biased herbivory in dioecious species is hampering our progress in this field.

2. Critique of Theory: Evolutionary Scenarios

While male-biased herbivory has been explained as a consequence of sex-specific selection of resistance determined by the cost of reproduction of each sex, few of the reviewed studies (Table 2) are actually placed within such evolutionary context. J. Lovett-Doust and L. Lovett-Doust [44] were the first to argue, citing Charnov [91], that an evolutionary divergence between sexes in resource allocation patterns could result in sex-biased resistance against herbivory. Danell et al. [57] based their expectation of male-biased herbivory on sexual selection: it may be more advantageous for males to invest less in resistance and more in reproduction when exposed to a cyclic herbivore compared to a noncyclic herbivore because males will lose fitness only once every four years, and their reproductive output during years of little or no herbivory will more than compensate for the fitness lost to herbivores on the heavy herbivory years. The above explanation would hold only to the extent that there are no carry-over effects from one year to another.

More recently, McCall [83] cited Bierzychudek and Eckhart [92] and Delph [87] in support of the claim that the reproductive output of females is more limited by resources than that of males. However, while there may be ample evidence that reproductive allocation is generally greater for the female function, it is possible that, upon the separation of sexes, physiological mechanisms involved in resource acquisition and allocation evolve in such way as to minimize the differences in reproductive effort between sexes. For example, in *Ilex glabra*, sexes do not differ in total reproductive biomass produced in a growing season because the greater unitary investment in pistillate flowers and fruit development is negated by the sevenfold greater flower production in males [75]. Given the importance of the tenet of greater female reproductive allocation for the expectation of male-biased herbivory, all studies of sex-biased herbivory

should test for intersexual differences in reproductive allocation or provide a reference to an empirical study that demonstrates such differences for the species in question. In the measurement of reproductive allocation, particular attention must be paid to obtaining reliable estimates of male reproductive output (pollen production), which presents its own logistical difficulties. In addition, resource expenditure in pollinator attraction needs to be considered, as this is another expenditure related to reproduction that may differ between sexes.

In essence, the presumed chain of evolutionary events that lead to male-biased herbivory in dioecious plants stems from the reallocation of those resources freed upon the loss of a sexual function in unisexual mutants towards defence. In most studies, defence has been equated to resistance. However, defence may also occur through tolerance [22, 93]. In comparison to resistance traits, tolerance traits have been more elusive. The capacity to store and mobilize carbohydrates, the presence of meristems and the capacity to activate them in response to damage have been proposed as tolerance traits [23]. Growth rate has also been proposed to influence tolerance, although it is controversial whether high or low growth rates favour tolerance [94–98]. A recent model shows that plants with low growth rates are more tolerant to herbivore damage [99]. This model also shows that plants that can change their growth rate positively in response to damage will tolerate damage better than those with a different response in growth rate. The activation of meristems and mobilization of resources in response to the loss of tissue are two well-documented responses [100, 101] that could contribute to increased plant growth rates in response to damage. Thus, according to this model, if females grow more slowly than males because of their greater allocation to reproduction, then females should be more tolerant to herbivory than males. This response was observed in *Urtica dioica* subjected to clipping of the stem apex [102]. Whether this prediction necessarily implies that females should be less resistant than males is not clear at this point since there is increasing evidence that these two modes of defence are not necessarily mutually exclusive [99, 103, 104].

There is one other possibility that has not been emphasized enough in the proposed models of the evolution of defence in dioecious species: while one possible consequence of a greater allocation of resources to reproduction in females is reduced allocation to growth, it is also possible that the main reduction in allocation is to defence. In this case, there would be no detectable detriment to growth. Consequently, female plants would suffer more damage (if they are less resistant; Figure 1, path c) or greater fitness losses (if they are less tolerant) compared to males. Greater damage in females could lead to fewer, more spaced reproductive events or greater interannual variability in reproductive output either directly through a decrease in the availability of resources for reproduction or indirectly through a decrease in pollinator visitation rates due to a lack of resources needed for floral display or nectar production [75, 105–108]. Fewer resources available for reproduction could also pose a selective pressure to become choosier about their mates, which may lead to increased fruit or seed abortion [75, 109, 110].

TABLE 2: List of dioecious species of angiosperms studied for dimorphic herbivore damage, and information on assessment of reproductive allocation, growth rate and resistance.

Species	Damage	Sex with greatest			Reference	Review
		Reproductive allocation	Growth rate	Resistance		
Alismatales						
Araceae						
Arisaema triphyllum	M	nm	nm	nd (N, C:N, leaf total phenolics)	[24]	3
		F	F (total dry mass)		[25]	
Arecales						
Arecaceae						
Chamaedorea alternans (= *C. tepejilote*)	nd	nm	nm	nm	[26]	1
		F	M (leaf production)		[27]	
					[28]	
Asterales						
Asteraceae						
Baccharis concinna	nd	nm	nm	nm	[29]	3
B. dracunculifolia	nd	nm	nd (leaf production) M (shoot length)	nm	[30]	3
			nm		[31]	3
			nd (shoot length)		[32, 33]	
B. halimifolia	M, F, nd, depends on herbivore	nd (flowers/shoot)	M	F (resin)	[34]	1
Brassicales						
Capparaceae						
Forchhammeria pallida	nd	nm	nm	nm	[35]	3
Caryophyllaceae						
Silene dioica	M	M (during flowering)	F	nm	[36]	1
	F	nm	nd (length of infected shoots)	nm	[37]	3
Chenopodiaceae						
Atriplex canescens	F	nm	nm	nm	[38]	3
	M	nm	nm; nd (height, width, fresh weight in spring), F (FW in winter)	nm	[39]	4
		(F)			[40]; the species includes hermaphrodites	
A. vesicaria	F	nm	nm	nm	[41, 42]	1
Nyctaginaceae						
Neea psychotrioides	M	M	nd (stem production)	nm	[43]	3
Polygonaceae						
Rumex acetosa	M	?	?	?	T. Elmqvist unpublished data	1

TABLE 2: Continued.

Species	Damage	Sex with greatest			Resistance	Reference	Review
		Reproductive allocation	Growth rate				
R. acetosella	M, F, nd	nd	nm		nm	[44]	1
			F (ramet production)			[45]	
Fagales							
Myricaceae							
Myrica gale	M, nd	nm	nm		F (1-digestibility), nd (phenolics, p-glycosides, tannins)	L. Ericson unpublished data	1
	nd	nm	nm		nd	[46]	
Laurales							
Lauraceae							
Lindera benzoin	nd	M (flowers/shoot), F (N and biomass)	nd, M, depending on year		nm	[47]	1
			M (plant volume)		F (phenolics on leaves, but nd on stems)	[48]	
Malpighiales							
Salicaceae							
Populus tremula	M	nm	nm		M (phenolics), nd (p-glycosides, tannins, digestibility)	[46]	1
Salix caprea	M,	nm	nm		F (1-digestibility), nd (phenolics, p-glycosides, tannins)	[46]	1
	nd	nm	nd			[49]	4
	nd	nm	nm		nm	[50]	1
S. cinerea	M, nd, varies by year	nm	nd		nm	[51]	3
						[52]	4
			nd		nd (phenolic glycosides)	[53]	3
S. eleagnos	M	nm	nm		nm	[50]	3
S. fragilis	nd	nm	nm		nm	[50]	4
S. lanata	nd	nm	nm		nm	[54]	3
S. lasiolepis	M (4 of 5 spp. of sawflies)	nm	M (shoot length)		F (phenols, marginally significant)	[55]	1
S. myrsinifolia-phylicifolia	nd (miners, gallers)	nm	nm		nm	[56]	1
	M (at high plant density)	nm	nm		nm	[36]	1
	M	nm	nd (new shoots)		nm	[57]	1
	M (in high productivity habitat; decreases at higher herbivore pressure)	nm	nd (biomass)		nm	[49]	1
S. pentandra	M	nm	nm		F (phenolics)	[46]	1
S. purpurea	nd	nm	nm		M (1-digestibility)	[50]	4
S. sericea	M marginal	nm	nm		nm	[58]	3

Table 2: Continued.

Species	Sex with greatest				Reference	Review
	Damage	Reproductive allocation	Growth rate	Resistance		
S. viminalis	nd	nm	nd	nd	[53]	3
	nd	nm	nd (regrowth after pruning)	nm	[59]	4
S. x rubens	nd	nm	nm	nm	[50]	4
		nm	nm	nm	[50]	4
Pandanales						
Pandanaceae						
Freycinetia arborea	M	nm	nm	nm	[60]	2
F. reineckei	M	nm	nm	nm	[60]	2, 3
Rosales						
Eleagnaceae						
Hippophae rhamnoides	M ?	?	?	?	L. Ericson unpublished data	1
			M		[61]	
Rhamnaceae						
Rhamnus alpinus	nd	F	nd if age < 10 y; M if age > 10 y	M (anthraquinones)	[62]; [63]	3
Rosaceae						
Rubus chamaemorus	M, nd	F but varies with fruit set	nm	nd	[64]; [65]	1
Urticaceae						
Urtica dioica	M ?	?	?	?	T. Elmqvist unpublished data	1
Sapindales						
Sapindaceae						
Acer negundo	M	nm	M (growth rings)	nd (astringency, total phenols, nitrogen, toughness), F (index of defence)	[66]	1
			variable: F near streams; M away from streams		[67]	
			nd		[68]	
			F		[69]	
Pistacia lentiscus	nd	nm	nm	nd (N)	[70]	1
Simaroubaceae						
Simarouba glauca	M	nm	nm	two flavonoid compounds on female flowers not present in male flowers	[71]	3

F: female, M: male, nd: no statistically significant intersexual differences, nm: not measured, CT: condensed tannins, TNC: total non-structural carbohydrates, N: nitrogen content (herbivores usually attracted to greater concentrations).

1: Ågren et al. 1999 [72], Table 2.
2: Ågren et al. 1999 [72], Table 3.
3: Cornelissen and Stiling 2005 [73].
4: Not mentioned in any of 1–3 above.

Evolutionary changes in the rate of resource acquisition in female individuals may occur through increased photosynthetic rates, canopy area, rates of mineral nutrient uptake, as well as greater branching of roots, and enhancement of mycorrhizal associations [19]. A greater rate of resource acquisition in females would decrease the relative differences in costs of reproduction between sexes: the sex with the greater resource demand for reproduction would have an increased capacity to garner resources. Under such scenario, the life-time cost of reproduction at the individual level would be equal between sexes, thus eliminating the source of inequalities in the patterns of allocation between males and females. In summary, nothing dictates that there is only one evolutionary pathway regarding changes in the patterns of resource allocation among reproduction, growth, and defence following the evolution of unisexuality (Figure 1).

Alternatively, a stage in which female individuals have heavier damage levels because of resource limitation for resistance may be a transient evolutionary stage prior to the invasion of mutants whose greater defence levels are attained at the expense of growth. In this case, we should observe female-biased herbivory in younger dioecious lineages and male-biased herbivory in those lineages in which there has been enough time for selection to reshape the patterns of resource allocation to reproduction, defence, and growth. We should be able to test this by means of relative dating of lineages with male- or female-biased herbivory.

Similarly, as long as there has not been selection on prereproductive growth rates following the evolution of unisexuality, we should not see differences in growth rates or other physiological vegetative traits between males and females before their first reproductive event. It is difficult to test this prediction without reliable morphological or genetic markers that allow juveniles to be sexed so that their performance can be compared on the basis of sex. Some sex-linked markers may be, effectively, sex-related traits expressed before the onset of reproduction. Whether the presence of these markers implies the existence of sex chromosomes is still an area in need of further investigation [84, 85, 111–113].

In short, without fitness gain curves for each sex, it is difficult to predict accurately which sex should evolve greater resistance against herbivores and whether we should expect or not male-biased herbivory in dioecious species [91]. In fact, we need fitness surfaces in order to include the effects of reductions in leaf area caused by herbivory. Moreover, the fitness surfaces should account for the short- and long-term responses of plants in terms of changes in photosynthetic rates, reallocation of resources to shoot or root, activation of meristems, and delays in phenology or shortening of life span brought about by herbivore damage [95, 99].

It has not escaped our attention that the evolution of defence in gynodioecious species can be approached from a similar perspective to the one presented above for dioecious species [12]. It is important to consider that some of gynodioecious species may be in evolutionary transition to dioecy while others are not [112, 114]. Another important difference with dioecious species is that, in gynodioecious plants, the morph that performs the male function—the

hermaphrodites—may have a greater cost of reproduction because of the expenditure of resources on two sexual functions. Does this mean that hermaphrodites would be less resistant (Figure 1, path C) or grow more slowly and evolve greater resistance (Figure 1, path B)? Clearly, making predictions with respect to gender dimorphism in defensive traits for bisexual conditions along the gradation from monoecy to dioecy is not straight forward.

3. Critique of Datasets Used to Conclude Male-Biased Herbivory

The collection of studies cited in the reviews of herbivory in dioecious species [72, 73] has, as a group, important shortcomings that weaken the conclusion of male-biased herbivory as a generality in dioecious species. The main shortcomings are (1) the taxonomic bias of the sample of species studied; and (2) failure to test for or provide references of empirical studies of intersexual differences in (a) resistance traits—the purported cause of the intersexual differences in herbivory levels; (b) growth rates—the purported cause of the intersexual differences in resistance to herbivore attack; and (c) reproductive effort—the purported cause of the aforementioned intersexual differences in growth rates. These deficiencies had been pointed out earlier [72, 89], but judging by statements included in the introduction or discussion of many of the papers published after the 1999 review, such caveats have not been considered to their full extent, and many authors take for granted either the generality of male-biased herbivory in dioecious species or its expectation without reference to any theoretical context.

3.1. Taxonomic Bias. Cornelissen and Stiling's meta-analysis of sex-biased herbivory includes 30 species, 28 of which are angiosperms. Focusing only on angiosperms, 13 of the 28 species were not considered previously in Ågren et al.'s review (Table 2). These 30 species represent a total of 20 genera, 18 families, and 10 orders. Nine of those species belong to the same genus: *Salix*. Adding to those *Populus tremula*, the species in the Salicaceae represent one-third of all species considered for the meta-analysis. Such distribution contrasts greatly with the taxonomic distribution of dioecy in 14,620 species, 959 genera, 157 families, and 36 orders [90]. Of the four dioecious genera (consisting of solely dioecious species) with most species (400), only *Salix* has been studied. *Pandanus*, *Diospyros*, and *Litsea*, with 700, 500+, and 400 species, respectively ([115]; S. Renner, University of Munich, unpublished data) have not been studied for sex-biased herbivory yet. Clearly, we need to direct our research efforts to the most understudied orders and families if we want to arrive at generalizations regarding the biology of dioecious species, and particularly the influence of herbivores in their ecology and evolution.

In addition to this taxonomic bias, a critical reexamination of that list of species casts serious doubt on the conclusion that male-biased herbivory is a rule in dioecious species: only 13 of those species were reported invariably to have male-biased herbivory. This list includes three *Salix*, two *Freycinetia*, and two species for which evidence of male-biased herbivory has not been published: *Hippophae rhamnoides* and *Rumex acetosa*. (In fact, male-biased herbivory in *Myrica gale*—not included in these 13 species—is also anecdotal.) Greater herbivore damage on females is reported for four species, while the rest show either no intersexual differences (16), or variation in the result, depending on different factors (species of herbivore, kind of herbivore, tissue damaged, time of year, phenological or ontogenetic stage, etc.). Moreover, it is possible that the results for those 13 species would show variation with either population or site, had these factors been studied.

3.2. Differential Growth between Sexes. Perhaps the most serious problem with several studies of herbivory and dioecy has been the failure to make the connection between sex-biased herbivore damage and intersexual differences in growth rate, precisely because the latter is the purported cause of the former. Of the 30 species of angiosperms in Table 2, either growth rate or a surrogate variable for growth (e.g., shoot length) was measured only in 21 species. Males grew faster in six species, females in two, no difference between sexes was detected in six species, and three species showed variable results. It must be noted that the same number of species shows no difference between genders as those that show greater growth rate in males. Considering solely the 13 species that invariably showed male-biased herbivory, only two show a greater growth rate in males: *Acer negundo* and *Hippophae rhamnoides*. However, as the evidence for male-biased herbivory in *H. rhamnoides* is anecdotal, we are left with only one species for which growth rate was measured in the same study as herbivore damage: *Acer negundo*.

3.3. Differential Reproductive Effort between Sexes. Only 12 of the 30 species listed were assessed for intersexual differences in reproductive allocation in terms of reproductive effort (the proportion of biomass or other currency devoted to reproductive structures relative to the total biomass or expenditure in the selected currency of an individual). Reproductive effort was greater in females of 10 species and in males for the other two species. In some species, reproductive effort was measured during flowering, but allocation to fruit production was not considered (e.g., *Silene dioica*). In those cases, we are left with an incomplete picture of reproductive allocation, and we can only join the authors in speculating whether species of the same genus have similar patterns of reproductive allocation.

The only species that have been assessed for foliar damage, growth rate, and reproductive allocation in the same study are *C. tepejilote*, *B. halimifolia*, *I. glabra*, *N. psychotrioides*, and *R. alpinus* (Table 2). These studies clearly made the chain of causal connections from sex bias in reproductive allocation all the way to sex bias in some resistance traits (except for *C. tepejilote* and *I. glabra*), and, as a consequence of the latter, sex bias in levels of damage.

The study on *R. acetosella* at least established the connection between damage and growth [44]. The study of *R. alpinus* went even further, comparing these attributes between pre- and postreproductive plants, and thus emphasizing that the root of the differences in growth rates, resistance traits, and leaf damage is in the patterns of reproductive resource expenditures [63, 116].

In some species, reproductive allocation, growth rate, and/or resistance were reported after the initial publication of sex-biased herbivory. However, even with these studies, the number of species for which we have a more complete picture of the causal links amongst these attributes remains low: nine more species (*C. alternans*, *B. dracunculifolia*. *A. canescens*, *R. acetosella*, *S. caprea*, *S. cinerea*, *S. lasiolepis*, *S. sericeae*, and *H. rhamnoides* [?]; Table 2) now have published data for damage and growth rate, bringing the number of species in this situation to 15. Two more species, for a total of three, now have data on damage, growth rate, and reproductive allocation (*C. alternans*, *S. dioica*, and *N. psychotrioides*). Two more species now have data on reproductive allocation, growth rate, and resistance apart from herbivore damage, for a total of four species with all four variables measured (*A. triphyllum*, *B. halimifolia*, *L. benzoin*, and *R. alpinus*).

In summary, the majority of studies on the topic of sex-biased herbivory have neglected the purported causal connections between bias in reproductive allocation, differential growth rate, resistance, and herbivore damage. Also, some authors seemed to confuse theoretical expectations with empirical evidence of greater female reproductive allocation: while Lloyd and Webb [20] argue convincingly for the expectation of greater reproductive effort in females, they provided empirical evidence only for *Rumex acetosella*, citing Putwain and Harper 1972 [117]. Therefore, Lloyd and Webb's excellent paper cannot be cited as solid empirical evidence of greater reproductive effort in females. Lastly, anecdotal evidence should be taken with great caution and always flagged as such until data are published (see entries marked "?" in Table 2).

Using the search terms herbiv* and dioec* for entries between January 1998 and May 2012 on the Web of Science, we found nine studies encompassing 14 species that were not included in either of the previous reviews of the topic. Of these, only the study on the three species of *Chamaedorea* palms measured reproductive allocation, growth rate, resistance, and herbivore damage (Table 3; N.B.: one of these species had been studied before: *C. alternans* = *C. tepejilote*). Only one other study measured damage and reproductive allocation (*Sclerocarya birrea*, Table 3). Similarly, growth rate was assessed in only one other species (*Salix arctica*). The taxonomic breadth of the studies of herbivory in dioecious species increased only by one family (in an unplaced order of the Euasterids I). The general lack of consistency in the level of detail and the variables that have been measured in all these studies could be addressed if researchers interested in this topic followed a minimally standardized protocol.

TABLE 3: Studies of defence on dioecious species published after 2004, or published earlier but not mentioned in Ågren et al or Cornelissen and Stiling's reviews.

Species	Damage	Sex with greatest Reproductive allocation	Growth rate	Resistance	Herbivores	Reference
Arecales						
Arecaceae						
Chamaedorea alternans (= *C. tepejilote*)	M	F	F	F	Chrysomelid beetles	[74]
C. pinnatifrons	M	F	M	F	Chrysomelid beetles	[74]
C. ernesti-augusti	M	F	M	F	Chrysomelid beetles	[74]
Aquifoliales						
Anacardiaceae						
Ilex glabra	nd; marginally F after flowering	nd	nd	nm	lepidopteran larvae and leaf spot (fungal pathogens)	[75]
Sapindales						
Anacardiaceae						
Sclerocarya birrea	F	nd (wood/reproductive shoot)	nm	nd (wood density, branch breakability)	Elephants	[76]
Spondias purpurea	F	nm	nm	M (N, TNC)	Cerambycid beetle	[77]
Malpighiales						
Salicales						
Salix discolor	nd	nm	nm	M (mortality of herbivore)	Leaf galler	[78]
S. polaris	nm	F	nd	nd (phenolics, CT)	Reindeer	[79]
S. arctica	nd	nm	nd	nm	Muskox	[80]
S. planifolia	nd	F	nm	nm	Insects	[81]
Laurales						
Lauraceae						
Lindera obtusiloba	nd	nm	nm	nm	Unspecified	[82]
L. praecox	nd	nm	nm	nm	Unspecified	[82]
L. umbellata	nd	nm	nm	nm	Unspecified	[82]
L. erythrocarpa	nd	nm	nm	nm	Unspecified	[82]
Unplaced (Euasterids I)						
Hydrophyllaceae						
Nemophila menziesii	nd	nm	nm	nm	Larvae of lepidoptera (2 spp.) and coleoptera (1 sp.)	[83]

F: female, M: male, nd: no statistically significant intersexual differences, nm: not measured, CT: condensed tannins, TNC: total non-structural carbohydrates, N: nitrogen content (herbivores usually attracted to greater concentrations).

TABLE 4: Total number of species, number of dioecious species, proportion of dioecious species, and estimated 2% of dioecious species in the top 30 most species-rich families with a proportion of dioecious species greater than 0.5 (from unpublished data from S. Renner, University of Munich).

Family	Total species	Dioecious species	Proportion of dioecious species	2% of dioecious species
Arecaceae	815	778	0.955	16
Pandanaceae	777	777	1.000	16
Lauraceae	1123	776	0.691	16
Menispermaceae	577	577	1.000	12
Ebenaceae	487	487	1.000	10
Anacardiaceae	594	439	0.739	9
Salicaceae	436	435	0.998	9
Myristicaceae	367	365	0.995	7
Clusiaceae	590	365	0.619	7
Restionaceae	387	364	0.941	7
Aquifoliaceae	400	300	0.750	6
Smilacaceae	215	205	0.953	4
Cucurbitaceae	390	197	0.505	4
Flacourtiaceae	209	192	0.919	4
Burseraceae	234	175	0.748	4
Cecropiaceae	184	174	0.946	3
Thymelaeaceae	236	119	0.504	2
Vitaceae	155	118	0.761	2
Loranthaceae	147	114	0.776	2
Meliaceae	181	105	0.580	2
Theaceae	155	94	0.606	2
Proteaceae	84	84	1.000	2
Hydrocharitaceae	123	75	0.610	2
Monimiaceae	108	74	0.685	1
Rhamnaceae	140	71	0.507	1
Nepenthaceae	70	70	1.000	1
Siparunaceae	93	68	0.731	1
Myricaceae	52	51	0.981	1
Chloranthaceae	57	51	0.895	1
Casuarinaceae	96	51	0.531	1
Total	9482	7751		155

4. Future Directions: Standardized Protocol and Broadening of Species Studied

New studies must clearly allude to the theoretical framework from which the prediction of sex-biased herbivory levels (resistance) stems—resource allocation theory, in particular, sex allocation. The claim that male-biased herbivory is expected because it has been reported as a pattern, whether implicit or explicit, lacks heuristic value because it does not address the causes of such pattern. Moreover, a plethora of factors may modify the expected pattern, as shown above.

Clearly, we need to increase the taxonomic breadth of the studies of herbivory in dioecious species. There are several ways to achieve greater taxonomic representation. We could direct our attention to those families with the greatest number of dioecious species or those with the greatest proportion of dioecious species. The first alternative will miss families with low species richness that may have a high proportion of dioecious species. The second method will miss families with high species richness but low proportion of dioecious species. One possible compromise is to focus our studies on the families with the greatest number of dioecious species among those with a large proportion of dioecious species, for instance, 50% or more (Table 4). So far, we have studied only 2% of the dioecious species in the most studied family (Salicaceae). If we took that as a target, we would have to study about 155 species for the 30 most dioecious species-rich families of angiosperms. However, by this method we would include only one species per family for many families, thus failing to achieve adequate representation of those families. In addition, we must consider that the conditions that determine sex-biased herbivory can change with habitat, and therefore some species may need to be studied in several habitats.

In addition to the taxonomic bias, there is a preponderance of studies of woody plants. While this is understandable because most dioecious species are woody, we should strive for representation of the herbaceous component. With increased research on herbaceous dioecious species, we can address the influence of life history traits on the evolution of dioecy and defence.

Lastly, we propose that all studies aimed at assessing whether herbivory levels differ between sexes and whether these differences are a consequence of differential growth rates (in turn resulting from differential allocation to reproduction) should conduct, at least, the following measurements and observations: (1) levels of herbivory, measured as precisely as possible (preferably for more than one growing season in perennials); (2) species of herbivores responsible for most of the damage; (3) growth rates, measured either as RGR for whole individuals or from increments in branch length or leaf production; (4) reproductive allocation, measured both as the number of reproductive structures (flowers and pollen production for males, flowers, and fruits for females), and also as reproductive effort (the proportion of individual or shoot biomass allocated to reproduction, and when possible N and P allocation to reproductive structures); and finally (5) the most important resistance characters that could be influencing the levels of herbivory and measure them quantitatively. In addition, these studies could add an experimental component in which plants are damaged at least at the highest rate seen in the surveys of natural damage, so as to measure tolerance to herbivory as well as resistance [75, 102]. Ideally, damage should be performed by placing natural herbivores on the plants because mechanical damage does not necessarily elicit the same physiological responses as herbivore damage [118]. Also, these studies should consider that resistance and tolerance may vary both with ontogeny and with respect to other reproductive phenology because the acquisition and expenditure of resources vary at different stages of development and life history [119–121]. We must reiterate that other authors have emphasized the need to address several of the points outlined above. It is our hope that future studies take these recommendations seriously so that we have to assume and speculate less, and we have empirical data to further our understanding of the evolution of defence in dioecious species.

5. Conclusions

The study of the evolution of sex-biased herbivory is hampered by the notion that male-biased herbivory in dioecious species is a rule. We have shown that the evidence used to support this conclusion has important shortcomings. We have presented other possible evolutionary outcomes with regards to sex-biased herbivory in the transition from hermaphroditic populations to dioecious ones. We have also discussed how these different outcomes can be predicted under different theoretical assumptions. Therefore, future studies of herbivory in dioecious species should be based on a clear theoretical framework. In particular, we urge that all new studies of herbivory in dioecious species include assessments of reproductive allocation, growth rates, and resistance traits deemed to differ between sexes and, therefore, determine sex-biased herbivory. In addition, tolerance should also be considered as a potentially important defence mode that can vary between sexes. In this manner, we should be able to explain better the results of any given study. The advancement of our knowledge about sex-related defence in plants should help us gain a better understanding of the evolution of sex-related traits in general.

Acknowledgments

The authors are grateful to César Domínguez, Mauricio Quesada, Nicholas Buckley, and Susana Magallón for their insight on the topic and the fruitful discussions. Thanks are due to Susan Renner for allowing them to use her data on the taxonomic distribution of dioecy across the angiosperms, to Caroline Tucker for sharing her expertise on Proteaceae, and to two anonymous reviewers for helpful suggestions to improve this paper. They are grateful to Alberto Civetta for inviting them to submit this paper.

References

[1] A. J. Richards, *Plant Breeding Systems*, Chapman & Hall, London, UK, 1997.

[2] A. K. Sakai and S. G. Weller, "Gender and sexual dimorphism in flowering plants: a review of terminology, biogeographic patterns, ecological correlates and phylogenetic approaches," in *Gender and Sexual Dimorphism in Flowering Plants*, M. A. Geber, T. E. Dawson, and L. F. Delph, Eds., pp. 1–31, Springer, Berlin, Germany, 1999.

[3] C. Darwin, *The Different Forms of Flowers of the Same Species*, John Murray, London, UK, 1877.

[4] D. G. Lloyd, "The maintenance of gynodioecy and androdioecy in angiosperms," *Genetica*, vol. 45, no. 3, pp. 325–339, 1975.

[5] R. B. Spigler and T. L. Ashman, "Gynodioecy to dioecy: are we there yet?" *Annals of Botany*, vol. 109, pp. 531–543, 2012.

[6] M. L. Van Etten, L. B. Prevost, A. C. Deen, B. V. Ortiz, L. A. Donovan, and S. M. Chang, "Gender differences in reproductive and physiological traits in a gynodioecious species, *Geranium maculatum* (Geraniaceae)," *International Journal of Plant Sciences*, vol. 169, no. 2, pp. 271–279, 2008.

[7] B. Charlesworth and D. Charlesworth, "A model for the evolution of dioecy and gynodioecy," *The American Naturalist*, vol. 112, pp. 975–997, 1978.

[8] D. Lewis, "Male sterility in natural populations of hermaphrodite plants," *The New Phytologist*, vol. 40, pp. 56–63, 1941.

[9] T. L. Ashman, "The role of herbivores in the evolution of separate sexes from hermaphroditism," *Ecology*, vol. 83, no. 5, pp. 1175–1184, 2002.

[10] T. L. Ashman, "Constraints on the evolution of males and sexual dimorphism: field estimates of genetic architecture of reproductive traits in three populations of gynodioecious *Fragaria virginiana*," *Evolution*, vol. 57, no. 9, pp. 2012–2025, 2003.

[11] T. L. Ashman, "The limits on sexual dimorphism in vegetative traits in a gynodioecious plant," *American Naturalist*, vol. 166, pp. S5–S16, 2005.

[12] T. L. Ashman, "The evolution of separate sexes: a focus on the ecological context," in *Ecology and Evolution of Flowers*, L. D. Harder and S. C. H. Barrett, Eds., pp. 204–222, Oxford University Press, Oxford, UK, 2006.

[13] G. Avila-Sakar and C. A. Domínguez, "Parental effects and gender specialization in a tropical heterostylous shrub," *Evolution*, vol. 54, no. 3, pp. 866–877, 2000.

[14] D. Charlesworth, "A further study of the problem of the maintenance of females in gynodioecious species," *Heredity*, vol. 46, pp. 27–39, 1981.

[15] D. Charlesworth, "Theories of the evolution of dioecy," in *Gender and Sexual Dimorphism in Flowering Plants*, M. A. Geber, T. E. Dawson, and L. F. Delph, Eds., pp. 33–60, Springer, Berlin, Germany, 1999.

[16] D. Couvet, O. Ronce, and C. Gliddon, "The maintenance of nucleocytoplasmic polymorphism in a metapopulation: the case of gynodioecy," *American Naturalist*, vol. 152, no. 1, pp. 59–70, 1998.

[17] S. A. Frank, "The evolutionary dynamics of cytoplasmic male sterility," *The American Naturalist*, vol. 133, pp. 345–376, 1989.

[18] P. Saumitou-Laprade, J. Cuguen, and P. Vernet, "Cytoplasmic male sterility in plants: molecular evidence and the nucleo-cytoplasmic conflict," *Trends in Ecology and Evolution*, vol. 9, no. 11, pp. 431–435, 1994.

[19] T. E. Dawson and M. A. Geber, "Dimorphism in physiology and morphology," in *Gender and Sexual Dimorphism in Flowering Plants*, M. A. Geber, T. E. Dawson, and L. F. Delph, Eds., pp. 175–215, Springer, Berlin, Germany, 1999.

[20] D. G. Lloyd and C. J. Webb, "Secondary sex characters in plants," *The Botanical Review*, vol. 43, no. 2, pp. 177–216, 1977.

[21] K. Mooney, A. Fremgen, and W. Petry, "Plant sex and induced responses independently influence herbivore performance, natural enemies and aphid-tending ants," *Arthropod-Plant Interactions*. In press.

[22] R. Karban and I. T. Baldwin, *Induced Responses to Herbivory*, The University of Chicago Press, Chicago, Ill, USA, 1997.

[23] S. Y. Strauss and A. A. Agrawal, "The ecology and evolution of plant tolerance to herbivory," *Trends in Ecology and Evolution*, vol. 14, no. 5, pp. 179–185, 1999.

[24] I. C. Feller, H. Kudosh, C. E. Tanner et al., "Sex-biased herbivory in Jack-in-the-pulpit (*Arisaema triphyllum*) by a specialist thrips (*Heterothrips arisaemae*)," in *Proceedings of the Royal 7th International Symposium on Thysanoptera*, pp. 163–172, 2001.

[25] J. Lovett-Doust and P. B. Cavers, "Sex and gender dynamics in jack-in-the-pulpit, *Arisaema triphyllum* (Araceae)," *Ecology*, vol. 63, no. 3, pp. 797–808, 1982.

[26] K. Oyama and R. Dirzo, "Ecological aspects of the interaction between *Chamaedorea tepejilote*, a dioecious palm and *Calyptocephala marginipennis*, a herbivorous beetle, in a Mexican rain forest," *Principes*, vol. 35, pp. 86–93, 1991.

[27] K. Oyama and R. Dirzo, "Biomass allocation in the dioecious palm *Chamaedorea tepejilote* and its life history consequences," *Plant Species Biology*, vol. 3, pp. 27–33, 1988.

[28] K. Oyama, "Variation in growth and reproduction in the neotropical dioecious palm *Chamaedorea tepejilote*," *Journal of Ecology*, vol. 78, no. 3, pp. 648–663, 1990.

[29] F. G. Madeira, T. G. Cornelissen, M. L. Faria et al., "Insect herbivore preference for plant sex and modules in *Baccharis concinna* Barroso (Asteraceae)," in *Ecology and Evolution of Plant-Feeding Insects in Natural and Man-Made Environments*, A. Raman, Ed., pp. 135–143, International Scientific, New Dehli, India, 1997.

[30] M. A. A. Carneiro, G. W. Fernandes, O. F. F. De Souza, and W. V. M. Souza, "Sex-mediated herbivory by galling insects on *Baccharis concinna* (Asteraceae)," *Revista Brasileira de Entomologia*, vol. 50, no. 3, pp. 394–398, 2006.

[31] M. M. Espírito-Santo and G. W. Fernandes, "Abundance of *Neopelma baccharidis* (Homoptera: Psyllidae) galls on the dioecious shrub *Baccharis dracunculifolia* (Asteraceae)," *Environmental Entomology*, vol. 27, no. 4, pp. 870–876, 1998.

[32] M. L. Faria and G. W. Fernandes, "Vigour of a dioecious shrub and attack by a galling herbivore," *Ecological Entomology*, vol. 26, no. 1, pp. 37–45, 2001.

[33] H. N. Ribeiro-Mendes, E. S. A. Marques, I. M. Silva, and G. W. Fernandes, "Influence of host-plant sex and habitat on survivorship of insect galls within the geographical range of the host-plant," *Tropical Zoology*, vol. 15, no. 1, pp. 5–15, 2002.

[34] V. A. Krischik and R. F. Denno, "Patterns of growth, reproduction, defense, and herbivory in the dioecious shrub *Baccharis halimifolia* (Compositae)," *Oecologia*, vol. 83, no. 2, pp. 182–190, 1990.

[35] B. J. Hendricks and B. D. Collier, "Effects of sex and age of a dioecious tree, *Forchhammeria pallida* (Capparaceae) on the performance of its primary herbivore, *Murgantia varicolor* (Hemiptera: Pentatomidae)," *Ecological Research*, vol. 18, no. 3, pp. 247–255, 2003.

[36] T. Elmqvist and H. Gardfjell, "Differences in response to defoliation between males and females of *Silene dioica*," *Oecologia*, vol. 77, no. 2, pp. 225–230, 1988.

[37] J. A. Lee, "Variation in the infection of *Silene dioica* (L.) Clairv. by *Ustilago violacea* (Pers.) Fuckel in north west England," *New Phytologist*, vol. 87, pp. 81–89, 1981.

[38] A. F. Cibils, D. M. Swift, and R. H. Hart, "Female-biased herbivory in fourwing saltbush browsed by cattle," *Journal of Range Management*, vol. 56, no. 1, pp. 47–51, 2003.

[39] D. Maywald, E. D. McArthur, G. L. Jorgensen, R. Stevens, and S. C. Walker, "Experimental evidence for sex-based palatability variation in fourwing saltbush," *Journal of Range Management*, vol. 51, no. 6, pp. 650–654, 1998.

[40] D. C. Freeman, E. D. McArthur, S. C. Sanderson, and A. R. Tiedemann, "The influence of topography on male and female fitness components of *Atriplex canescens*," *Oecologia*, vol. 93, no. 4, pp. 538–547, 1993.

[41] R. D. Graetz, "The influence of grazing by sheep on the structure of a saltbush (*Atriplex vesicaria* Hew ex Benth.) population," *The Australian Rangeland Journal*, vol. 1, pp. 117–125, 1978.

[42] D. G. Williams, D. J. Anderson, and K. R. Slater, "The influence of sheep on pattern and process in *Atriplex vesicaria* populations of the Riverine Plain of New South Wales," *Australian Journal of Botany*, vol. 26, pp. 381–392, 1978.

[43] L. M. Wolfe, "Differential flower herbivory and gall formation on males and females of *Neea psychotrioides*, a dioecious tree," *Biotropica*, vol. 29, no. 2, pp. 169–174, 1997.

[44] J. Lovett-Doust and L. Lovett-Doust, "Sex ratios, clonal growth and herbivory in *Rumex acetosella*," in *Studies on Plant Demography: A Festschrift for John L Harper*, J. White, Ed., pp. 327–341, Academic Press, London, UK, 1985.

[45] L. Lovett-Doust and J. Lovett-Doust, "Leaf demography and clonal growth in female and male *Rumex acetosella*," *Ecology*, vol. 68, pp. 2056–2058, 1987.

[46] J. Hjältén, "Plant sex and hare feeding preferences," *Oecologia*, vol. 89, no. 2, pp. 253–256, 1992.

[47] R. Niesenbaum, "The effects of light environment on herbivory and growth in the dioecious shrub *Lindera benzoin* (Lauraceae)," *American Midland Naturalist*, vol. 128, pp. 270–275, 1992.

[48] M. L. Cipollini and D. F. Whigham, "Sexual dimorphism and cost of reproduction in the dioecious shrub *Lindera benzoin* (Lauraceae)," *American Journal of Botany*, vol. 81, no. 1, pp. 65–75, 1994.

[49] K. Danell, J. Hjältén, L. Ericson, and T. Elmqvist, "Vole feeding on male and female willow shoots along a gradient of plant productivity," *Oikos*, vol. 62, no. 2, pp. 145–152, 1991.

[50] J. P. Kopelke, J. Amendt, and K. Schönrogge, "Patterns of interspecific associations of stem gallers on willows," *Diversity and Distributions*, vol. 9, no. 6, pp. 443–453, 2003.

[51] M. C. Alliende, "Demographic studies of a dioecious tree. II. The distribution of leaf predation within and between trees," *Journal of Ecology*, vol. 77, no. 4, pp. 1048–1058, 1989.

[52] M. C. Alliende and J. L. Harper, "Demographic studies of a dioecious tree. I. Colonization, sex and age structure of a population of *Salix cinerea*," *Journal of Ecology*, vol. 77, no. 4, pp. 1029–1047, 1989.

[53] C. M. Nichols-Orians, R. S. Fritz, and T. P. Clausen, "The genetic basis for variation in the concentration of phenolic glycosides in *Salix sericea*: clonal variation and sex-based differences," *Biochemical Systematics and Ecology*, vol. 21, no. 5, pp. 535–542, 1993.

[54] M. Predavec and K. Danell, "The role of lemming herbivory in the sex ratio and shoot demography of willow populations," *Oikos*, vol. 92, no. 3, pp. 459–466, 2001.

[55] W. J. Boecklen, P. W. Price, and S. Mopper, "Sex and drugs and herbivores: sex-biased herbivory in arroyo willow (*Salix lasiolepis*)," *Ecology*, vol. 71, no. 2, pp. 581–588, 1990.

[56] W. J. Boecklen, S. Mopper, and P. W. Price, "Sex-biased herbivory in arroyo willow: are there general patterns among herbivores?" *Oikos*, vol. 71, no. 2, pp. 267–272, 1994.

[57] K. Danell, T. Elmqvist, L. Ericson, and A. Salomonson, "Sexuality in willows and preference by bark-eating voles: defence or not?" *Oikos*, vol. 44, no. 1, pp. 82–90, 1985.

[58] R. S. Fritz, "Direct and indirect effects of plant genetic variation on enemy impact," *Ecological Entomology*, vol. 20, no. 1, pp. 18–26, 1995.

[59] I. Åhman, "Growth, herbivory and disease in relation to gender in *Salix viminalis* L.," *Oecologia*, vol. 111, no. 1, pp. 61–68, 1997.

[60] P. A. Cox, "Vertebrate pollination and the maintenance of dioecism in *Freycinetia*," *American Naturalist*, vol. 120, no. 1, pp. 65–80, 1982.

[61] C. Li, G. Xu, R. Zang, H. Korpelainen, and F. Berninger, "Sex-related differences in leaf morphological and physiological responses in Hippophae rhamnoides along an altitudinal gradient," *Tree Physiology*, vol. 27, no. 3, pp. 399–406, 2007.

[62] M. J. Bañuelos, M. Sierra, and J. R. Obeso, "Sex, secondary compounds and asymmetry. Effects on plant-herbivore interaction in a dioecious shrub," *Acta Oecologica*, vol. 25, no. 3, pp. 151–157, 2004.

[63] M. J. Bañuelos and J. R. Obeso, "Resource allocation in the dioecious shrub *Rhamnus alpinus*: the hidden costs of reproduction," *Evolutionary Ecology Research*, vol. 6, no. 3, pp. 397–413, 2004.

[64] J. Ågren, "Intersexual differences in phenology and damage by herbivores and pathogens in dioecious *Rubus chamaemorus* L.," *Oecologia*, vol. 72, no. 2, pp. 161–169, 1987.

[65] J. Ågren, "Sexual differences in biomass and nutrient allocation in the dioecious *Rubus chamaemorus*," *Ecology*, vol. 69, no. 4, pp. 962–973, 1988.

[66] S. W. Jing and P. D. Coley, "Dioecy and herbivory: the effect of growth rate on plant defense in *Acer negundo*," *Oikos*, vol. 58, no. 3, pp. 369–377, 1990.

[67] T. E. Dawson and J. R. Ehleringer, "Gender-specific physiology, carbon isotope discrimination, and habitat distribution in boxelder, *Acer negundo*," *Ecology*, vol. 74, no. 3, pp. 798–815, 1993.

[68] M. F. Willson, "On the costs of reproduction in plants: *Acer negundo*," *American Midland Naturalist*, vol. 115, no. 1, pp. 204–207, 1986.

[69] P. F. Ramp and S. N. Stephenson, "Gender dimorphism in growth and mass partitioning by box-elder (*Acer negundo* L.)," *American Midland Naturalist*, vol. 119, no. 2, pp. 420–430, 1988.

[70] J. Hjältén, M. Astrom, E. Aberg et al., "Biased sex ratios in Spanish populations of *Pistacia lentiscus* (Anacardiaceae): the possible role of herbivory," *Anales del Jardín Botánico de Madrid*, vol. 51, pp. 49–53, 1993.

[71] K. S. Bawa and P. A. Opler, "Why are pistillate inflorescences of *Simarouba glauca* eaten less than staminate inflorescences?" *Evolution*, vol. 32, pp. 673–676, 1978.

[72] J. Ågren, K. Danell, T. Elmqvist et al., "Sexual dimorphism and biotic interactions," in *Gender and Sexual Dimorphism in Flowering Plants*, M. A. Geber, T. E. Dawson, and L. F. Delph, Eds., pp. 217–246, Springer, Berlin, Germany, 1999.

[73] T. Cornelissen and P. Stiling, "Sex-biased herbivory: a meta-analysis of the effects of gender on plant-herbivore interactions," *Oikos*, vol. 111, no. 3, pp. 488–500, 2005.

[74] V. Cepeda-Cornejo and R. Dirzo, "Sex-related differences in reproductive allocation, growth, defense and herbivory in three dioecious neotropical palms," *PloS ONE*, vol. 5, no. 3, Article ID e9824, 2010.

[75] N. E. Buckley and G. Avila-Sakar, "Trade-offs among reproduction, growth and defense vary with gender and reproductive allocation in an evergreen, dioecious shrub," *American Journal of Botany*. In press.

[76] A. M. Hemborg and W. J. Bond, "Do browsing elephants damage female trees more?" *African Journal of Ecology*, vol. 45, no. 1, pp. 41–48, 2007.

[77] C. A. Uribe-Mú and M. Quesada, "Preferences, patterns and consequences of branch removal on the dioecious tropical tree *Spondias purpurea* (Anacardiaceae) by the insect borer *Oncideres albomarginata chamela* (Cerambycidae)," *Oikos*, vol. 112, no. 3, pp. 691–697, 2006.

[78] R. S. Fritz, B. A. Crabb, and C. G. Hochwender, "Preference and performance of a gall-inducing sawfly: plant vigor, sex, gall traits and phenology," *Oikos*, vol. 102, no. 3, pp. 601–613, 2003.

[79] C. F. Dormann and C. Skarpe, "Flowering, growth and defence in the two sexes: consequences of herbivore exclusion for *Salix polaris*," *Functional Ecology*, vol. 16, no. 5, pp. 649–656, 2002.

[80] A. Tolvanen, J. Schroderus, and G. H. R. Henry, "Age- and stage-based bud demography of *Salix arctica* under contrasting muskox grazing pressure in the High Arctic," *Evolutionary Ecology*, vol. 15, no. 4–6, pp. 443–462, 2001.

[81] J. Turcotte and G. Houle, "Reproductive costs in *Salix planifolia* ssp. *planifolia* in subarctic Québec, Canada," *Ecoscience*, vol. 8, no. 4, pp. 506–512, 2001.

[82] Y. L. Dupont and M. Kato, "Sex ratio variation in dioecious plant species: a comparative ecological study of six species of

Lindera (Lauraceae)," *Nordic Journal of Botany*, vol. 19, no. 5, pp. 529–540, 1999.

[83] A. C. McCall, "Leaf damage and gender but not flower damage affect female fitness in *Nemophila menziesii* (Hydrophyllaceae)," *American Journal of Botany*, vol. 94, no. 3, pp. 445–450, 2007.

[84] B. Charlesworth, "The evolution of sex chromosomes," *Science*, vol. 251, no. 4997, pp. 1030–1033, 1991.

[85] R. Ming, A. Bendahmane, and S. S. Renner, "Sex chromosomes in land plants," *Annual Review of Plant Biology*, vol. 62, pp. 485–514, 2011.

[86] B. Vyskot and R. Hobza, "Gender in plants: sex chromosomes are emerging from the fog," *Trends in Genetics*, vol. 20, no. 9, pp. 432–438, 2004.

[87] L. F. Delph, "Sexual dimorphism in life history," in *Gender and Sexual Dimorphism in Flowering Plants*, M. A. Geber, T. E. Dawson, and L. F. Delph, Eds., pp. 149–173, Springer, Berlin, Germany, 1999.

[88] P. D. Coley, J. P. Bryant, and F. S. Chapin, "Resource availability and plant antiherbivore defense," *Science*, vol. 230, no. 4728, pp. 895–899, 1985.

[89] M. A. Watson, "Sexual differences in plant developmental phenology affect plant-herbivore interactions," *Trends in Ecology and Evolution*, vol. 10, no. 5, pp. 180–182, 1995.

[90] S. S. Renner and R. E. Ricklefs, "Dioecy and its correlates in the flowering plants," *American Journal of Botany*, vol. 82, pp. 596–606, 1995.

[91] E. L. Charnov, *The Theory of Sex Allocation*, Princeton University Press, Princeton, NJ, USA, 1982.

[92] P. Bierzychudek and V. Eckhart, "Spatial segregation of the sexes of dioecious plants," *American Naturalist*, vol. 132, no. 1, pp. 34–43, 1988.

[93] P. Tiffin, "Mechanisms of tolerance to herbivore damage: what do we know?" *Evolutionary Ecology*, vol. 14, no. 4–6, pp. 523–536, 2000.

[94] D. W. Hilbert, D. M. Swift, J. K. Detling, and M. I. Dyer, "Relative growth rates and the grazing optimization hypothesis," *Oecologia*, vol. 51, no. 1, pp. 14–18, 1981.

[95] C. B. Marshall, G. Avila-Sakar, and E. G. Reekie, "Effects of nutrient and CO_2 availability on tolerance to herbivory in *Brassica rapa*," *Plant Ecology*, vol. 196, no. 1, pp. 1–13, 2008.

[96] A. E. Weis, E. L. Simms, and M. E. Hochberg, "Will plant vigor and tolerance be genetically correlated? Effects of intrinsic growth rate and self-limitation on regrowth," *Evolutionary Ecology*, vol. 14, no. 4–6, pp. 331–352, 2000.

[97] T. G. Whitham, J. Maschinski, K. C. Larson et al., "Plant responses to herbivory: the continuum from negative to positive and underlying physiological mechanisms," in *Plant-Animal Interactions: Evolutionary Ecology in Tropical and Temperate Regions*, pp. 227–256, John Wiley & Sons, New York, NY, USA, 1991.

[98] M. J. Wise and W. G. Abrahamson, "Effects of resource availability on tolerance of herbivory: a review and assessment of three opposing models," *American Naturalist*, vol. 169, no. 4, pp. 443–454, 2007.

[99] G. Avila-Sakar and A. Laarakker, "The shape of the function of tolerance to herbivory," *The Americas Journal of Plant Science*, vol. 5, pp. 76–82, 2011.

[100] K. Lehtila and A. S. Larsson, "Meristem allocation as a means of assessing reproductive allocation," in *Reproductive Allocation in Plants*, E. G. Reekie and F. A. Bazzaz, Eds., pp. 51–75, Elsevier, Boston, Mass, USA, 2005.

[101] K. N. Paige and T. G. Whitham, "Overcompensation in response to mammalian herbivory: the advantage of being eaten," *American Naturalist*, vol. 129, no. 3, pp. 407–416, 1987.

[102] P. Mutikainen, M. Walls, and A. Ojala, "Sexual differences in responses to simulated herbivory in *Urtica dioica*," *Oikos*, vol. 69, no. 3, pp. 397–404, 1994.

[103] R. Mauricio, M. D. Rausher, and D. S. Burdick, "Variation in the defense strategies of plants: are resistance and tolerance mutually exclusive?" *Ecology*, vol. 78, no. 5, pp. 1301–1311, 1997.

[104] J. Núñez-Farfán, J. Fornoni, and P. L. Valverde, "The evolution of resistance and tolerance to herbivores," *Annual Review of Ecology, Evolution, and Systematics*, vol. 38, pp. 541–566, 2007.

[105] R. S. Freeman, A. K. Brody, and C. D. Neefus, "Flowering phenology and compensation for herbivory in *Ipomopsis aggregata*," *Oecologia*, vol. 136, no. 3, pp. 394–401, 2003.

[106] J. A. Lau and S. Y. Strauss, "Insect herbivores drive important indirect effects of exotic plants on native communities," *Ecology*, vol. 86, no. 11, pp. 2990–2997, 2005.

[107] R. J. Marquis, "The selective impact of herbivores," in *Plant Resistance to Herbivores and Pathogens: Ecology, Evolution and Genetics*, R. S. Fritz, Ed., pp. 301–325, University of Chicago Press, Chicago, Ill, USA, 1992.

[108] S. Y. Strauss and R. E. Irwin, "Ecological and evolutionary consequences of multispecies plant-animal interactions," *Annual Review of Ecology, Evolution, and Systematics*, vol. 35, pp. 435–466, 2004.

[109] A. G. Stephenson, "Flower and fruit abortion: proximate causes and ultimate functions," *Annual Review of Ecology and Systematics*, vol. 12, pp. 253–279, 1981.

[110] M. F. Willson, *Plant Reproductive Ecology*, John Wiley & Sons, New York, NY, USA, 1983.

[111] D. Charlesworth, "Plant sex determination and sex chromosomes," *Heredity*, vol. 88, no. 2, pp. 94–101, 2002.

[112] S. L. Dellaporta and A. Calderon-Urrea, "Sex determination in flowering plants," *Plant Cell*, vol. 5, no. 10, pp. 1241–1251, 1993.

[113] R. B. Spigler, K. S. Lewers, and T. L. Ashman, "Genetic architecture of sexual dimorphism in a subdioecious plant with a proto-sex chromosome," *Evolution*, vol. 65, no. 4, pp. 1114–1126, 2011.

[114] D. Charlesworth and F. R. Ganders, "The population genetics of gynodioecy with cytoplasmic-genic male-sterility," *Heredity*, vol. 43, pp. 213–218, 1979.

[115] P. F. Stevens, Angiosperm Phylogeny Website. Version 9, 2008.

[116] M. J. Bañuelos, M. Sierra, and J. R. Obeso, "Sex, secondary compounds and asymmetry. Effects on plant-herbivore interaction in a dioecious shrub," *Acta Oecologica*, vol. 25, no. 3, pp. 151–157, 2004.

[117] P. D. Putwain and J. L. Harper, "Studies in the dynamics of plant populations. V. Mechanisms governing the sex ratio in *Rumex acetosa* and *R. acetosella*," *Journal of Ecology*, vol. 60, pp. 113–129, 1972.

[118] I. T. Baldwin, "Herbivory simulations in ecological research," *Trends in Ecology and Evolution*, vol. 5, no. 3, pp. 91–93, 1990.

[119] E. K. Barto and D. Cipollini, "Testing the optimal defense theory and the growth-differentiation balance hypothesis in *Arabidopsis thaliana*," *Oecologia*, vol. 146, no. 2, pp. 169–178, 2005.

[120] K. Boege and R. J. Marquis, "Facing herbivory as you grow up: the ontogeny of resistance in plants," *Trends in Ecology and Evolution*, vol. 20, no. 8, pp. 441–448, 2005.

[121] C. Tucker and G. Avila-Sakar, "Ontogenetic changes in tolerance to herbivory in *Arabidopsis*," *Oecologia*, vol. 164, no. 4, pp. 1005–1015, 2010.

19

Deep Phylogenetic Divergence and Lack of Taxonomic Concordance in Species of *Astronotus* (Cichlidae)

Olavo Pinhatti Colatreli,[1] **Natasha Verdasca Meliciano,**[1] **Daniel Toffoli,**[1,2] **Izeni Pires Farias,**[1] **and Tomas Hrbek**[1]

[1] *Laboratório de Evolução e Genética Animal (LEGAL), Universidade Federal do Amazonas (UFAM), 69077-000 Manaus, AM, Brazil*
[2] *Departamento de Genética e Evolução, Universidade Federal de São Carlos (UFSCar), 18052-780 São Carlos, SP, Brazil*

Correspondence should be addressed to Tomas Hrbek, hrbek@evoamazon.net

Academic Editor: Martin J. Genner

The neotropical cichlid genus *Astronotus* currently comprises two valid species: *A. ocellatus* Agassiz, 1831 and *A. crassipinnis* Heckel, 1840. The diagnosis is based on color pattern and meristics counts. However, body color pattern is highly variable between regions and the meristic counts show a considerable overlap between populations differing in color patterning. They do not represent true synapomorphies that diagnose species. Purportedly the only truly diagnostic character is the presence or absence of one or more ocelli at the base of the dorsal fin, diagnosing *A. ocellatus* and *A. crassipinnis*, respectively. Using the 5′ portion of the mitochondrial COI gene and EPIC nuclear markers, the validity of the dorsal ocelli as diagnostic character was tested in individuals sampled from ten localities in the Amazon basin. Analyses rejected the hypothesis that dorsal ocelli are diagnostic at the species level. However, they revealed the existence of five hypothetical, largely allopatrically distributed morphologically cryptic species. The phylogeographic structure is not necessarily surprising, since species of the genus *Astronotus* have sedentary and territorial habits with low dispersal potential. The distribution of these hypothetical species is coincident with patterns observed in other Amazonian aquatic fauna, suggesting the role of common historical processes in generating current biodiversity patterns.

1. Introduction

The neotropical cichlid genus *Astronotus* currently comprises two valid species: *A. ocellatus* Agassiz, 1831 and *A. crassipinnis* Heckel, 1840 [1]. Kullander [1] reports a number of diagnostic characters, however, with the exception of the presence of ocelli at the base of the dorsal fin in *A. ocellatus* and their absence in *A. crassipinnis*, all other characters show considerable overlap in their statistical distributions. The two species are characterized by differences in the modal number of lateral line scales (35 to 40 in *A. crassipinnis* versus. 33 to 39 in *A. ocellatus*), and the number of rays and spines of the dorsal fin (modal XIII.20 in *A. ocellatus* versus. modal XII.21-22 in *A. crassipinnis*). There are also reported differences in color hue and patterning where *A. crassipinnis* is darker than *A. ocellatus*, the first light vertical bar is above the anal fin base in *A. ocellatus* versus more anteriorly in *A. crassipinnis*, and *A. crassipinnis* has two more or less well-separated dark vertical bars in the position of the first light bar in

A. ocellatus. Although proposed as diagnostic characters, the position of the vertical bars and body color appears highly variable between localities and individuals (authors' obs.), and the meristic counts are not truly diagnostic (are not synapomorphies) since they represent modal values and overlap between species.

While the presence of ocelli on the dorsal fin is considered a diagnostic character of *A. ocellatus*, Kullander ([1]; see http://www2.nrm.se/ve/pisces/acara/astronot.shtml), only individuals from Peru were analyzed by Kullander [1] in his reanalysis of the genus. Moreover, Kullander [1] raises the possibility that ocelli are unique to specimens of western Amazonia, requiring a possible reinstatement or reclassification of species considered synonyms of *A. ocellatus*. The geographic distribution of *A. ocellatus* spans the whole Amazon basin and the Oyapock and Approuague drainages. It does not include the Bolivian basin which is a subbasin of the Amazon basin.

The quantity and size of ocelli further appear to be influenced by reproductive state. In a study by Queiroz and Barcelos [2] of *Astronotus ocellatus* (diagnosed as such by the presence of ocelli) from the Mamirauá Sustainable Development Reserve located in the western Amazon north of the city of Tefé, the authors demonstrated that the number of ocelli and their size are positively and linearly correlated with gonadal development in both males and females. These potential difficulties do not prevent, however, the common acceptance of ocelli as strictly diagnostic character of the two species (e.g., [3]).

Of the type series of *A. crassipinnis*, only two syntypes from the Guaporé River are known. Other type material reported from the Negro and Branco Rivers according to Kullander [1] likely represents *A. ocellatus* or some undescribed species. *Astronotus crassipinnis* is therefore restricted to the upper Paraguay River and the Bolivian Amazon including the Guaporé, Mamoré, and Madre de Dios rivers. However, pending designation of a lectotype from the Guaporé River, Kullander [1] considers the classification of Paraguayan and Bolivian Amazonian specimens as *A. crassipinnis* provisory. Kullander [1] also recognizes that *A. ocellatus* could be restricted to the western Amazon and that *Astronotus ocellatus* var. *zebra* Pellegrin, 1904 and *Astronotus orbiculatus* Haseman, 1911 both described from Santarem and currently considered junior synonyms of *A. ocellatus* could represent valid species or may be synonyms of *A. crassipinnis*. Kullander [1, 4] further mentions the existence of an *Astronotus* species from the Orinoco basin but does not recommend any kind of classification of these specimens.

Phenotypic variation of *A. ocellatus* at the scale of the Amazon basin would not be surprising given the extent of geographic distribution of the species and the biology of cichlids. Both species of the genus *Astronotus* inhabiting lentic environments are sedentary. Males have strong territorial behavior, and both sexes build nests and exhibit parental care. First gonadal maturation occurs between 15 and 24 months, and reproduction may occur more than once a year. Both species are also relatively large for fishes of the family Cichlidae (up to 35 cm SL and 1.5 kg). The geographic distribution of species of *Astronotus* as well as the species themselves may therefore carry signatures of climatic and geological events.

While phenotypic variation is evident in the species of *Astronotus*, it is not clear if the currently used sets of characters are fully diagnostic. An alternative approach to species diagnosis may be through the use of DNA barcoding [5]. DNA barcoding has rapidly expanded in the last years, and already the fish faunas of several countries have been barcoded (e.g., [6–9]). One of the objectives of the DNA barcoding initiative is to generate a curated database of reference material. The usefulness of this database depends on the quality of the reference specimens and the quality of the underlying taxonomic information. For example, recently diverged species may share DNA barcodes (COI haplotypes), or multiple species may be subsumed within the same morphospecies, and both cases will lower the quality of the database. Identifying these instances is the first step in generating a reliable biodiversity database.

TABLE 1: Number of *Astronotus* specimens sampled at each site. We have no information about the phenotype (*Astronotus ocellatus*/ *Astronotus crassipinnis*, presence/absence of dorsal ocelli, resp.) for specimens identified as *Astronotus* sp., but, in each of the Careiro do Castanho and Araguari River localities, both species of *Astronotus* occurred, were sampled, and were included in the analyses.

| Localities | Specimen identification | | | All |
	A. crassipinnis	*A. ocellatus*	*Astronotus* sp.	
Tabatinga		4		4
Tefé/Mamirauá		4		4
Eirunepé		4		4
Guajará-Mirim	5			5
Borba	5	3		8
Barcelos		10		10
Sta Isabel do rio Negro		3		3
Careiro do Castanho			6	6
Oriximiná	3	2		5
Araguari river			8	8
Total	13	30	14	57

Many neotropical fish species have broad geographic distributions, often occurring allopatrically in the tributaries of the Amazon River, or are even shared between the Amazon and other South American basins (see [10]). While some species truly appear to be biological species with weak or nearly nonexistent population structuring across its distributional range (e.g., [11–14]), others probably comprise morphologically cryptic species complexes, recently diverged groups, or complexes of hybridizing groups (e.g., [15–18]).

The goal of this study was to assess population structuring and reassess the taxonomy of the genus *Astronotus* based on an analysis of molecular data and assess the utility of a traditionally used diagnostic character for the species *A. ocellatus* and *A. crassipinnis*.

2. Material and Methods

2.1. Sampling. Tissue samples (dorsal muscle or pectoral fins) were collected from specimens purchased directly from artisanal fishermen and from fishes sampled with 50 mm mesh gillnets. The tissues were deposited in the tissue collection of the Laboratory of Animal Genetics and Evolution, Federal University of Amazonas. Most individuals were photographed, and vouchers are being deposited at the ichthyological collection of the Instituto Nacional de Pesquisas da Amazonia (INPA).

We sampled 10 localities in the Amazon basin (Figure 1), and individuals were classified as *A. ocellatus* or *A. crassipinnis* based on the presence/absence of at least one ocellus or dark spot on the posterior part of the dorsal fin (Table 1). We do not have exact information about the state of ocelli for the Tabatinga and Mamirauá/Tefé specimens; however, based on field identification, fishes from Tabatinga and Mamirauá/Tefé were classified as *A. ocellatus*. Several studies [2, 19] also only report *A. ocellatus* from Mamirauá. Similarly although the presence/absence of ocelli was not recorded for

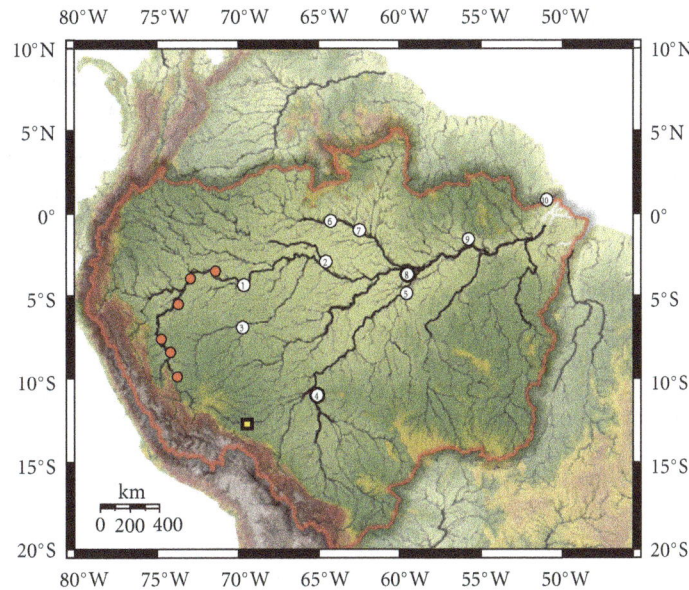

FIGURE 1: Sampling localities of species of *Astronotus* in the Brazilian Amazon. Base map was obtained from WWF (http://assets.panda.org/ img/original/hydrosheds_amazon_large.jpg). Numbers correspond to sampling localities: (1) Tabatinga; (2) Mamirauá; (3) Juruá; (4) Guajará Mirim; (5) Borba; (6) Santa Isabel; (7) Barcelos; (8) Careiro Castanho; (9) Oriximiná; (10) Araguari. Red circles and yellow squares are localities of *A. ocellatus* and *A. crassipinnis*, respectively, studied by Kullander [1]. Reddish-brown line delimits the periphery of the Amazon basin.

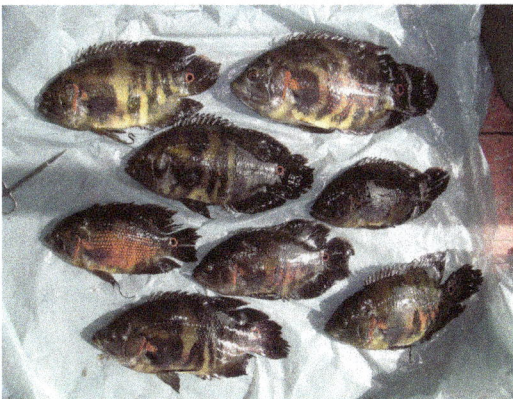

FIGURE 2: Photograph of fishes of the genus *Astronotus* collected in the Araguari River and showing the presence and absence of dorsal ocelli in the same locality. In addition to the Araguari locality, both *A. ocellatus* and *A. crassipinnis* phenotypes were collected in Oriximiná, Careiro Castanho, and Borba. Photo by S. C. Willis.

individual specimens at the time of collection at the localities of Careiro do Castanho and the Araguari River, both *A. ocellatus* and *A. crassipinnis* phenotypes were observed and sampled (Figure 2).

2.2. Polymerase Chain Reaction (PCR) and Sequencing.

We amplified and sequenced one mitochondrial and two nuclear gene regions. All PCR reactions were carried out in a final volume of 15 μL containing 7.0 μL of ddH$_2$O, 1.5 μL of MgCl$_2$ (25 mM), 1.5 μL of dNTPs (10 mM), 1.2 μL of 10x PCR buffer (100 mM Tris-HCl, 500 mM KCl), 1.2 μL of each primer (2 μM), 0.3 μL of Taq DNA Polymerase (1 U/μL), and 1 μL of DNA (concentration varied between 50 ng and 100 ng).

We amplified the COI barcode region with the primers COIFishF.2 (5′-CGACTAATCATAAAGATATCGGCAC-3′) and COIFishR.1 (5′-TTCAGGGTGACCGAAGAATCAGAA-3′), and the EPIC region primers 18049E2 (18049E2f2— 5′-GTGGTGGAGATGCAYGAYGTGAC-3′; 18049E2r2—5′-TAGTAAAGGTCYCCRTGGATGGTGAG-3′), and 14867E4 (14867E4f2—5′-TGTGATCAGGGGACAGAGRAAAGGTG-3′; 14867E4r2—5′-CAGTARATGAACTGBCCGGTGTGG-3′) obtained from the online supplement of Li and Riethoven [20]. PCR reaction consisted of 35 cycles of denaturation at 93°C for 5 seconds, primer annealing at 50°C; 50°C and 56°C, respectively, for 35 seconds, and primer extension at 72°C for 90 seconds, followed by a final extension at 72°C for 5 minutes. PCR products were purified using the polyethylene glycol/ethanol precipitation [21] and subjected to cycle sequencing reaction using both amplification primers following the manufacturer's recommended protocol for BigDye sequencing chemistry (Applied Biosystems). Subsequent to the cycle sequencing reaction, the products were precipitated with 100% ethanol/125 mM EDTA solution, resuspended in Hi-Di formamide, and resolved on an ABI 3130xl automatic sequencer (Applied Biosystems). Base calls were verified by viewing electropherograms in the program Bioedit [22], sequences were aligned in the program Clustal W [23], and alignment was verified by eye. Sequences of nuclear genes were separated into alleles prior to analyses. Sequences were deposited in Genbank (JQ965997-JQ966020).

2.3. DNA Barcode Analysis (COI mtDNA).

Genetic distances between individuals were calculated using the JC69 model of molecular evolution [24], and individuals were clustered using the BIONJ algorithm [25]. The analyses were implemented in the online version of the ABGD software [26] whose objective is to automatically and in an unbiased way delimit clades. Clade delimitation was done assuming a range of possible intraclade θs from 0.001 to 0.1. Once clades were identified, we also estimated average divergences between and within clades using the JC69 model of molecular evolution [24] in the program MEGA 5 [27]. Although the K2P model of molecular evolution [28] is the recommended [29] and has become the *defacto* model in DNA barcoding studies, it poorly fits the data at the species level divergence [30]. Collins et al. [30] recommend the use of uncorrected divergences or simplest models possible. Further, intraspecific divergences—employed in DNA barcoding threshold and barcoding gap methods, and pairwise divergences between sister taxa—employed in DNA barcoding gap methods, normally need no correction for multiple mutational hits and saturation due to their inherently shallow phylogenetic divergences.

We also performed an individual level Population Aggregation Analysis (PAA) [31] to identify clades. In the DNA barcoding literature, the use of molecular synapomorphies to delimit clades has been described by Rach et al. [32] under the acronym CAOS.

2.4. Phylogenetic Inference and Hypothesis Testing. Maximum likelihood topology for the mtDNA dataset was inferred in the program Treefinder [33], and the robustness of the tree topology was assessed using the nonparametric bootstrap with 1,000 replicates. The most appropriate model of molecular evolution for the mtDNA dataset was inferred as HKY85 [34] with a portion of the sites considered invariable in the program Treefinder [33]. Model selection criterion was the corrected Akaike Information Criterion [35]. Association of lineages and phenotypes was tested by comparing the constrained topology (phenotypes are monophyletic) with the most likely unconstrained topology. Significance was tested using the approximately unbiased test of Shimodaira [36]. A test of phylogenetic distribution of ocelli was performed using the CAPER package [37] in the statistical program R (http://www.cran.r-project.org/). A test of genetic structuring at nuclear loci, assuming the existence of groups identified in the ABGD [26] analysis of the COI barcode region, was performed in the software Arlequin 3.5.1 [38].

2.5. Phylogenetic Networks. Due to the low number of variable sites, phylogenetic relationships of nuclear haplotypes were inferred as a haplotype network using the PEGAS package [39] in the statistical program R (http://www.cran.r-project.org/).

3. Results

We sequence data for one mitochondrial and two nuclear DNA regions. We collected 664 bp of the mtDNA COI barcode region, representing 19 haplotypes separated by 31 mutations. No stop codons were observed in the COI barcode region. We also collected 397 bp of the nDNA 18049E2

TABLE 2: Mean intra- and interspecific distances and their standard errors estimated between COI haplotypes using the Jukes Cantor model of molecular evolution [24]. Hypothetical species were inferred using the ABGD [26] algorithm.

Average divergence between groups (below diagonal), and associated standard errors (above diagonal)					
	East	West	Bolivia	Jurua	Negro
East		0.56%	0.36%	0.55%	0.36%
West	2.17%		0.55%	0.33%	0.57%
Bolivia	0.98%	2.20%		0.57%	0.42%
Jurua	2.08%	0.86%	2.42%		0.49%
Negro	0.97%	2.20%	1.31%	1.80%	

Average divergence within groups (left column), and standard errors (right column)		
East	0.03%	0.02%
West	0.06%	0.03%
Bolivia	0.10%	0.07%
Jurua	0.13%	0.12%
Negro	0.09%	0.05%

EPIC regions, representing three haplotypes separated by three mutations. We further collected 248 bp of the nDNA 14867E4 EPIC region, resulting in two haplotypes separated by one mutation.

Using the ABGD software, we were able to infer five clades potentially representing species. Minimal divergence between these clades is 0.9% (Table 2). Individuals from all localities but Borba, a locality in the lower Madeira River, belong to just one clade. In the case of Borba, one individual is part of a clade that otherwise has a distribution in the Bolivian basin (upper Madeira River), while the remaining individuals are members of a clade found in the western Amazon basin. All five groups, with the exception of the Jurua group, are supported by at least one molecular synapomorphy (Table 3). For the sake of convenience, these clades will be referred to as East, Bolivia, Negro, West, and Jurua groups (Figure 3).

The 18049E2 nDNA gene was represented by three haplotypes (Figure 4), with the most common haplotype being present in all localities but Tabatinga-western-most locality of the West clade, the second most common haplotype not occurring in the Negro River and upper Madeira River, corresponding to the Negro and Bolivia groups, and the third haplotype being restricted to the upper Madeira River—Bolivia group. The 14867E4 nDNA gene was represented by only two haplotypes (Figure 5), one common haplotype not found in western localities corresponding to the West and Jurua groups and another restricted to the central Amazonian localities. Both nDNA gene regions show strong structuring, that is, alleles are not randomly distributed among the five groups identified in ABGD analysis. Analysis of molecular variance of the 18049E2 nDNA gene was significant ($F_{ST} = 0.4163$, $P < 0.001$) as was that of the 14867E4 nDNA gene ($F_{ST} = 0.8099$, $P < 0.001$).

Ocelli were not phylogenetically clustered (Figure 3). A constrained topology where individuals with and without ocelli were forced into reciprocal monophyly, that is an

TABLE 3: Matrix of molecular synapomorphies of the hypothetical species inferred using the ABGD [26] algorithm. Molecular synapomorphies are in bold. Column numbers indicate position within the sequenced COI fragment.

	89	98	131	143	152	209	215	227	236	248	260	305	443	447	464	539	578	590	596	662
East	G	C	G	G	T	T	C	A	A	T	T	C	T	C	A	A	**A**	T	C	T
Bolivia	G	C	G	A	T	T	C	A	A	T	C	C	T	C	A	**G**	G	T	**A**	**C**
Negro	**A**	C	G	G	T	T	C	G	A	C	C	T	T	C	A	A	G	T	C	T
West	G	T	**T**	A	C	C	A	A	G	T	C	T	C	**T**	G	A	G	A	C	T
Jurua	G	T	G	A	C	C	A	G	G	C	T	T	C	C	G	A	G	A	C	T

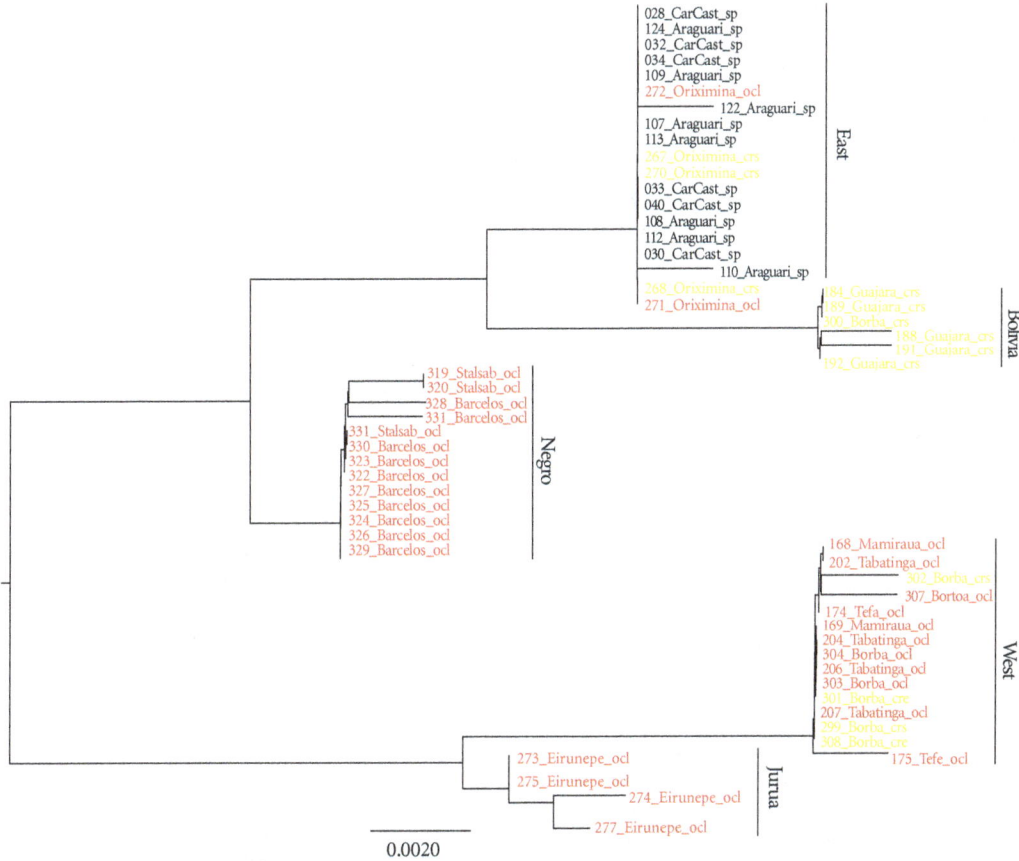

FIGURE 3: Maximum likelihood phylogenetic hypothesis ($-\ln = 1166.583$) of relationships of individuals of *Astronotus* sampled throughout Brazilian Amazônia based on the mtDNA COI barcode region. The topology is significantly different ($P = 0.003$) from constrained topology enforcing monophyly of *A. ocellatus* and *A. crassipinnis*. red—*A. ocellatus* (ocelli present); yellow—*A. crassipinnis* (ocelli absent); black—unknown.

explicit phylogenetic test of the usefulness of the presence/absence of ocelli as a diagnostic character, resulted in a significantly less likely topology ($P = 0.003$) and thus a rejection of the null hypothesis. However, analyses in CAPER indicated that ocelli were not distributed randomly across the ML topology (Fritz and Purvis' $D = 0.3862$, $P < 0.001$) but also were not clumped ($P = 0.021$).

4. Discussion

DNA barcode analyses revealed five, largely geographically restricted clades. Each clade with the exception of the Jurua group was supported by at least one molecular synapomorphy in the mtDNA dataset. While having less phylogenetic information, patterns of geographic distribution of nuclear

DNA haplotype distribution did not contradict the mtDNA results and supported certain phylogeographic divisions observed in the mtDNA phylogeny. The Bolivia group had a private allele of the 18049E2 nDNA gene, while the second most common haplotype of this gene was absent in the Bolivia and Negro groups. Of the two 14867E4 nDNA alleles, the more common allele do not occur in the West and Jurua groups, while the rarer allele occurred infrequently in the group East.

The five groups predicted with Automatic Barcode Gap Discovery (ABGD) [26] and supported by the analyses of nuclear DNA loci can be taken as a first set of species hypotheses that need to be tested with other data. The algorithm is based on the statistical properties of the coalescent, and baring recent radiations, will identify evolutionary entities compatible with the coalescent. Other methods of

EPIC 18049E2

EPIC 14867E4

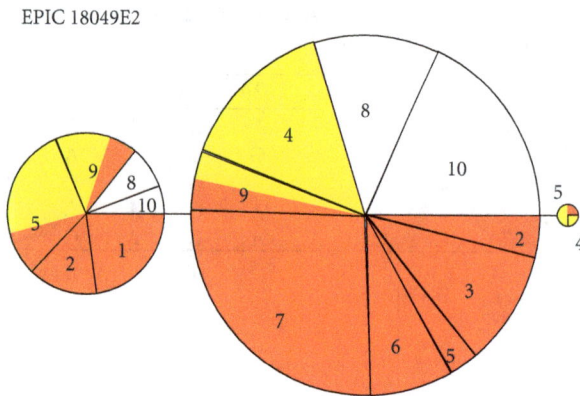

FIGURE 4: Haplotype network of the E0PIC 18049E2 nDNA region. Colors correspond to phenotypes: red—*A. ocellatus* (ocelli present); yellow—*A. crassipinnis* (ocelli absent); white—unknown. Numbers correspond to sampling localities: (1) Tabatinga; (2) Mamirauá; (3) Juruá; (4) Guajará Mirim; (5) Borba; (6) Santa Isabel; (7) Barcelos; (8) Careiro Castanho; (9) Oriximiná; (10) Araguari.

FIGURE 5: Haplotype network of the EPIC 14867E4 nDNA region. Colors correspond to phenotypes: red—*A. ocellatus* (ocelli present); yellow—*A. crassipinnis* (ocelli absent); white—unknown. Numbers correspond to sampling localities: (1) Tabatinga; (2) Mamirauá; (3) Juruá; (4) Guajará Mirim; (5) Borba; (6) Santa Isabel; (7) Barcelos; (8) Careiro Castanho; (9) Oriximiná; (10) Araguari.

identifying species from DNA barcode data are generally subjective or not generalizable across a broad range of organisms. The commonly used criterion of delimiting species such as the 3% interspecific divergence criterion, DNA barcodes differing by more than 3% belonging to different taxa [40], or the 10x rule, interspecific divergences that are 10x or larger than intraspecific divergences [41], fails to generalize for a number of taxonomic groups (e.g., [15, 42, 43]). Similarly, the interspecific and intraspecific divergences often overlap among closely related taxa (e.g., [15, 44–46]).

While it is clear that clades identified by ABGD [26] as potential species are geographically structured, the same cannot be said of the presence/absence of ocelli. Ocelli are not randomly distributed on the mtDNA phylogeny nor the nDNA haplotype networks; however, they also do not form monophyletic groups. Individuals of the Bolivia group do not have dorsal ocelli, while dorsal ocelli characterize all individuals of the Negro and Jurua groups. With the exception of individuals from the Borba locality, all other individuals pertaining to the group West are also characterized by the presence of ocelli. The group East is, on the other hand, characterized by a mix of individuals exhibiting both phenotypes (Figures 2 and 3). It should be noted that the Borba locality in the lower Madeira River is geographically intermediate between the Bolivia and the East groups. Thus, while some groups are monomorphic with respect to the presence/absence of ocelli, this character is not diagnostic and cannot be used to delimit species. Thus, currently, there are no morphological characters that can be used to diagnose and delimit species of *Astronotus*. On the other hand, ocelli are not randomly distributed throughout the phylogeny and do retain some phylogenetic information. In effect, specimens sampled from the vicinity of the main stream of the Amazon River (groups East and West) show both phenotypes, while specimens sampled from major affluents show either one or the other phenotype.

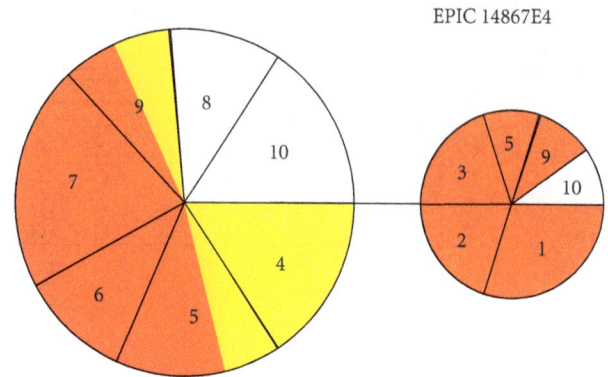

Broadly, however, the biodiversity patterns observed in the genus *Astronotus* are consistent with Kullander's [1] analysis. The group Bolivia is likely to be *Astronotus crassipinnis*, and one of its characteristics is lack of dorsal ocelli. What is currently considered *Astronotus ocellatus* harbors multiple species, a possibility also raised by Kullander, and while not diagnostic, specimens in the western Amazon basin have ocellated dorsal fin. Additional potential species currently subsumed under *A. ocellatus* include the groups from Jurua and Negro Rivers and from the central and eastern Amazon (group East).

The strong phylogeographic structure and the discovery of potentially new species of *Astronotus* are not necessarily surprising. *Astronotus* species are sedentary and territorial, have low power of dispersion, and therefore are likely to be influenced by climatic and geomorphological events. Perhaps the most interesting observation is that the division between the group East and West (not considering the Borba locality) parallels the division between the cichlid fishes *Symphysodon* sp. 2 (phenotype blue) and *Symphysodon tarzoo* (phenotype green) [16, 17, 47]. Also intriguing is that all but one specimen from the Borba locality in the lower Madeira River share haplotypes with the group West, which again parallels haplotype sharing between lower Madeira River and western Amazon observed by Ready et al. [47] in *Symphysodon*. The differentiation of the Bolivia group from all other *Astronotus* is potentially explained by the presence of the series of rapids on the Madeira River. These series of rapids are thought to delimit the geographic distributions of such diverse taxa as *Inia geoffrensis* and *I. boliviensis* [48, 49], *Cichla monoculus* and *C. pleiozona* [50], or they act as barriers, restricting gene flow in *Colossoma macropomum* [13] and *Podocnemis expansa* [51]. The physiochemical composition of the Negro River has also been suggested to act as a barrier between and within species [16, 17, 52, 53]. The patterns observed in *Astronotus* are likely to be general, implying that multiple additional species in broadly distributed Amazonian taxa are almost inevitably to be discovered.

Acknowledgments

This research was supported by grants from CNPq (554057/2006-9 and 557004/2009-8). I. P. Farias and T. Hrbek were supported by a *Bolsa de Pesquisa* scholarship from CNPq during the study. Permits for field collection and DNA accession were given by IBAMA/CGEN no. 045/DIFAP/2005 and 148/DIFAP/2006. S. Willis participated in fieldwork. This paper is based on an undergraduate monograph of O. P. Colatreli.

References

[1] S. O. Kullander, *Cichlid Fishes of the Amazon River Drainage of Peru*, Museum of Natural History, Stockholm, Sweden, 1986.

[2] H. L. Queiroz and M. Barcelos, *Seminário Anual de Pesquisas do IDSM*, Instituto do Desenvolvimento Sustentavel Mamirauá, Tefé, AM, Brazil, 2005.

[3] G. M. D. Santos, E. J. G. Ferreira, and J. A. S. Zuanon, *Peixes Comerciais de Manaus*, IBAMA/Pró-Várzea, Manaus, AM, Brazil, 2006.

[4] S. O. Kullander, "Family Cichlidae," in *Check List of the Freshwater Fishes of South and Central America*, R. E. Reis, S. O. Kullander, and C. J. Ferraris, Eds., pp. 605–654, EDIPUCRS, Porto Alegre, Brazil, 2003.

[5] P. D. N. Hebert, A. Cywinska, S. L. Ball, and J. R. DeWaard, "Biological identifications through DNA barcodes," *Proceedings of the Royal Society B*, vol. 270, no. 1512, pp. 313–321, 2003.

[6] M. Valdez-Moreno, N. V. Ivanova, M. Elías-Gutiérrez, S. Contreras-Balderas, and P. D. N. Hebert, "Probing diversity in freshwater fishes from Mexico and Guatemala with DNA barcodes," *Journal of Fish Biology*, vol. 74, no. 2, pp. 377–402, 2009.

[7] N. Hubert, R. Hanner, E. Holm et al., "Identifying Canadian freshwater fishes through DNA barcodes," *PLoS ONE*, vol. 3, no. 6, Article ID e2490, 2008.

[8] R. D. Ward, T. S. Zemlak, B. H. Innes, P. R. Last, and P. D. N. Hebert, "DNA barcoding Australia's fish species," *Philosophical Transactions of the Royal Society B*, vol. 360, no. 1462, pp. 1847–1857, 2005.

[9] J. April, R. L. Mayden, R. H. Hanner, and L. Bernatchez, "Genetic calibration of species diversity among North America's freshwater fishes," *Proceedings of the National Academy of Sciences of the United States of America*, vol. 108, no. 26, pp. 10602–10607, 2011.

[10] R. E. Reis, S. O. Kullander, and C. J. Ferraris, Eds., *Check List of the Freshwater Fishes of South and Central America*, EDIPUCRS, Porto Alegre, Brazil, 2003.

[11] M. C. F. Santos, M. L. Ruffino, and I. P. Farias, "High levels of genetic variability and panmixia of the tambaqui *Colossoma macropomum* (Cuvier, 1816) in the main channel of the Amazon River," *Journal of Fish Biology*, vol. 71, pp. 33–44, 2007.

[12] T. Hrbek, I. P. Farias, M. Crossa, I. Sampaio, J. I. R. Porto, and A. Meyer, "Population genetic analysis of *Arapaima gigas*, one of the largest freshwater fishes of the Amazon basin: implications for its conservation," *Animal Conservation*, vol. 8, no. 3, pp. 297–308, 2005.

[13] I. P. Farias, J. P. Torrico, C. García-Dávila, M. D. C. F. Santos, T. Hrbek, and J. F. Renno, "Are rapids a barrier for floodplain fishes of the Amazon basin? A demographic study of the keystone floodplain species *Colossoma macropomum* (Teleostei:

Characiformes)," *Molecular Phylogenetics and Evolution*, vol. 56, no. 3, pp. 1129–1135, 2010.

[14] J. S. Batista and J. A. Alves-Gomes, "Phylogeography of *Brachyplatystoma rousseauxii* (Siluriformes-Pimelodidae) in the Amazon Basin offers preliminary evidence for the first case of "homing" for an Amazonian migratory catfish," *Genetics and Molecular Research*, vol. 5, no. 4, pp. 723–740, 2006.

[15] D. Toffoli, T. Hrbek, M. L. G. de Araújo, M. P. de Almeida, P. Charvet-Almeida, and I. P. Farias, "A test of the utility of DNA barcoding in the radiation of the freshwater stingray genus *Potamotrygon* (Potamotrygonidae, Myliobatiformes)," *Genetics and Molecular Biology*, vol. 31, no. 1, pp. 324–336, 2008.

[16] I. P. Farias and T. Hrbek, "Patterns of diversification in the discus fishes (*Symphysodon* spp. Cichlidae) of the Amazon basin," *Molecular Phylogenetics and Evolution*, vol. 49, no. 1, pp. 32–43, 2008.

[17] M. V. Amado, T. Hrbek, and I. P. Farias, "A molecular perspective on systematics, taxonomy and classification Amazonian discus fishes of the genus *Symphysodon*," *International Journal of Evolutionary Biology*, vol. 2011, Article ID 360654, 16 pages, 2011.

[18] S. C. Willis, M. S. Nunes, C. G. Montaña, I. P. Farias, and N. R. Lovejoy, "Systematics, biogeography, and evolution of the Neotropical peacock basses *Cichla* (Perciformes: Cichlidae)," *Molecular Phylogenetics and Evolution*, vol. 44, no. 1, pp. 291–307, 2007.

[19] R. A. Chaves, Universidade Federal do Pará (UFPA), EmpresaBrasileira de Pesquisa Agropecuária—Amazônia Oriental (EMPBRAPA-Oriental), and Universidade Federal Rural da Amazônia (UFRA), M.Sc., 2007.

[20] C. Li and J. J. M. Riethoven, "Exon-primed intron-crossing (EPIC) markers for non-model teleost fishes," *BMC Evolutionary Biology*, vol. 10, no. 1, article 90, 2010.

[21] K. R. Paithankar and K. S. N. Prasad, "Precipitation of DNA by polyethylene glycol and ethanol," *Nucleic Acids Research*, vol. 19, no. 6, p. 1346, 1991.

[22] T. Hall, "BioEdit: a user-friendly biological sequence alignment editor and analysis program for Windows 95/98/NT," *Nucleic Acids Symposium Series*, vol. 41, no. 1, pp. 95–98, 1999.

[23] J. D. Thompson, D. G. Higgins, and T. J. Gibson, "CLUSTAL W: improving the sensitivity of progressive multiple sequence alignment through sequence weighting, position-specific gap penalties and weight matrix choice," *Nucleic Acids Research*, vol. 22, no. 22, pp. 4673–4680, 1994.

[24] T. H. Jukes and C. R. Cantor, "Evolution of protein molecules," in *Mammalian Protein Metabolism*, H. N. Munro, Ed., pp. 21–123, Academic Press, New York, NY, USA, 1969.

[25] O. Gascuel, "BIONJ: an improved version of the NJ algorithm based on a simple model of sequence data," *Molecular Biology and Evolution*, vol. 14, no. 7, pp. 685–695, 1997.

[26] N. Puillandre, A. Lambert, S. Brouillet, and G. Achaz, "ABGD, Automatic Barcode Gap Discovery for primary species delimitation," *Molecular Ecology*, vol. 21, no. 8, pp. 1864–1877, 2012.

[27] S. Kumar, M. Nei, J. Dudley, and K. Tamura, "MEGA: a biologist-centric software for evolutionary analysis of DNA and protein sequences," *Briefings in Bioinformatics*, vol. 9, no. 4, pp. 299–306, 2008.

[28] M. Kimura, "A simple method for estimating evolutionary rates of base substitutions through comparative studies of nucleotide sequences," *Journal of Molecular Evolution*, vol. 16, no. 2, pp. 111–120, 1980.

[29] P. D. N. Hebert, E. H. Penton, J. M. Burns, D. H. Janzen, and W. Hallwachs, "Ten species in one: DNA barcoding reveals

cryptic species in the neotropical skipper butterfly *Astraptes fulgerator*," *Proceedings of the National Academy of Sciences of the United States of America*, vol. 101, no. 41, pp. 14812–14817, 2004.

[30] R. A. Collins, L. M. Boykin, R. H. Cruickshank, and K. F. Armstrong, "Barcoding's next top model: an evaluation of nucleotide substitution models for specimen identification," *Methods in Ecology and Evolution*, vol. 3, no. 3, pp. 457–465, 2012.

[31] J. I. Davis and K. C. Nixon, "Populations, genetic variation, and the delimitation of phylogenetic species," *Systematic Biology*, vol. 41, no. 4, pp. 421–435, 1992.

[32] J. Rach, R. DeSalle, I. N. Sarkar, B. Schierwater, and H. Hadrys, "Character-based DNA barcoding allows discrimination of genera, species and populations in Odonata," *Proceedings of the Royal Society B*, vol. 275, no. 1632, pp. 237–247, 2008.

[33] G. Jobb, A. Von Haeseler, and K. Strimmer, "TREEFINDER: a powerful graphical analysis environment for molecular phylogenetics," *BMC Evolutionary Biology*, vol. 4, article 18, 2004.

[34] M. Hasegawa, H. Kishino, and T. Yano, "Dating of the human-ape splitting by a molecular clock of mitochondrial DNA," *Journal of Molecular Evolution*, vol. 22, no. 2, pp. 160–174, 1985.

[35] H. Akaike, "A new look at the statistical model identification," *IEEE Transactions on Automatic Control*, vol. 19, no. 6, pp. 716–723, 1974.

[36] H. Shimodaira, "An approximately unbiased test of phylogenetic tree selection," *Systematic Biology*, vol. 51, no. 3, pp. 492–508, 2002.

[37] C. D. L. Orme, R. P. Freckleton, G. H. Thomas, T. Petzoldt, S. A. Fritz, and N. J. B. Isaac, "CAPER: comparative analyses of phylogenetics and evolution in R," *Methods in Ecology and Evolution*, vol. 3, no. 1, pp. 145–151, 2012.

[38] L. Excoffier and H. E. L. Lischer, "Arlequin suite ver 3.5: a new series of programs to perform population genetics analyses under Linux and Windows," *Molecular Ecology Resources*, vol. 10, no. 3, pp. 564–567, 2010.

[39] E. Paradis, "Pegas: an R package for population genetics with an integrated-modular approach," *Bioinformatics*, vol. 26, no. 3, pp. 419–420, 2010.

[40] M. A. Smith, B. L. Fisher, and P. D. N. Hebert, "DNA barcoding for effective biodiversity assessment of a hyperdiverse arthropod group: the ants of Madagascar," *Philosophical Transactions of the Royal Society B*, vol. 360, no. 1462, pp. 1825–1834, 2005.

[41] P. D. N. Hebert, M. Y. Stoeckle, T. S. Zemlak, and C. M. Francis, "Identification of birds through DNA barcodes," *PLoS Biology*, vol. 2, no. 10, article e310, 2004.

[42] A. Gómez, P. J. Wright, D. H. Lunt, J. M. Cancino, G. R. Carvalho, and R. N. Hughes, "Mating trials validate the use of DNA barcoding to reveal cryptic speciation of a marine bryozoan taxon," *Proceedings. Biological sciences / The Royal Society*, vol. 274, no. 1607, pp. 199–207, 2007.

[43] R. Meier, G. Zhang, and F. Ali, "The use of mean instead of smallest interspecific distances exaggerates the size of the "barcoding gap" and leads to misidentification," *Systematic Biology*, vol. 57, no. 5, pp. 809–813, 2008.

[44] C. P. Meyer and G. Paulay, "DNA barcoding: error rates based on comprehensive sampling," *PLoS biology*, vol. 3, no. 12, article e422, 2005.

[45] M. Elias, R. I. Hill, K. R. Willmott et al., "Limited performance of DNA barcoding in a diverse community of tropical butterflies," *Proceedings of the Royal Society B*, vol. 274, no. 1627, pp. 2881–2889, 2007.

[46] M. Wiemers and K. Fiedler, "Does the DNA barcoding gap exist? A case study in blue butterflies (Lepidoptera: Lycaenidae)," *Frontiers in Zoology*, vol. 4, article 8, 2007.

[47] J. S. Ready, E. J. G. Ferreira, and S. O. Kullander, "Discus fishes: mitochondrial DNA evidence for a phylogeographic barrier in the Amazonian genus *Symphysodon* (Teleostei: Cichlidae)," *Journal of Fish Biology*, vol. 69, pp. 200–211, 2006.

[48] G. Pilleri and M. Gihr, "Observations on the Bolivian, *Inia boliviensis*, (D'Orbigny, 1834) and the Amazonian bufeo, *Inia geoffrensis* (Blainville, 1817), with a description of a new subspecies (*Inia geoffrensis humboldtiana*)," *Experientia*, vol. 24, no. 9, pp. 932–934, 1968.

[49] E. Banguera-Hinestroza, H. Cárdenas, M. Ruiz-García et al., "Molecular identification of evolutionarily significant units in the Amazon river dolphin *Inia* sp. (Cetacea: Iniidae)," *Journal of Heredity*, vol. 93, no. 5, pp. 312–322, 2002.

[50] S. O. Kullander and E. J. G. Ferreira, "A review of the South American *cichlid* genus *Cichla*, with descriptions of nine new species (Teleostei: Cichlidae)," *Ichthyological Exploration of Freshwaters*, vol. 17, no. 4, pp. 289–398, 2006.

[51] D. E. Pearse, A. D. Arndt, N. Valenzuela, B. A. Miller, V. Cantarelli, and J. W. Sites Jr., "Estimating population structure under nonequilibrium conditions in a conservation context: continent-wide population genetics of the giant Amazon river turtle, *Podocnemis expansa* (Chelonia; Podocnemididae)," *Molecular Ecology*, vol. 15, no. 4, pp. 985–1006, 2006.

[52] B. de Thoisy, T. Hrbek, I. P. Farias, W. R. Vasconcelos, and A. Lavergne, "Genetic structure, population dynamics, and conservation of Black caiman (*Melanosuchus niger*)," *Biological Conservation*, vol. 133, no. 4, pp. 474–482, 2006.

[53] N. V. Meliciano, Instituto Nacional de Pesquisas da Amazônia (INPA) and Universidade Federal do Amazonas (UFAM), M.Sc., 2008.

20

Genomic Structure and Evolution of Multigene Families: "Flowers" on the Human Genome

Hie Lim Kim,[1,2] **Mineyo Iwase,**[1] **Takeshi Igawa,**[3] **Tasuku Nishioka,**[1] **Satoko Kaneko,**[4] **Yukako Katsura,**[5] **Naoyuki Takahata,**[6] **and Yoko Satta**[5]

[1] *Center for the Promotion of Integrated Sciences, The Graduate University for Advanced Studies (SOKENDAI), Hayama, Kanagawa 240-0193, Japan*
[2] *Department of Biochemistry and Molecular Biology, Pennsylvania State University, 312 Wartik Laboratory, University Park, PA 16802, USA*
[3] *Institute for Amphibian Biology, Graduate School of Science, Hiroshima University, Higashihiroshima, Hiroshima 739-8526, Japan*
[4] *Laboratory of Plant Genetics, Graduate School of Agriculture, Kyoto University, Kyoto 606-8502, Japan*
[5] *Department of Evolutionary Studies of Biosystems, The Graduate University for Advanced Studies (SOKENDAI), Hayama, Kanagawa 240-0193, Japan*
[6] *The Graduate University for Advanced Studies (SOKENDAI), Hayama, Kanagawa 240-0193, Japan*

Correspondence should be addressed to Yoko Satta, satta@soken.ac.jp

Academic Editor: Hirohisa Kishino

We report the results of an extensive investigation of genomic structures in the human genome, with a particular focus on relatively large repeats (>50 kb) in adjacent chromosomal regions. We named such structures "Flowers" because the pattern observed on dot plots resembles a flower. We detected a total of 291 Flowers in the human genome. They were predominantly located in euchromatic regions. Flowers are gene-rich compared to the average gene density of the genome. Genes involved in systems receiving environmental information, such as immunity and detoxification, were overrepresented in Flowers. Within a Flower, the mean number of duplication units was approximately four. The maximum and minimum identities between homologs in a Flower showed different distributions; the maximum identity was often concentrated to 100% identity, while the minimum identity was evenly distributed in the range of 78% to 100%. Using a gene conversion detection test, we found frequent and/or recent gene conversion events within the tested Flowers. Interestingly, many of those converted regions contained protein-coding genes. Computer simulation studies suggest that one role of such frequent gene conversions is the elongation of the life span of gene families in a Flower by the resurrection of pseudogenes.

1. Introduction

A genomic structure is a region of repeats located on adjacent chromosomal regions and consists of combinations of tandem or/and inverted repeats and palindromes. Genomic structures are generated by genomic rearrangements such as duplications, deletions, and inversions in a genome. If the genes are located within a rearrangement such as a duplication or deletion, the number of genes would vary between individuals, and consequently, expression levels of those genes could also vary [1, 2]; this may thus affect phenotypic traits. Well-known inherited diseases, such as Prader-Willi and Williams-Beurens syndromes, are caused by

variation in the gene numbers and, in particular, by deletions in 15q11–q13 and 7q11.23, respectively [3]. On the other hand, inversions do not result in changes in the copy number of genes, but could affect recombination frequency between an intact and inverted segment (haplotype). Recombination is suppressed, and therefore, both haplotypes accumulate specific mutations. This accumulation enhances genetic differentiation between genomes within a species. If genes in the recombination-suppressed region are involved in mating or adaptation to environmental changes, the inversion might affect reproductive isolation or speciation [4].

Genomic structures can be detected by comparing nucleotide sequences within an individual genome. Such structures are not distributed uniformly in a genome and sometimes are located on particular regions in a chromosome. One well-known example is a cluster of palindromes on the Y chromosome [5]. There are eight palindromes of various sizes, and six of them carry protein-coding genes. The palindromes have been maintained via genetic exchanges between arms within a palindrome or between different palindromes [6]. These palindromes are important in keeping a large number of copies of male-specific Y chromosomal genes identical [7].

Duplication occurring in the genome is often called "segmental duplication," which is distinct from whole genome duplication. There have been several studies involving screening of segmental duplications collectively for the entire human genome [8, 9]. Within human populations, copy number variations (CNVs) in segmental duplications or genomic structures have been found [8, 9]. Massive structural variations were reported [8, 9], but the biological significance of the frequent genomic rearrangements during evolution remains unclear. It is thus necessary to shed light on each region to understand the evolution of genomic rearrangements.

The aim of this study was to reveal how rearrangements in the genome or genomic structures have been maintained and to determine the biological significance of their retention, from an evolutionary point of view. Large-scale duplicated sequences are amongst the most difficult sequences to assemble. For our study, the quality of assembly is most important. In this regard, the human genome sequence is the most reliable. Thus, we have studied the evolution of the genomic structures on the human genome.

2. Materials and Methods

2.1. Identification and Statistical Analyses of Flowers. Human genome sequence data (build 36; 2,858,142,293 bp) and corresponding gene information was obtained from NCBI (http://www.ncbi.nlm.nih.gov/). To detect genomic structures (such as an inversions, tandem repeats, or deletions/insertions), dot-matrix and BLAST analyses using the software package *GenomeMatcher* [10] were applied to the genome sequence. *GenomeMatcher* was used to search for homologous sequences between two input sequences and to visualize the results as dot plots. To detect regions where at least two duplications are located in close proximity, we first divided genomic sequences into 1-Mb nonoverlapping regions. In each of these regions, we compared sequences by itself using BLAST with a command line "*blastn-F T-W 40-e 0.01-D 1.*"

The results of "BLAST hits" for each region are alignments of pairs of the query and subject sequences. Our method focused on identifying genomic structures, which are duplication-concentrated regions generated by genomic rearrangements. A "Flower" is composed of an aligned sequence of more than 1 kb, excluding the concentrated regions of repetitive elements, such as LINEs, SINEs, and microsatellites. The size of a Flower, that is, the length of the region containing the alignments, was selected as longer than 50 kb. In the case of duplications located across neighboring

windows, we ran BLAST for extended windows in order to detect large-sized Flowers. If the location of two duplications was not within 1 Mb in length, our detection method could miss such regions, but they were not of interest to us.

One example of a Flower is shown in Figure 1. Several measurements for each Flower were defined (Figure 1, Table 1). The start and end points of a Flower were determined by BLAST analysis. In the example in Figure 1, *Flower length* was defined as the consecutive segment size from the start to the end point of a Flower. The *D length* for a Flower was defined as the sum of lengths of duplicated regions in a Flower. The effective number of copy units was calculated by the ratio of the total length of aligned sequences to the D length. *Copy unit length* was defined by the D length/the effective number of copy units. *Length of inverted copy* was defined as the total length of sequences matched to the query sequences in an opposite direction. The *minimum and maximum identity* were defined as the minimum and maximum values among identities of all alignments in a single Flower, respectively. The *gene region length over D length* stands for the proportion of gene regions in a repeated region, defined as the *gene number* in a Flower. When the number is more than 0.1, the gene was determined to be a "Flower gene".

2.2. Association of Flowers with Genes. To examine the relationship between Flowers and the genes in them, we compared gene density (measured by the number of genes per 1 Mb) in the duplicated regions of Flowers with the density of genes in randomly selected regions of the genome. Among 291 Flowers, there were 277 Flowers whose *D* lengths were unique. We sampled 277 regions of same sizes from the human genome, with the exception of Flower regions and sequencing gaps. Based on the annotation data of NCBI build 36, the number of genes in non-Flower and nongap regions was counted. Thousands of sampling repeats were performed to obtain the distribution of the gene number in such regions.

To investigate the known or reported functions of Flower genes, we extracted the Gene Ontology (GO) ID and obtained information on the GO categories from data of human genes at NCBI (http://www.ncbi.nlm.nih.gov/) and the GO database (http://www.geneontology.org/; The Gene Ontology Consortium).

2.3. Gene Conversion Detection Test. To detect gene conversion events within Flowers, we used the CHAP package [12]. To run the package, at least one orthologous sequence for each Flower must be determined as an outgroup. We performed BLAST with queries of Flower sequences and subjects of chimpanzee or gorilla genome assemblies. The identification of orthologs was difficult for large-sized Flowers or Flowers on sex chromosomes because of the quality of those assemblies. For accurate detection of gene conversion, we analyzed only Flowers for which orthologs were clearly determined. Finally, we succeeded in obtaining 189 Flowers with their orthologous sequences. We counted only conversion events occurring between human paralogous sequences after split with chimpanzees or gorilla.

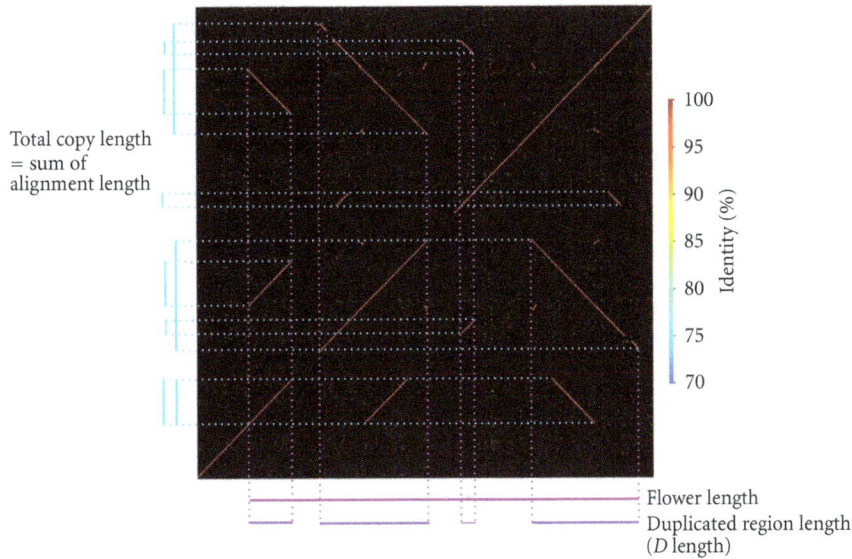

FIGURE 1: Example of a Flower. A dot plot shows an example of a Flower located on chromosome 10. Diagonal lines in the plot indicate the positions of blast hits, and the colors of the lines represent the identity of alignments as shown in the right-hand side of the plot. A blue bar indicates a detected copy unit, and the sum of alignments of blast hits equals the sum of the copy length. A pink bar indicates a Flower region, and the length of the bar represents the length of the Flower. A purple bar indicates a region called the duplicated region that contains the copy.

TABLE 1: Statistical data for 291 Flowers. Definitions of each term are described in Section 2.

$N = 291$	Sum	Mean	S.D.	Median	Min.	Max.
Flower length (Kb)	179,197	615	1,349	210	51	11,695
D length/Flower length		0.33	0.22	0.27	0.01	0.96
Effective number of copy units		3	3	3	2	22
Copy unit length (Kb)		50	95	23	1	1,122
Length of inverted copy/D length		0.4	0.4	0.5	0.0	1.0
Minimum identity (%)		88.5	5.0	88.0	77.7	100.0
Maximum identity (%)		98.0	2.6	99.2	89.8	100.0
Gene region length/D length		0.4	0.3	0.4	0.0	1.0
Gene number[a]	2,844	10	19	4	0	133
Pseudogene number/gene number[a]		0.3	0.3	0.3	0.0	1.0

[a] The number of Flower genes.

2.4. Simulation Analyses. Our simulation model assumed two loci in a diploid population with an effective size of $N = 50$. Although $N = 50$ could be considered as being too small given the size of the human population ($N = 10^4$), under neutrality, the simulation can be scaled while keeping the population parameters $N\mu$ and Nc constant.

We set generation 1 as the time when a duplicated identical gene is fixed in a population. Mutations occurred at a rate of $N\mu = 0.5$ per locus per generation, and each mutation caused pseudogenization. A backward mutation was not allowed and infinite sites model for mutations was used. Gene conversion occurred from one gene to another at a rate of $Nc = 0$–20 per gene per generation. Since we assume two loci with four genes in a diploid individual, the conversion can occur in any of three different schemes: allelic trans, allelic cis, and nonallelic trans (Figure 4). The number

of pseudogenes (n_ψ) in a diploid individual ranges from 0 to 4. When n_ψ was 0 to 3, no natural selection against an individual having the pseudogene(s) was considered ($Ns = 0$, where s is a selection coefficient), whereas when n_ψ in an individual becomes 4, there is negative selection against the individual with $Ns = 2$–5. In one generation, simulations were performed according to the following order: mutation in a gamete $\rightarrow$ gene conversion between gametes in an individual $\rightarrow$ random sampling of gametes $\rightarrow$ selection against a progeny (zygote). A simulation counted the number of generations from generation 1 until all genes at both loci in a population became pseudogenes (fixation of pseudogenes). We carried out 10,000 repeats of this simulation and calculated the mean and standard error of this fixation time of 10,000 replications for each set of parameters (i.e., Nc and Ns).

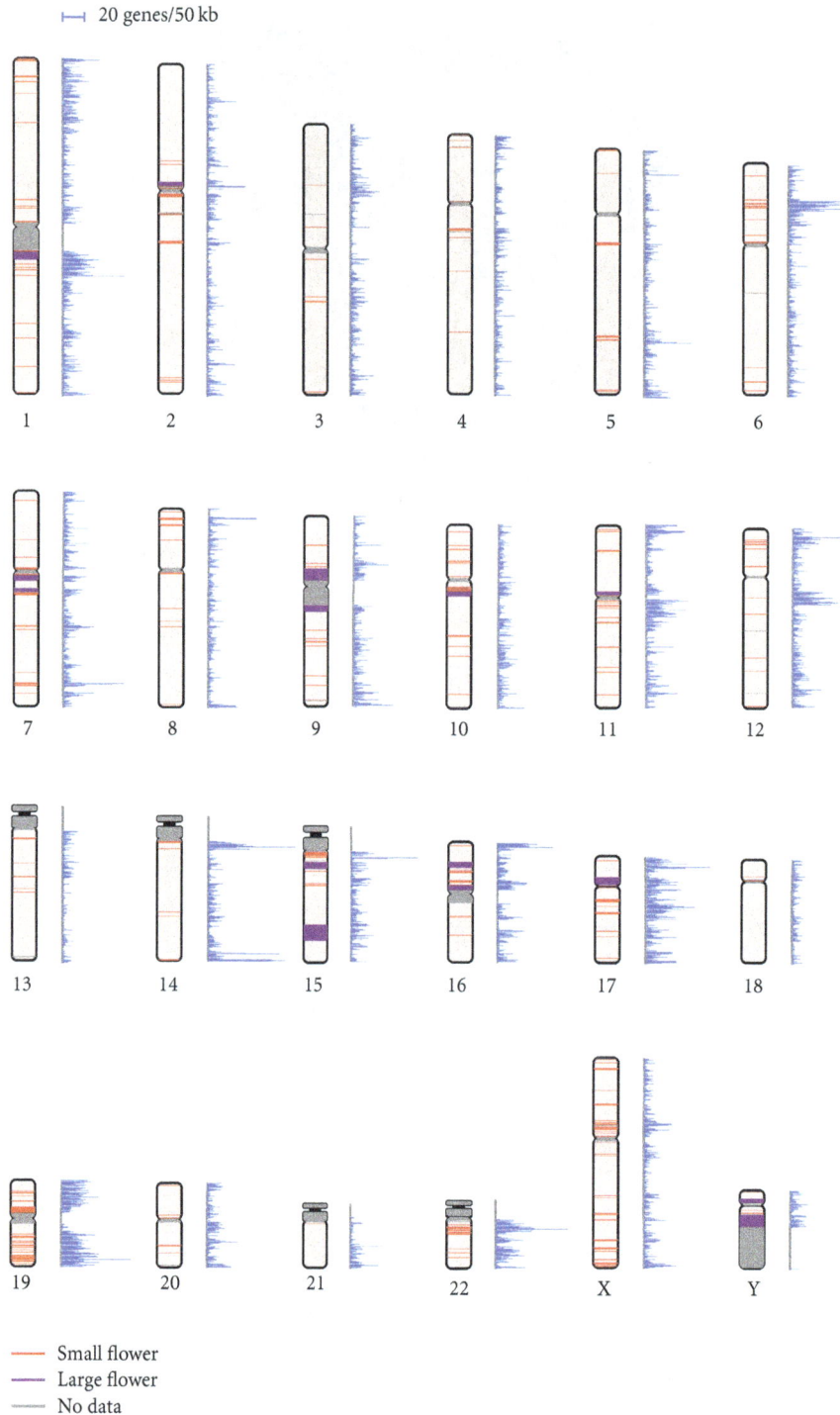

FIGURE 2: Distribution of Flowers and gene density on chromosomes. A bar on a chromosome denotes the position of a Flower and the width of the bar represents the Flower length. The color of the bar represents the patterns of Flowers: small (red) and large (purple) Flowers. Gray bars stand for no genomic sequence data. Mapping of the bars was accomplished using *ColoredChromosomes* [11]. Gene density is represented alongside each chromosome. The scale for the gene density is placed at the top of the figure.

3. Results

3.1. 291 Flowers. We scanned the entire human genome and, based on our definitions, detected 291 Flowers (Figure 2). The Flowers were composed of at least two duplications with some degree of similarity and were more than 50 kb in length (see Section 2). Summary statistics for the 291 Flowers are shown in Table 1 (detailed information is presented in Table S1 in Supplementary Material available online at doi:10.1155/2012/917678).

The sum of the 291 Flower lengths was approximately 179 Mb, which comprised 6% of the genome. The mean length of a Flower was 615 kb and the median length was 210 kb (Table 1). The mean number of copy units in a Flower was approximately three and the mean length of a copy unit was 50 kb. Most Flowers (87%) were shorter than 1 Mb, and only 5% of Flowers exceeded 3 Mb (Figure S1). We categorized Flowers into two groups to distinguish exceptionally large Flowers and then fixed the 5% cutoff point (at 3 Mb) in the Flower length distribution (large Flowers (≥3 Mb, 15 Flowers) and small Flowers (<3 Mb, 276 Flowers). Compared with small Flowers, large Flowers were composed of more complicated structures. The mean number of copy units was 7.3, more than twice that of small Flowers (3.1). Interestingly, 10 of the 15 large Flowers were located in pericentromeric regions or near heterochromatin regions (Figure 2). This finding is consistent with the previous observation of numerous segmental duplications in pericentromeric regions [8].

The largest Flower (11.7 Mb), however, was located on a euchromatic region on chromosome 15. The structure of the Flower was extremely complicated, with many duplication units, as represented in Figure S2. A part of this region is likely responsible for a recurrent microdeletion syndrome [13]. Four unrelated patients with this syndrome shared breakpoints for de novo 1.7 to 3.9 Mb deletions in the region. The deletions could be the result of recurrent nonallelic homologous recombination between highly similar duplicates. Additionally, a polymorphic 1.2-Mb inversion has also been mapped to the same region [14]. Even though such dynamic rearrangements occur, this region contains 46 protein-coding genes.

Two classes of Flowers might be different types of genomic structures in mechanisms for generating or maintaining Flowers. Large Flowers were usually located in pericentromeric regions, and small Flowers were located in euchromatic regions. Since Large Flowers showed a more complicated structure than small Flowers, they have likely experienced more dynamic rearrangements compared to small Flowers.

3.2. Flower Distribution and Gene Density.

Flowers were distributed nonuniformly over the human genome (Figure 2 and Table 2). The pattern of distribution was different not only in the location on a chromosome, but also between the 24 chromosomes. The arithmetic mean of Flower density (i.e., the number of Flowers per 1 Mb) for a chromosome was 0.12 ± 0.10 (Table 2). Most chromosomes showed a relatively similar density to one another (Table 2), with the exception of chromosome 19. The gene density of chromosome 19 (0.47) was the largest, and about 20% of the length of the chromosome was occupied by 26 Flowers.

Previous studies suggested that genomic structures, such as segmental duplications or CNVs, are often observed in gene-rich regions [15–18]. A possible reason for the high Flower density in chromosome 19 might be related to the fact that the density of genes in the chromosome is high. In fact, the association between Flowers and their gene content was also observed. The gene density of each chromosome seemed to be associated with the location of Flowers (Figure 2). To evaluate this association, we compared the gene density in 277 Flowers (D length < 1 Mb) with the gene density in randomly selected regions from the genome (Table 3). The mean gene density in a Flower was 0.61 per 10 kb, more than five times the value for the randomly selected regions (i.e., 0.12; $P < 10^{-5}$, Z-test). Also, the density of protein-coding genes, 0.34, was 3.4 times larger than that of the randomly selected regions (0.09; $P < 10^{-5}$, Z-test). The gene density of 14 large Flowers (D length > 1 Mb) was 0.35, larger than that of the randomly selected regions (data not shown). Therefore, we concluded that Flowers are generich relative to other regions of the genome.

3.3. Flower Genes.

Among all Flowers, 82% included at least one gene. We found 2,844 genes in these Flowers, and these genes comprised 8% of the human genes annotated in NCBI build 36. Among these, there were 1,417 protein-coding genes (50%), 1,085 pseudogenes (38%), and 116 RNA genes (4%) (Table S2). The remaining 226 (8%) were "unknown" or "other" categories in the NCBI annotation. To examine the functions of the genes in Flowers, we classified the genes according to the GO categories. To detect biases in the functions of these genes compared to the entire set of human genes, we compared the observed gene number of Flowers in each given category with the expected number. The expected gene numbers were calculated based on the proportion of human genes in each category (Table S3). Table 4 shows GO categories that were significantly overrepresented in Flowers ($P < 10^{-4}$, hypergeometric test).

The most overrepresented category was the alpha-amylase multigene family (AMY), which is located on chromosome 1. The family consists of AMY1 and AMY2. Amylases catalyze the breakdown of starch and glycogen into disaccharides or trisaccharides, and the genes are highly expressed in the salivary gland and pancreas. AMY1 showed extensive CNVs that are known to be present among several ethnic populations. The copy number, however, did not depend on the geographic distribution of populations, but instead was associated with differences in diets. Populations with high-starch diets have a larger number of AMY1 gene copies than those with low-starch diets [19, 20]. The number of AMY1 gene copies correlates with the amount of AMY1 proteins in saliva [20].

The second most frequent category in Table 4 was glucuronosyltransferase activity, which includes the UDP glucuronosyltransferase (UGT) multigene family. Members of the UGT gene family are divided into two subfamilies, UGT1 and UGT2. They are located on 2q37 and 4q13-13.2 and in one and two Flowers, respectively. UGTs encode enzymes that catabolize small lipophilic molecules, such as steroids, bilirubins, hormones, drugs, environmental toxicants, and carcinogens, into water-soluble glucuronides [21]. A mature UGT1 mRNA is composed of five exons. However, depending on the substrate, several distinct mRNAs are observed. Interestingly, this variety in mRNAs is caused by alternative splicing. The region for the UGT1 subfamily on chromosome 2 encodes 13 sets of exon 1 and a single set of exon 2 to exon 5. Splicing of each variable exon (exon 1) to the four constant

TABLE 2: Statistics of Flowers in each chromosome.

Chr.	Flower number	Flower number /1Mb	Flower length[a] (Kb)	Copy unit length[a] (Kb)	Effective number of copy units[a]	Min. identity[a] (%)	Max. identity[a] (%)	Gene number[a,b]	Pseudogene number/Gene number [a,b]
1	27	0.12	472	80	3.3	89.8	98.7	12	0.3
2	10	0.04	1,110	102	2.8	86.5	98.8	19	0.3
3	6	0.03	271	17	2.3	88.0	96.4	4	0.7
4	10	0.05	346	47	3.5	88.2	97.6	6	0.4
5	9	0.05	556	39	5.7	88.0	96.5	9	0.2
6	15	0.09	238	39	2.5	88.1	97.2	6	0.3
7	17	0.11	789	54	3.0	88.8	97.6	12	0.4
8	11	0.08	373	40	3.4	87.6	97.2	14	0.3
9	12	0.10	1,308	129	3.2	88.3	97.2	17	0.3
10	14	0.11	707	74	2.7	90.3	97.5	8	0.2
11	18	0.14	405	26	4.0	87.5	96.8	8	0.5
12	12	0.09	223	18	2.4	88.7	96.4	2	0.1
13	6	0.06	264	24	2.1	90.2	97.3	2	0.4
14	6	0.07	449	75	2.6	86.8	99.1	20	0.5
15	7	0.09	2,911	141	3.8	87.1	99.9	27	0.4
16	10	0.13	1,082	85	3.0	89.1	99.3	13	0.4
17	11	0.14	1,031	73	2.7	91.4	99.2	12	0.3
18	3	0.04	268	16	4.1	87.5	99.0	1	0.0
19	26	0.47*	422	23	3.9	86.5	97.0	7	0.2
20	5	0.08	374	35	2.7	86.1	98.5	2	0.0
21	1	0.03	94	5	2.0	95.6	98.2	1	1.0
22	10	0.29*	351	31	2.4	90.3	98.3	11	0.3
X	39	0.26	270	28	3.8	89.1	99.2	5	0.2
Y	6	0.23	2,319	110	6.1	88.6	99.9	30	0.6
Mean	12	0.12	693	52	3.5	88.7	98.0	10	0.3
S.D.	9	0.10	680	37	1.2	2.0	1.1	8	0.2
Mean w/o XY	11	0.11	638	51	3.3	88.7	97.9	10	0.3
S.D. w/o XY	7	0.10	607	37	0.9	2.1	1.0	7	0.2

[a] Mean values of each chromosome, [b] Flower genes.
*$P < 0.05$ by Z-test.

exons (exon 2 to 5) generates diverse functional *UGT1* mRNA. Nine of the 13 first exons encode the specific N-terminal domains, conferring the substrate specificity of the enzyme [22], whereas the remaining four are pseudogenized. Similar to *UGT1*, the region for the protocadherin beta (*PCDHB*) gene family encodes 16 different proteins with variable N-termini. In contrast to *UGT1*, each PCDHB protein is encoded by a single exon, and in total, 16 independent exons exist in this region. *PCDHB* is also in an overrepresented GO category in Flowers (see Table 4).

Except for *AMY* and *PCDHB*, most overrepresented GO categories in Table 4 are related to immune responses (multigene families of immunoglobulin, major histocompatibility complex (*MHC*), and defensin) and detoxifications (multigene families of *UGT*, glutathione S-transferase, and cytochrome P450). This is consistent with previous results for segmental duplications and CNVs showing that these regions are rich in genes that can interact with their environment [23, 24].

3.4. Gene Conversion within a Flower. We calculated the minimum and maximum identities between copies of a gene in a Flower to investigate the age of Flowers. We can estimate the time of occurring of the oldest and most recent duplications, from the minimum and maximum identities, respectively. For example, the lowest minimum identity was 77.7% (Table 1), suggesting that the duplication likely occurred prior to the placental mammal radiation. In the distributions of minimum and maximum identities, the minimum identity was evenly distributed from low to high identity (78% ~ 100%), whereas the maximum identity was concentrated at 100% identity (Figure S3). One hundred Flowers (34%) had exactly identical copies and 55 Flowers (19%) had almost identical copies (i.e., 99% to <100%).

TABLE 3: Comparison of gene density of Flowers with randomly selected regions. Except Flowers with D length ≥ 1 Mb, the number of Flower genes was compared to that of 1,000 randomly selected regions on the human genome.

		Gene number in 277 Flowers	Gene number/10 Kb	Protein-coding gene number	Protein-coding gene number/10 kb
Flower		1,841	0.61*	1,030	0.34*
Randomly selected region ($n = 1,000$)	Mean	371	0.12	282	0.09
	S.D.	39	0.01	30	0.01
	Min.	272	0.09	204	0.07
	Max.	561	0.19	439	0.15

$^*P = 0$, Z-test.

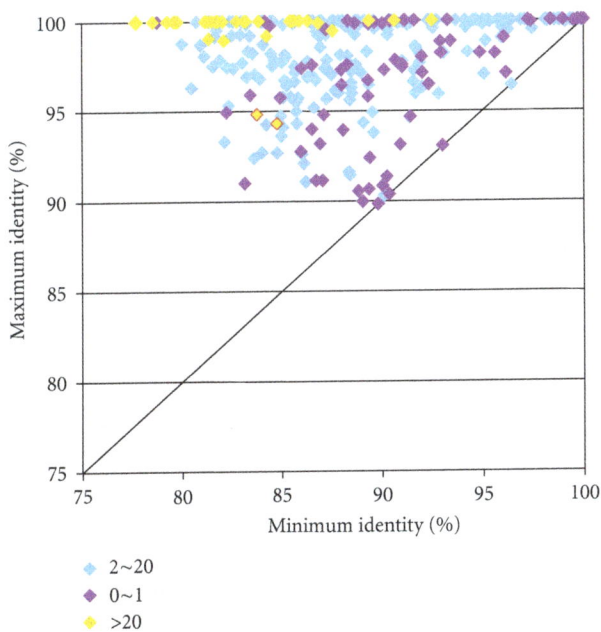

FIGURE 3: Scatter plot of maximum and minimum identity. The X and Y axes show the maximum and minimum percent identity of blast hits in a Flower. A dot represents one Flower. The 291 Flowers were classified into three groups, based on the number of Flower genes: $0 \sim 1$, $2 \sim 20$, and more than 20, are colored purple, blue, and yellow, respectively. The two yellow dots outlined in red are two exceptional Flowers with low maximum identity.

When we considered all the BLAST hits in Flowers, the number of hits with $\geq 99\%$ identity was larger than that with $<99\%$ identity (Figure S4).

Interestingly, the maximum identity did not depend upon the values of the minimum identities. Many old Flowers, which had rather low minimum identities, showed 100% maximum identity (Figure 3). In addition, we found that the maximum identity depended upon the number of genes included in a Flower. Flowers containing up to 20 genes showed variety in maximum identities, whereas Flowers containing more than 20 genes showed almost 100% maximum identity (Figure 3). This suggests that the number of gene copies is associated with the genetic distance between paralogs within a Flower.

This observation suggested one hypothesis to explain the association of gene copy numbers with the large number of highly similar copies in Flowers: gene conversions occurred between gene copies, generating identical sequences within a Flower. The conversion is likely to be enriched in the case of a large number of gene copies. To examine the possibility of gene conversions within a Flower, we performed a gene conversion detection test, using the CHAP package [12]. This method identifies the orthologous and paralogous sequences within input sequences and detects regions showing significantly higher identity between the paralogs rather than between the orthologs. This pipeline was developed to analyze gene cluster regions and was useful for the detection of gene conversions within Flowers. We applied this method and obtained results for 189 Flowers. There were some technical difficulties in producing results for all Flowers (see Section 2).

Among the 189 Flowers, gene conversion events that occurred in the human lineage were detected in 157 Flowers (83% of the tested Flowers). In the entire duplicated regions in the 157 Flowers, the average number of gene conversion events was 21, which corresponds to 20% of the duplicated regions (Table 5). Furthermore, in 798 genes of the 189 tested Flowers, 67% (533 genes) had experienced conversion. On average, 49% of the converted regions in a Flower had overlapped gene regions. In the 533 genes with conversions, the average number of conversions per gene was 6, and the proportion of converted region in the gene region was 21% (Table 5). These results indicate that gene conversions have been occurring frequently and recently within a Flower, especially in the gene regions, but also in the intergenic regions. Gene conversion events could play an important role in the evolution of Flower genes.

3.5. Simulations of Gene Conversions. To understand the effects of gene conversion on the evolution of Flower genes, we performed simulation studies (for details, see Section 2). The null hypothesis was that the fixation time of a pseudogene at a locus does not depend upon the rates of gene conversion. We measured the fixation time of a pseudogene in a population under neutrality or negative selection against fixation of pseudogenes in a diploid individual population. We assumed that gene conversion between alleles at two loci can take place in *any* of three different schemes: allelic-trans, allelic-cis, and nonallelic-trans (Figure 4). In all cases,

TABLE 4: Functions of genes in Flowers. For a GO category, we tested significance of frequency of Flower genes compared to total number of the human genes. This table represents GO categories showing $P < 10^{-4}$, the observed number of Flower genes ≥ 3, and the ratio of observation to expectation ≥ 3.

GO: ID	Detail	Catalog	Observation[a]	Obs/Exp[b]	Multigene families on Flowers
GO: 0004556	Alpha-amylase activity	Function	3	21.1	Amylase alpha
GO: 0015020	Glucuronosyltransferase activity	Function	15	15.0	UDP glucuronosyltransferase
GO: 0019864	IgG binding	Function	7	14.7	Fc fragment of IgG
GO: 0016339	Calcium-dependent cell-cell adhesion	Process	12	10.5	Protocadherin beta
GO: 0003823	Antigen binding	Function	19	10.0	Immunoglobulin, leukocyte immunoglobulin-like receptor, killer cell immunoglobulin-like receptor
GO: 0004364	Glutathione transferase activity	Function	9	9.0	Glutathione S-transferase
GO: 0019882	Antigen processing and presentation	Process	13	8.6	Major histocompatibility complex, class I, MHC class I polypeptide-related sequence, retinoic acid early transcript, UL16-binding protein, C-type lectin domain family
GO: 0006805	Xenobiotic metabolic process	Process	10	8.4	UDP glucuronosyltransferase, defensin, alpha, aldo-keto reductase family
GO: 0006952	Defense response	Process	17	4.9	Interferon, alpha, leukocyte immunoglobulin-like receptor, pregnancy-specific beta-1-glycoprotein, major histocompatibility complex, class I, SP140 nuclear body protein
GO: 0032312	Regulation of ARF GTPase activity	Process	7	4.8	ArfGAP with GTPase domain, centaurin, gamma-like family
GO: 0020037	Heme binding	Function	24	4.4	Cytochrome P450, nitric oxide synthase, HECT domain and RLD
GO: 0007565	Female pregnancy	Process	10	4.3	Pregnancy-specific beta-1-glycoprotein
GO: 0005792	Microsome	Component	33	3.9	UDP glucuronosyltransferase, cytochrome P450, flavin-containing monooxygenase, hydroxy-delta-5-steroid dehydrogenase
GO: 0042742	Defense response to bacterium	Process	12	3.2	Defensin, alpha, defensin, beta, MHC class I polypeptide-related sequence
GO: 0009615	Response to virus	Process	12	3.1	Defensin, alpha, interferon, alpha, chemokine (C-C motif) ligand, leukocyte immunoglobulin-like receptor

[a] The observed number of Flower genes for a GO category.
[b] The ratio of the observed number of Flower genes to the expected number of genes.

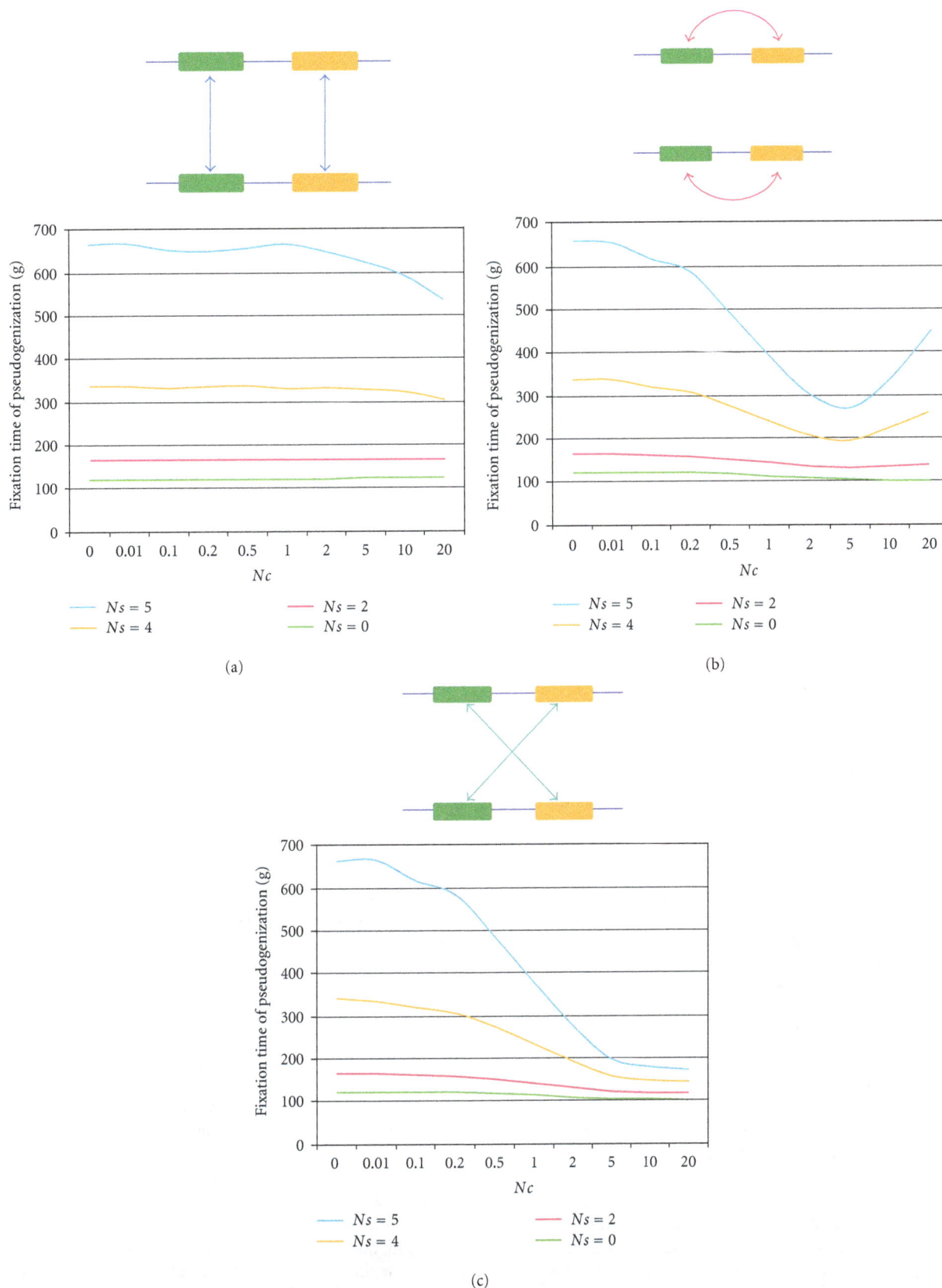

FIGURE 4: Fixation time of pseudogenes. We tested three types of gene conversion (double-headed arrows): (a) allelic-*trans*, (b) *cis*, and (c) nonallelic-*trans* gene conversion. The three graphs show that the pseudogene fixation time (in units of a generation) in a population ($2N = 100$) depends on the gene conversion rate ($Nc = 0–20$). The mutation rate is assumed to be constant ($2N\mu = 1$) for the three cases shown here.

TABLE 5: Statistics of detected gene conversion events within a Flower. The average and standard deviation of several values in the 157 Flowers and 533 genes, experiencing gene conversion events in the human lineage.

157 Flowers	No. of events /Flower	Prop. of converted /D region	Prop. of gene regions in the converted region	No. of genes /Flower	No. of converted genes/Flower
Average	21	0.25	0.49	5	3
S.D.	29	0.31	0.35	5	3
533 Genes	No. of events/gene		Prop. of converted/gene		
Average	6		0.21		
S.D.	20		0.30		

under neutrality, the fixation time of a pseudogene did not appear to depend upon the rate of gene conversion ($Ns = 0$, Figure 4). Introduction of the purifying selection ($Ns \geq 2$) generally increased the fixation time with a constant Nc, and if Nc increases the fixation time gradually decreased. Nevertheless, in the case of cis gene conversion, when the conversion rate was lower than the mutation rate ($N\mu = 0.5$ and $Nc \leq 5$), the fixation time decreased when Nc increased, but when $Nc \geq 5$, the fixation time increased. The life span of a multigene family in a Flower seems to be extended by gene conversion (in cis), probably because frequent gene conversion enhances homogenization of functional paralogs and thus counteracts pseudogenization. Frequent conversion also helps to convert pseudogenes to functional genes.

4. Discussion

In the human genome, we detected 291 regions with particular genomic structures that we termed "Flowers." Based on our characterization of the Flowers, we can draw two main conclusions. First, genes appear to be enriched in Flowers. Second, there is evidence of frequent gene conversion between duplicates within Flowers. These two findings may sometimes be contradictory, because gene conversion from a pseudogene has the potential to be deleterious for functional genes. Genomic rearrangements associated with several inherited diseases could be caused by this kind of gene conversion from paralogous pseudogenes [25]. In our observations, 30% of Flower genes were pseudogenes (Table 2 and Table S1).

The results of this study, on the other hand, showed that frequent gene conversion could play an important role in the preservation of functionality of multigene families. In Figure 3, the 32 Flowers that had over 20 genes showed higher maximum identity than Flowers with fewer genes. Flowers containing a large copy number of genes have likely experienced more frequent gene conversions than other Flowers, and the gene copies had been homogenized. However, there were two exceptional Flowers (yellow dots outlined in red in Figure 3), which contained a large number of genes, but showed a maximum identity of less than 95%. These exceptions included immunoglobulin lambda variable (*IGLV*) and major histocompatibility complex class I (*MHC*) genes. These families need to maintain genetic diversity,

which is important for their functions. This result implies that the *MHC* and immunoglobulin gene families have evolved under purifying selection against homogenization between members of a multigene family [26].

A similar example of purifying selection was observed in melanoma antigen family A (*MAGE-A*) in a Flower on the X chromosome [27]. Members of this multigene family are expressed in cancer cells, encode epitopes recognized by *MHC*, and are associated with cancer immunity [28]. The genetic diversity between the *MAGE-A3* and *A6* genes has likely been preserved by purifying selection against homogenization to maintain association with a particular MHC molecule.

In this study, we characterized several biologically significant features of genomic structures, called Flowers, in the human genome. First, large-sized and complex Flowers were usually located in pericentromeric regions. In primates, the genomic structures of pericentromeric regions resulted from numerous segmental duplications of euchromatin regions [8, 29]. Recent reports suggest that complex rearrangements played a role in the deterioration of functional genes during the generation of novel centromeres [30, 31].

Second, we suggested an evolutionary mechanism for the preservation of multigene families within Flowers. Simulation studies showed that frequent gene conversion, probably at a higher rate than the mutation rate, could extend the age (lifespan) of a gene family. Negative selection against pseudogenization was a likely driving force in the maintenance of a multigene family.

Third, we provided evidence for the interrelationship of Flowers with the functions of multigene families. Multigene families related to immune responses and detoxification were overrepresented in the Flowers. These gene families have evolved under purifying selection opposing gene conversion to maintain the genetic variety between paralogous genes. Frequent genomic rearrangements in Flowers could drive the duplication of genes and increase the number of genes. This evolutionary mode was likely to be more favorable for multigene families related to immune responses and detoxification, which need to adapt to environmental changes.

In conclusion, a Flower has an important role in the evolution of multigene families. Future studies will be extended to the genomes of other organisms to further understand the evolution of multigene families contained in genomic structures.

Acknowledgments

The authors thank Dr. Masafumi Nozawa for his critical comments on an early version of this paper. This work was supported in part by a Grant (no. 17018032) from the Ministry of Education, Culture, Sports, Science and Technology (MEXT) of Japan.

References

[1] L. Huminiecki and K. H. Wolfe, "Divergence of spatial gene expression profiles following species-specific gene duplications in human and mouse," *Genome Research A*, vol. 14, no. 10, pp. 1870–1879, 2004.

[2] R. Blekhman, A. Oshlack, and Y. Gilad, "Segmental duplications contribute to gene expression differences between humans and chimpanzees," *Genetics*, vol. 182, no. 2, pp. 627–630, 2009.

[3] H. C. Mefford and E. E. Eichler, "Duplication hotspots, rare genomic disorders, and common disease," *Current Opinion in Genetics and Development*, vol. 19, no. 3, pp. 196–204, 2009.

[4] F. J. Ayala and M. Coluzzi, "Chromosome speciation: humans, Drosophila, and mosquitoes," *Proceedings of the National Academy of Sciences of the United States of America*, vol. 102, no. 1, pp. 6535–6542, 2005.

[5] H. Skaletsky, T. Kuroda-Kawaguchi, P. J. Minx et al., "The male-specific region of the human Y chromosome is a mosaic of discrete sequence classes," *Nature*, vol. 423, no. 6942, pp. 825–837, 2003.

[6] B. K. Bhowmick, Y. Satta, and N. Takahata, "The origin and evolution of human ampliconic gene families and ampliconic structure," *Genome Research*, vol. 17, no. 4, pp. 441–450, 2007.

[7] S. Rozen, H. Skaletsky, J. D. Marszalek et al., "Abundant gene conversion between arms of palindromes in human and ape Y chromosomes," *Nature*, vol. 423, no. 6942, pp. 873–876, 2003.

[8] X. She, J. E. Horvath, Z. Jiang et al., "The structure and evolution of centromeric transition regions within the human genome," *Nature*, vol. 430, no. 7002, pp. 857–864, 2004.

[9] E. V. Linardopoulou, E. M. Williams, Y. Fan, C. Friedman, J. M. Young, and B. J. Trask, "Human subtelomeres are hot spots of interchromosomal recombination and segmental duplication," *Nature*, vol. 437, no. 7055, pp. 94–100, 2005.

[10] Y. Ohtsubo, W. Ikeda-Ohtsubo, Y. Nagata, and M. Tsuda, "GenomeMatcher: a graphical user interface for DNA sequence comparison," *BMC Bioinformatics*, vol. 9, article 376, 2008.

[11] S. Böhringer, R. Gödde, D. Böhringer, T. Schulte, and J. T. Epplen, "A software package for drawing ideograms automatically," *Online Journal of Bioinformatics*, vol. 1, pp. 51–61, 2002.

[12] G. Song, C.-H. Hsu, C. Riemer et al., "Conversion events in gene clusters," *BMC Evolutionary Biology*, vol. 11, article 226, 2011.

[13] A. J. Sharp, R. R. Selzer, J. A. Veltman et al., "Characterization of a recurrent 15q24 microdeletion syndrome," *Human Molecular Genetics*, vol. 16, no. 5, pp. 567–572, 2007.

[14] F. Antonacci, J. M. Kidd, T. Marques-Bonet et al., "Characterization of six human disease-associated inversion polymorphisms," *Human Molecular Genetics*, vol. 18, no. 14, pp. 2555–2566, 2009.

[15] L. Zhang, H. H. S. Lu, W. Y. Chung, J. Yang, and W. H. Li, "Patterns of segmental duplication in the human genome," *Molecular Biology and Evolution*, vol. 22, no. 1, pp. 135–141, 2005.

[16] X. She, G. Liu, M. Ventura et al., "A preliminary comparative analysis of primate segmental duplications shows elevated substitution rates and a great-ape expansion of intrachromosomal duplications," *Genome Research*, vol. 16, no. 5, pp. 576–583, 2006.

[17] D. Q. Nguyen, C. Webber, J. Hehir-Kwa, R. Pfundt, J. Veltman, and C. P. Ponting, "Reduced purifying selection prevails over positive selection in human copy number variant evolution," *Genome Research*, vol. 18, no. 11, pp. 1711–1723, 2008.

[18] D. Q. Nguyen, C. Webber, and C. P. Ponting, "Bias of selection on human copy-number variants," *Plos Genetics*, vol. 2, no. 2, p. e20, 2006.

[19] P. C. Groot, W. H. Mager, and R. R. Frants, "Interpretation of polymorphic DNA patterns in the human α-amylase multigene family," *Genomics*, vol. 10, no. 3, pp. 779–785, 1991.

[20] G. H. Perry, N. J. Dominy, K. G. Claw et al., "Diet and the evolution of human amylase gene copy number variation," *Nature Genetics*, vol. 39, no. 10, pp. 1256–1260, 2007.

[21] R. H. Tukey and C. P. Strassburg, "Human UDP-glucuronosyltransferases: metabolism, expression, and disease," *Annual Review of Pharmacology and Toxicology*, vol. 40, pp. 581–616, 2000.

[22] Q.-H. Gong, J. W. Cho, T. Huang et al., "Thirteen UDPglucuronosyltransferase genes are encoded at the human UGT1 gene complex locus," *Pharmacogenetics*, vol. 11, no. 4, pp. 357–368, 2001.

[23] J. A. Bailey, Z. Gu, R. A. Clark et al., "Recent segmental duplications in the human genome," *Science*, vol. 297, no. 5583, pp. 1003–1007, 2002.

[24] G. M. Cooper, D. A. Nickerson, and E. E. Eichler, "Mutational and selective effects on copy-number variants in the human genome," *Nature Genetics*, vol. 39, no. 1, pp. S22–S29, 2007.

[25] J. M. Chen, D. N. Cooper, N. Chuzhanova, C. Férec, and G. P. Patrinos, "Gene conversion: mechanisms, evolution and human disease," *Nature Reviews Genetics*, vol. 8, no. 10, pp. 762–775, 2007.

[26] M. Nei and A. P. Rooney, "Concerted and birth-and-death evolution of multigene families," *Annual Review of Genetics*, vol. 39, pp. 121–152, 2005.

[27] Y. Katsura and Y. Satta, "Evolutionary history of the cancer immunity antigen MAGE gene family," *Plos ONE*, vol. 6, no. 6, Article ID e20365, 2011.

[28] P. Van der Bruggen, C. Traversari, P. Chomez et al., "A gene encoding an antigen recognized by cytolytic T lymphocytes on a human melanoma," *Science*, vol. 254, no. 5038, pp. 1643–1647, 1991.

[29] J. E. Horvath, C. L. Gulden, R. U. Vallente et al., "Punctuated duplication seeding events during the evolution of human chromosome 2p11," *Genome Research*, vol. 15, no. 7, pp. 914–927, 2005.

[30] M. Ventura, F. Antonacci, M. F. Cardone et al., "Evolutionary formation of new centromeres in macaque," *Science*, vol. 316, no. 5822, pp. 243–246, 2007.

[31] M. Lomiento, Z. Jiang, P. D'Addabbo, E. E. Eichler, and M. Rocchi, "Evolutionary-new centromeres preferentially emerge within gene deserts," *Genome Biology*, vol. 9, no. 12, article R173, 2008.

21

Drosophila melanogaster Selection for Survival of *Bacillus cereus* Infection: Life History Trait Indirect Responses

Junjie Ma,[1] Andrew K. Benson,[1] Stephen D. Kachman,[2] Zhen Hu,[3] and Lawrence G. Harshman[3]

[1] Department of Food Science and Technology, University of Nebraska Lincoln, Lincoln, NE 68583, USA
[2] Department of Statistics, University of Nebraska Lincoln, Lincoln, NE 68583, USA
[3] School of Biological Sciences, University of Nebraska Lincoln, Lincoln, NE 68588, USA

Correspondence should be addressed to Lawrence G. Harshman, lharshman1@unl.edu

Academic Editor: Alberto Civetta

To study evolved resistance/tolerance in an insect model, we carried out an experimental evolution study using *D. melanogaster* and the opportunistic pathogen *B. cereus* as the agent of selection. The selected lines evolved a 3.0- to 3.3-log increase in the concentration of spores required for 50% mortality after 18–24 generations of selection. In the absence of any treatment, selected lines evolved an increase in egg production and delayed development time. The latter response could be interpreted as a cost of evolution. Alternatively, delayed development might have been a target of selection resulting in increased adult fat body function including production of antimicrobial peptides, and, incidentally, yolk production for oocytes and eggs. When treated with autoclaved spores, the egg production difference between selected and control lines was abolished, and this response was consistent with the hypothesis of a cost of an induced immune response. Treatment with autoclaved spores also reduced life span in some cases and elicited early-age mortality in the selected and wound-control lines both of which were consistent with the hypothesis of a cost associated with induction of immune responses. In general, assays on egg production yielded key outcomes including the negative effect of autoclaved spores on egg production.

1. Introduction

Genetic selection in the laboratory provides a powerful tool for evolutionary analysis of complex traits [1]. It has been used to study many phenomena at different levels of biological organization including life histories, physiology, demography and population dynamics, behavior, form, sex, whole-genome evolution, altruism, and speciation [2]. Selection results in amplification of genetic differences between selected and control lines which is the basis of phenotypic differentiation. Often, correlated (indirect) responses to selection are of particular interest in these experiments as they can suggest tradeoffs between traits. For example, selection for increased *D. melanogaster* life span and late-age reproduction resulted in decreased early-age reproduction [3, 4]. The nature of tradeoffs between traits is an important topic in life history evolution [5].

In the present study, the insect model *D. melanogaster* has been used in selection experiments for increased survival after bacterial infection. A previous study of responses in a laboratory selection experiment using *Pseudomonas aeruginosa* has examined the impact of *D. melanogaster* resistance on life history traits [6]. This study showed considerable costs in life span and larval survival as correlated responses to selection. While fly survival increased from 15% to 70% within ten generations in the selected lines, adult and larval viability decreased markedly relative to the control lines. In this selection experiment, microarray data indicated that a greater number of cellular immunity genes changed expression in the selected lines compared to the number of humoral immunity genes that changed, suggesting the relative importance of cellular immunity for resistance to *P. aeruginosa*.

A series of laboratory selection studies has also been conducted in which *Drosophila* evolved resistance against parasitoids [7, 8]. After 5 generations of selection the encapsulation frequency, important against parasitoids, increased from 5% to 60% in response to *A. tabida* and 0.5% to 45% in response to *L. boulardi*. Increased resistance to parasitoids was accompanied by a correlated evolutionary response in a number of traits, including doubling of the number of circulating haemocytes, smaller adult size, lower fecundity, reduced larval competitive ability, and increased pupal susceptibility [9, 10].

One area of tremendous interest is the selection response of host organisms to infectious or zoonotic diseases. These diseases have significant impact on human and animal health, and understanding the evolutionary underpinnings of responses in humans may provide keys to alternative methods of prevention or intervention [11–16].

To further understand the evolutionary implications of infectious disease resistance, and the potential for novel interventions, we have exploited *D. melanogaster* to study the effects of selection for resistance/tolerance to a spore-forming bacterial species (*Bacillus cereus*) which is closely related to the pathogenic spore-forming bioterrorism agent *Bacillus anthracis*. Because the spore is the most frequently encountered form of this organism (natural or otherwise), we used the spore form as a basis for selection. A strong response to selection for resistance to infection by spores was obtained, observed by a 3.3-log change in the number of spores required for approximately 50% mortality, within 24 generations. Here, we now demonstrate that life history traits were also affected as a consequence of selection and introduction of autoclaved spores. In wound-control and selected lines, exposure to autoclaved spores decreased life span. Selection was strongly positively correlated with egg production. When treated with autoclaved (dead) spores, the large difference between selected and control lines was abolished suggesting a cost of activating an immune response. There was a difference between untreated selected and control lines in progeny development time; the selected line progeny developed relatively slowly. After a series of matings were conducted to separate female and male effects, it was documented that exposure of selected and wound-control line males to autoclaved (dead) spores resulted in relatively rapid progeny development time. Finally, the selected and wound-control lines evolved heightened early-age mortality in response to the autoclaved spore treatment, which was interpreted as being consistent with the hypothesis of a cost associated with inducing an immune response.

2. Materials and Methods

2.1. Fly Populations. The procedure for establishing the base population as well as subpopulations used for selected and control lines in the present study was described in Schwasinger-Schmidt et al. [17]. Briefly, a large base population was maintained at approximately 10,000 individuals in an overlapping generation regime for approximately two years before being subdivided into 9 subpopulations (lines) for 5 generations in a similar population maintenance regime to that designed for the selected and control lines. Each rearing vial was seeded with 100 eggs to standardize density during the 5 discrete generations before the initiation of the selection experiment. In this artificial laboratory

selection experiment, there were 9 lines which were separate outbred populations that evolved independently: selected lines, wound-control lines (punctured only with sterile H_2O), and no-treatment lines. There were three replicates of each line type; replicates lines were independent populations that were subject to essentially the same conditions. The different sets of lines (selected, wound-control, and no perturbation) are referred to as "line types" in the present study. In *D. melanogaster* experimental evolution studies, "lines" are commonly used as a term to describe selected and control populations (e.g., see Rose 1984 and many subsequent *Drosophila* laboratory selection experiments).

2.2. Selection and Control Lines. Selection was conducted in a specific manner. *B. cereus* spores were used for selection. Spores were introduced into adults of the selected lines using a tungsten needle dipped into spores suspended in H_2O. A concentration of spores that killed approximately 50% of the females and males was determined before the first generation of selection (2×10^6 per mL). Every generation of selection the goal was to attain approximately 50% mortality after introduction of live spores. Each generation of selection, one thousand virgin females and the same number of virgin males were infected for each of three selected lines. After three days at room temperature, the number of survivors was determined for each selected line (S1, S2, and S3). Typically, there were 500 surviving females and 500 males per line. Surviving males and females within each line were counted and randomly placed in bottles to mate at a density of approximately 80 flies per bottles. Flies were kept in bottles for 24 hours to mate and lay eggs. The next day, eggs were collected from the bottles to initiate the following generation. Approximately, one hundred eggs were collected for each vial used to propagate the next generation. The vials with eggs were placed in 18°C for development. A temporal synopsis of the selected and control lines, and treatments, is presented in Figure 1.

There were two types of control lines. For the wound-control lines (CP for control punctured), females and males were punctured with sterile water without spores. Another set of control lines (CN) was used, and, in this case, there was no perturbation (no infection, no puncture).

The number of breeders used for each generation was the same for each set of matched lines of each type (selected, wound-control, and no perturbation). For example, the number of survivors after selection in selected line S2 was matched to the numbers of the control-punctured line CP2 and the no-perturbation line CN2. There were approximately 1000 flies (500 females and 500 males) used as breeders for each line each generation.

2.3. Treatments. All flies used from all lines were subjected to one of three treatments prior to life history assays. The treatments were similar, or the same, as used for selection and in the control lines (Figure 1). The autoclaved spore treatment (AS) was designed to induce an immune response with dead spores; it was analogous to administration of live spores. Live spores could not be used before assays at the level used for selection as they would cause excessive mortality.

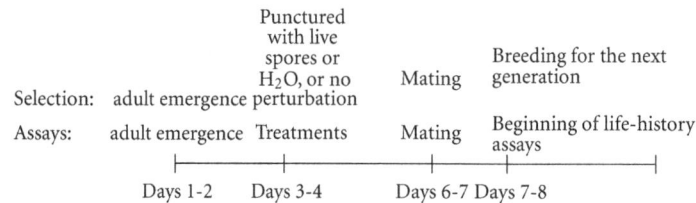

FIGURE 1: Timeline of the selection experiment and life-history assays. The life-history assays were designed to be conducted in parallel with the process used for the selected and control lines. Days were recorded as days posteclosion.

The treatment with sterile H_2O was the same as used for the wound-control lines. Also, no treatment (NON) was the same condition as used for the no perturbation control lines.

2.4. Bacterial Culture and Spore Isolation. *Bacillus cereus* ATCC 10987 was used for spore purification using a step gradient of Renografin [18]. A single colony of *B. cereus* was inoculated in 25 mL of Difco Sporulation Medium (DSM) and incubated at 37°C on a rotary shaker (150 rpm) until mid-log phase. Inoculation of this culture into 2 L of DSM generated a 1 : 10 dilution which was followed by incubation at 37°C on the rotary shaker for 48 hours. The culture was centrifuged for 10 min at 10,000 ×g, and the supernatant gently discarded. The pellet was resuspended in 200 mL of 4°C sterile water followed by the same centrifugation procedure. The pellet was again resuspended in 200 mL of sterile water and stored overnight at 4°C. After repeating the centrifuge-resuspension-centrifuge procedure, greater than 90% of bright-field spores were observed under phase contrast microscopy. The pellet was resuspended in 20% Renografin and the suspension transferred to a 30 mL glass core tube with 15 mL of 50% Renografin. The spore suspension was centrifuged for 30 min at 4°C at 10,000 ×g. All layers containing vegetative cells were removed and the spore pellet retained. The pellet was resuspended in 10 mL of 4°C sterile water in an Oak Ridge tube. The spore suspension was centrifuged for 10 min at 10,000 ×g at 4°C. Trace amounts of Renografin were removed, by three washes with 4°C sterile water as described immediately above. The spore pellet was suspended in 2 mL of 4°C sterile water. The concentration of spores was determined by serial dilution and spread-plating.

Spores were isolated twice during the portion of the selection experiment presented in this study. They were isolated the first time for selection generations 1–11. Also, they were isolated a second time for selection generations 12–24. Each preparation had very similar effects on mortality.

2.5. Life History Assays. Conditions were standardized prior to life history assays. There was no selection for two generations prior to life history assays to minimize the impact of any effects that carry over from generation to generation such as maternal effects. There were nine lines and each was subject to three conditions prior to a life history assay (see Treatments section). Life span and egg production assays were conducted on flies two generations after relaxing selection at generation 18. Development time

was conducted on flies derived from selection generation 24 after two generations of relaxed selection.

The age of flies assayed for life histories was designed to conform to the conditions of the selection experiment (Figure 1). The start of the life-history assays corresponded to the age of the flies used for the start of selection which was 7–9 days old (days posteclosion). Virgin flies (days 1-2) were collected from the breeding vials. At 3–5 days of adult life, the virgin flies were treated (punctured with autoclaved spores, punctured with water, and untreated). After three days, the level of survival was tabulated. Then, males and females were combined and allowed to mate and lay eggs for 24 hours before life-history assays were initiated.

2.6. Life Span. Life span, and all other life-history assays, were conducted with populations of flies held under standard conditions (25°C, 12 : 12 L : D). The cages used for the life span assay were made out of quart-size plastic containers. The lid had mesh inserted for ventilation. There was a grommet in the side of each container with a tube allowing for replacement of used food vials with fresh food vials every three days. A rubber patch was sewn into the opposite side of each container, and it had a slash in it to allow insertion of a Pasteur pipette. The pipette was used to aspirate dead flies from the bottom of the cage allowing them to be removed and recorded every three days. Each cage initially received 30 flies of the same sex that had been allowed to mate for 24 hours prior to the assay. There were four replicate cages for each sex and treatment for all lines. The cages were monitored until all flies were dead.

2.7. Egg and Progeny Production. Egg production was recorded for all lines and treatment combinations. Twenty mated females (males were discarded after the 24-hour mating period) from each line and treatment were placed individually in vials at 25°C. Females were transferred to new vials each day for 49 days at which time almost all of the eggs were produced.

For determination of adult progeny numbers, after eggs were counted, the replicate vials for all lines and treatments were placed at 18°C. Emergent progeny was counted from all vials until all adults emerged.

2.8. Progeny Development Time. The design for development time was more complicated than for life span or fecundity as additional combinations of matings were used. For each line, five different F_1 crosses were employed to parse out

TABLE 1: Mating design for the progeny development time assay.

	Mating for progeny development time
One sex treated with autoclaved spores	A treated females × untreated males
	B treated males × untreated females
One sex types treated with sterile H_2O	C treated females × untreated males
	D treated males × untreated females
No treatment	E untreated females × untreated males

All line types (selected—S, puncture control—CP, and no perturbation—CN) were used for crosses A, B, C, D, and E. The average emergence (eclosion) time of progeny was determined for each cross (A–E).

male and female treatment effects on progeny development time (Table 1). For example, females treated with autoclaved spores were mated to untreated males which allows for assessment of the effect of female treatment on progeny development time. The reciprocal mating allows for assessment of the effect of the autoclaved spore treatment on males with progeny development time as the outcome. These were the first two matings. Similarly, the effect of puncturing females or males with sterile H_2O was evaluated by reciprocal crosses with progeny development time as the outcome. These were the third and fourth matings. The fifth mating allowed for assessment of progeny development time when neither male nor female was treated. There were six replicates of each of the five different F_1 matings. The five matings were used for all lines to investigate progeny development time. Approximately 100 eggs were collected from each cross after the 24-hour mating period and placed in each vial to control larval density. All of the vials (matings, lines) were randomized with respect to the order that matings were initiated. The time of first emergence of adults was t_0 for the progeny development time assay. The counting period of emerged progeny continued well beyond the time when no additional adults eclosed.

2.9. Survival after Administration of Autoclaved Spores. There was a three-day period after treatments, when survival was monitored before the beginning of life-history assays. In Figure 1, this period is shown as occurring after the administration of treatments and before mating (posteclosion ages 3-4 to 6-7). Mortality during this period was considered separately from life span or other life-history assays. This data was tabulated and statistically analyzed.

2.10. Statistical Analysis. The data was analyzed by ANOVA using SAS version 9.3 (SAS, 2009). The data was treated as continuous. A mixed model analysis of variance was used with line types and treatments as fixed effects. Random effects were derived from variation among the three replicate lines of the same type. Variation among lines of the same type was nested within fixed effects for the analysis. For each sex, all lines and treatments were analyzed with one ANOVA for every life history trait. Any significant, or nonsignificant, interaction terms were derived from an analysis using the full model. Residuals were examined by QQ-plots and histograms in order to detect deviations from normality.

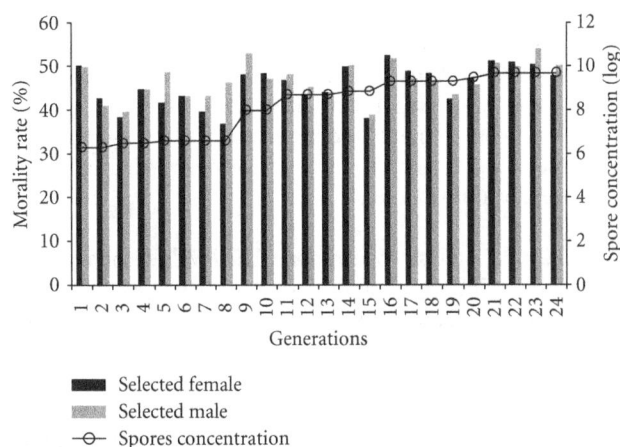

FIGURE 2: Direct response to selection for survival after *B. cereus* live spore infection. After 24 generations of approximately 50% mortality of females and males, the concentration of spores required for this level of mortality increased by 3.3log.

For life span, the average survival time, median survival time, first quartile survival (25% mortality), and time to third quartile survival (75% mortality) of females or males were used for statistical analysis. For development time, the average progeny eclosion time was used for statistical analysis. Total egg and progeny numbers were used for statistical analysis of reproduction. Any statistically significant interactions between treatments and lines were explicitly described in the present study in terms of the pattern of the data.

3. Results

3.1. Statistical Analyses. The statistical analysis was conducted using mixed model ANOVAs. The degrees of freedom and F values are presented in supplemental tables (see S1a—life span, S1b—egg and progeny production, and S1c—progeny developmental time in Supplementary material available online at doi:10.1155/2012/935970). Examination of QQ-plots and histograms for all of the data indicated no major deviations from normality.

3.2. Direct Response to Selection. In order to exert steady selective pressure across multiple generations, the spore concentration used for selection at each generation was increased to attain 50% mortality. As shown in Figure 2, this approach led to a steady incremental response to selection over 24 generations, producing a 3.3-log increase in spore concentration necessary for 50% mortality. The response to selection was almost log-linear. In some cases, there was an increase in survival in the next generation after relatively strong selection in the previous generations (generations 8 to 9 and generations 15 to 16). This might have resulted from selecting to greater degree in generations 8 and 15, hypothetically resulting in a genetically more resistance subset of the population, which could have responded in the next generation by elevated survival in generations 9 and 16, respectively.

For the S1 selected line, a late generation of selection (generation 36) was tested for resistance by introducing the spore concentration (2×10^6 per mL) used for the first generation of selection into 100 females and 100 males. There was no mortality after the standard three-day observation period. This result complements the observation of an incremental response over 24 generations and provides further evidence of a direct response to selection.

3.3. Indirect Responses to Selection. The P values for all of the line types (selected, wound-control, and no perturbation) and treatments (autoclaved spores introduction by puncturing, punctured with H_2O, and no treatment) are presented in Table 1.

For reporting results in the following text, there typically are separate subheadings for untreated samples and samples treated with autoclaved spores. However, the sterile H_2O treatment was sometimes also included. Untreated lines represent the consequences of evolution, whereas lines treated with autoclaved spores hypothetically represent the consequences of eliciting an induced immune response. In the parlance of McKean and Lazaro [19], untreated lines are analogous to "standing defense," and autoclaved spore-treated lines were designed to be analogous to "deployment."

3.4. Life Span

3.4.1. Untreated Lines. The average percent survival (life span) of untreated selected and control-line females was determined. Overall, there were no line effects for females or for males. The survival curves for both sexes when untreated are presented in supplemental figures (S2a and b).

3.4.2. Autoclaved Spore—Treated Lines. The average percent survival (life span) of the autoclaved spore treatment applied to selected and control females is presented in Figure 3(a). The average percent survival (life span) of the autoclaved spore treatment applied to selected and control males is presented in Figure 3(b). The average survival time after treatments is presented in Table 2(a) (females) and Table 2(b) (males). Overall, there was a statistically significant effect of treatments on females ($P < 0.0001$) and males ($P = 0.0003$). Treatment with autoclaved spores reduced mean female life span in the selected lines (27.65 days) and wound-control lines (27.92 days) compared to the no-perturbation lines (31.35 days). The decrease in selected female life span was approximately 10%. Autoclaved spores reduced mean

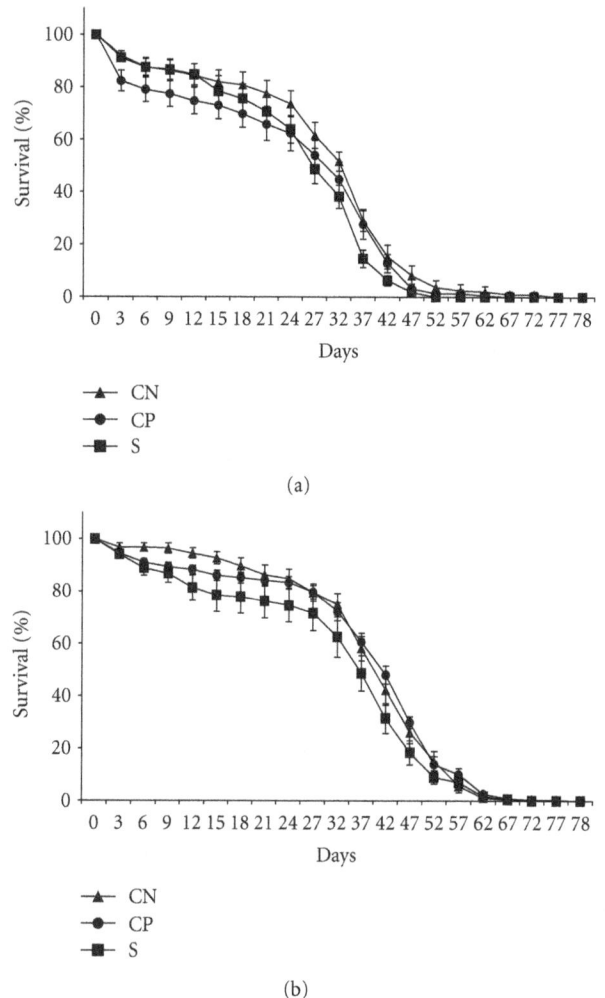

Figure 3: (a) Average percentage survival of adult females from the selected and control lines treated with autoclaved spores. The mean was determined from the replicate lines of the same type: S—selected lines, CP—lines punctured with H_2O (wound control), and CN—no perturbation lines. (b) Average percentage survival of adult males from the selected and control lines treated with autoclaved spores. The mean was determined from the replicate lines of the same type: S—selected lines, CP—lines punctured with H_2O (wound control), and CN—no perturbation lines.

male survival in the selected lines (35.17 days), compared to wound-control (40.10 days) and no-perturbation (40.50) lines. Autoclaved spores decreased the survival of selected males by 14.5%. In neither females nor males was there a statistically significant interaction between treatments and line types. In general, the administration of autoclaved spores reduced life span in the wound-control lines and for selected line males. This observation is consistent with the hypothesis of the cost of an induced immune response (deployment).

3.5. Egg and Progeny Production

3.5.1. Untreated Lines—Egg Production. Total average egg production number is shown in Table 3 for lines and per female per day in Figure 4(a). Overall, for total egg production, there were statistically significant line ($P = 0.0215$),

TABLE 2: *P* values from comparisons (average survival of females, average survival of males, total egg production, total progeny production, and average progeny development time) of treatments within lines of the same type and comparisons of line types for each treatment.

Overall effects	Lines/treatments	Comparisons	P value of female life span	P value of male life span	P value of egg production	P value of progeny production	P value of progeny development time
Lines			0.4444	0.5330	0.0215	0.0961	0.1457
Treatments			<0.0001	0.0003	0.0039	0.0005	0.0051
Lines* treatments			0.1027	0.1603	0.0280	0.1535	0.1790
	Lines						
	No perturbation control (CN)	Treatment[a]	0.1969	0.6687	0.4517	0.0022	0.8001
	Control punctured (CP)	Treatment[a]	<0.0001	0.0027	0.3270	0.1776	0.1773
	Selected (S)	Treatment[a]	0.4262	0.0019	0.0001	0.0354	0.0012
	Treatments						
	AS	Lines[b]	0.6967	0.1663	0.6729	0.0462	
	H$_2$O	Lines[b]	0.2074	0.2502	0.1727	0.0288	
	NON	Lines[b]	0.2717	0.8843	<0.0001	0.9560	
	A[c]	Lines[b]					0.0088
	B[c]	Lines[b]					0.8619
	C[c]	Lines[b]					0.3675
	D[c]	Lines[b]					0.4857
	E[c]	Lines[b]					0.0098

[a]Treatments: punctured with autoclaved spores (AS), punctured with sterile water (H$_2$O), and no treatment (NON).
[b]Line types: selected (S), punctured with H$_2$O each generation (CP), and no perturbation each generation (CN).
[c]A–E were F_1 matings for the progeny development time assay (see Table 1).

TABLE 3: Average survival time of female and male flies, average total number of eggs, and average progeny production: lines and treatments.

Lines	Treatments	Mean (S.E.) female survival	Mean (S.E.) male survival	Mean (S.E.) egg number	Mean (S.E.) progeny number
No perturbation control (CN)	Autoclaved spores	31.35 (1.84)	40.50 (1.63)	577 (45.43)	185 (18.7)
No perturbation control (CN)	H$_2$O	35.37 (0.84)	42.88 (0.64)	639 (47.48)	274 (17.6)
No perturbation control (CN)	NON	35.65 (0.84)	42.22 (1.64)	561 (37.49)	214 (18.1)
Control punctured (CP)	Autoclaved spores	27.92 (2.16)	40.10 (1.19)	636 (46.17)	225 (18.9)
Control punctured (CP)	H$_2$O	37.17 (1.24)	47.65 (1.62)	699 (38.21)	262 (17.3)
Control punctured (CP)	NON	38.20 (1.24)	40.65 (1.83)	737 (50.09)	219 (17.5)
Selected (S)	Autoclaved spores	27.65 (1.68)	35.17 (2.83)	588 (55.27)	161 (17.5)
Selected (S)	H$_2$O	29.37 (3.51)	43.44 (1.93)	763 (45.48)	210 (18.0)
Selected (S)	NON	30.64 (3.10)	41.57 (1.86)	874 (55.66)	222 (17.3)

Treatments: punctured with autoclaved spores (AS), punctured with sterile water (H$_2$O), and no treatment (NON).
Line types: selected (S), punctured with H$_2$O each generation (CP), and no perturbation each generation (CN).

treatment (P = 0.0039), and line by treatment interaction (P = 0.0280) effects (Table 2). A major difference between line types was observed when there was no treatment (P < 0.0001). Selected lines produced a markedly high number of eggs (874), wound-control lines were intermediate (737), and no-perturbation lines produced the lowest number eggs (561) (Table 3). The selected line produced 19% more eggs than the wound-control lines and 56% more eggs than the no-perturbation lines. The wound-control lines produced 31% more eggs than the no-perturbation lines.

3.5.2. Autoclaved Spore Treatment—Egg Production. Egg production for autoclaved spore-treated females (and males) for all line types is presented in Table 3 and Figure 4(b). The effect of treatment with autoclaved spores was to markedly reduce average egg production in the selected lines (588 eggs) and in the wound-control lines (636 eggs) (Table 3). The effect of treatment was statistically significant for the selected lines (P = 0.0001), but not the wound-control lines. There was no reduction of average total egg production in the no-perturbation control lines as a result of treatment

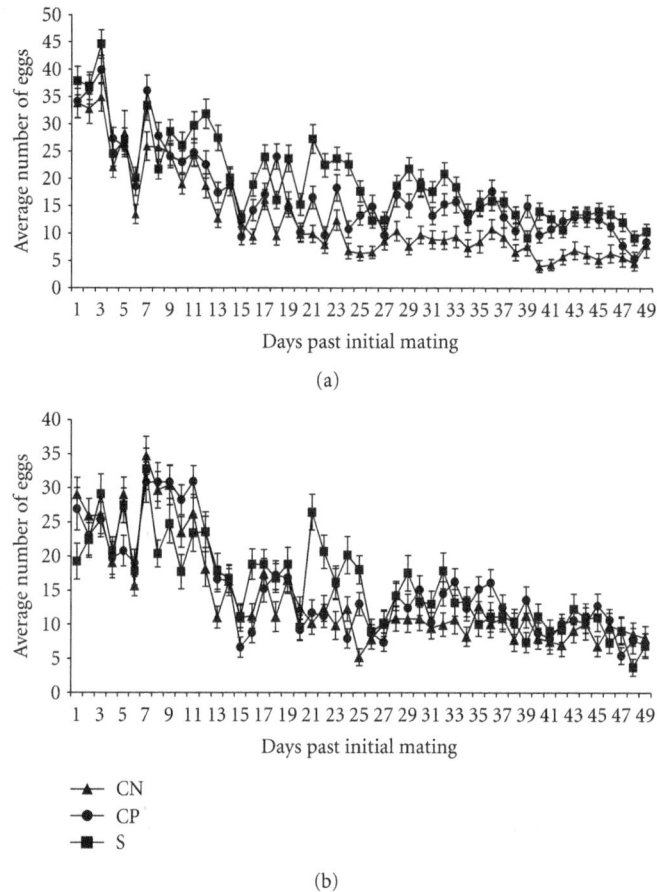

FIGURE 4: (a) Average daily egg production among line types when there was no treatment of adult females and males. The mean was determined from the replicate lines of the same type: S—selected lines, CP—lines punctured with H_2O, and CN—no perturbation lines. (b) Average daily egg production among lines when adult females and males from the same line were treated with autoclaved spores. The mean was determined from the replicate lines of the same type: S—selected lines, CP—lines punctured with H_2O, and CN—no perturbation lines.

with autoclaved spores (Table 3). The statistically significant interaction resulted from the decrease in egg production in the selected and wound-control lines, but not the no-perturbation lines. This reduction in total egg production after treatment with autoclaved spores, especially acute in the selected lines, is consistent with the hypothesis of a cost associated with induction of an immune response (deployment).

3.5.3. Untreated Lines—Progeny Production. Table 3 presents the progeny number for lines and treatments. There were only marginal statistically significant differences in progeny production among line types ($P = 0.0961$, Table 2). In dramatic contrast to egg production, the selected lines did not produce significantly more progeny (Table 2). Progeny production is dependent on the number of sperm transferred to females and stored after the 24-hour mating period, and this did not differ appreciably among line types.

3.5.4. Autoclaved Spore and Sterile H_2O Treatments—Progeny Production. Table 3 presents the progeny numbers after treatment with autoclaved spores or sterile H_2O. The treatment effects were statistically significant overall ($P = 0.0005$,

Table 2). Autoclaved spore treatment resulted in the fewest number of progeny (CN and S) and punctured with sterile H_2O resulted in the greatest number (CN and CP) (Table 3). There were no statistically significant interactions between line types and treatments.

3.6. Progeny Development Time. Progeny development time represents egg to adult emergence time. As described in the materials and methods and Table 4, four different crosses allowed us to separate the progeny development time effects of treatment (autoclaved spores or sterile H_2O) on progenitor females or males. The fifth mating allowed us to evaluate an absence of treatments (mating pattern E in Table 4). All adults from all lines were subjected to each of the five crosses (A–E in Table 4).

3.6.1. Untreated Adult Females and Males. Figure 5 presents the cumulative percentage progeny emergence per time period for all lines types when there was no adult fly treatment. The progeny development time for selected line flies was slowest (Table 4). Statistically significant effects on progeny development were observed when there was no adult treatment ($P = 0.0098$, Table 2). This observation is

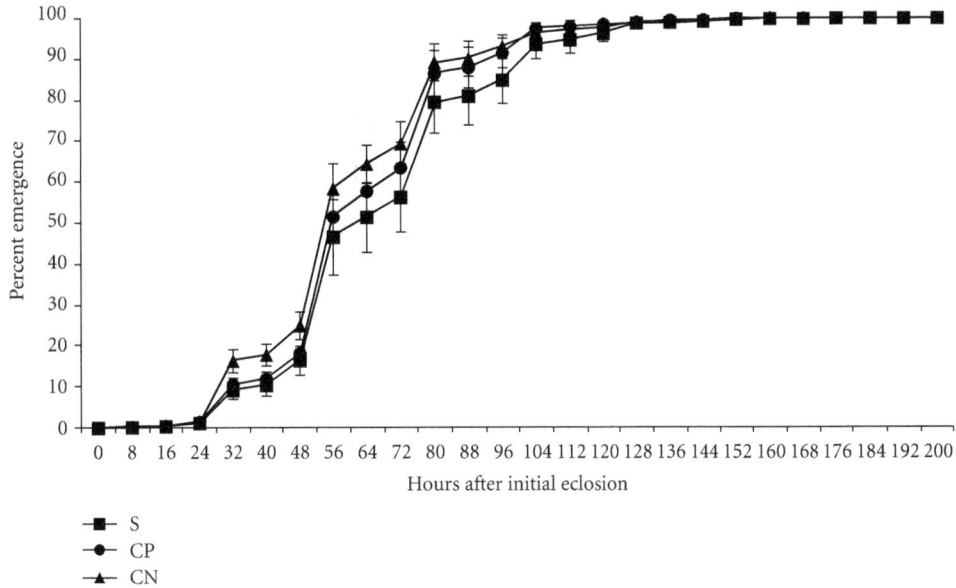

FIGURE 5: Average adult progeny emergence (eclosion) for line types when females and males were untreated. Line types: S—selected lines, CP—lines punctured with H_2O each generation, and CN—no perturbation lines.

TABLE 4: Average progeny emergence (eclosion) time from each of the following crosses applied to each line type.

Lines	Treatments	Mean (hours)	S.E.
No perturbation control (CN)	A	63.78	2.08
No perturbation control (CN)	B	63.17	2.78
No perturbation control (CN)	C	65.31	2.82
No perturbation control (CN)	D	64.14	2.58
No perturbation control (CN)	E	62.96	2.02
Control punctured (CP)	A	67.63	3.22
Control punctured (CP)	B	64.53	2.43
Control punctured (CP)	C	68.34	2.74
Control punctured (CP)	D	64.20	2.67
Control punctured (CP)	E	66.32	2.08
Selected (S)	A	71.60	3.61
Selected (S)	B	63.75	3.21
Selected (S)	C	68.45	3.50
Selected (S)	D	66.80	2.84
Selected (S)	E	70.68	3.08

Treatments: females punctured with autoclaved spores and mated with untreated males (A), males punctured with autoclaved spores and mated with untreated females (B), females punctured with sterile water and mated with untreated males (C), males punctured with sterile water and mated with untreated females (D), and untreated females mated with untreated males (E).
Line types: selected (S), punctured with H_2O each generation (CP), no perturbation each generation (CN).

consistent with the hypothesis of an evolved cost (standing defense) which is manifest as delayed progeny development time. However, an alternate hypothesis is presented in the discussion.

3.6.2. Treatment of Adult Females and Males. There was a statistically significant sex-dependent effect of adult treatment on progeny development time ($P = 0.0051$, Table 2). When females were treated with autoclaved spores, or sterile H_2O, progeny development time was similar to that observed when neither sex was treated (Figures 6(a) and 6(b)). The rank order of progeny development time was selected lines slowest, wound-control lines intermediate, and no-perturbation lines fastest. The grand mean value for male treatments (autoclaved spores and sterile H_2O, treatments B and D in Table 4) was 64.4. This value was similar to 66.7 which was the mean for untreated females and males (mating E in Table 4). Thus, when males were treated, the adult progeny emergence time dropped below the level of the untreated lines (Figures 6(a) and 6(b)). This male effect was more pronounced in the selected lines ($P = 0.0012$), again emphasizing the impact of evolution for resistance in these lines.

The number of adult progeny emerging was tabulated for each population and treatment to evaluate if there was a density effect on progeny development time. Comparing the number of progeny from treated males (treatments B and D) with treated females (treatments A and C), there was a 0.9% difference. The effect of treated males on progeny development time was not due to larval density as inferred from the number of adult progeny that eclosed.

3.7. Three-Day Survival after Administration of Autoclaved Spores. The selected lines and wound-control lines evolved an increase in mortality in the three-day period after autoclaved spores were introduced (Figure 7). Importantly, this three-day period occurred before the start of life-history assays (Figure 1). The no-perturbation control lines exhibited 2.11% mortality after introduction of the autoclaved

(a)

(b)

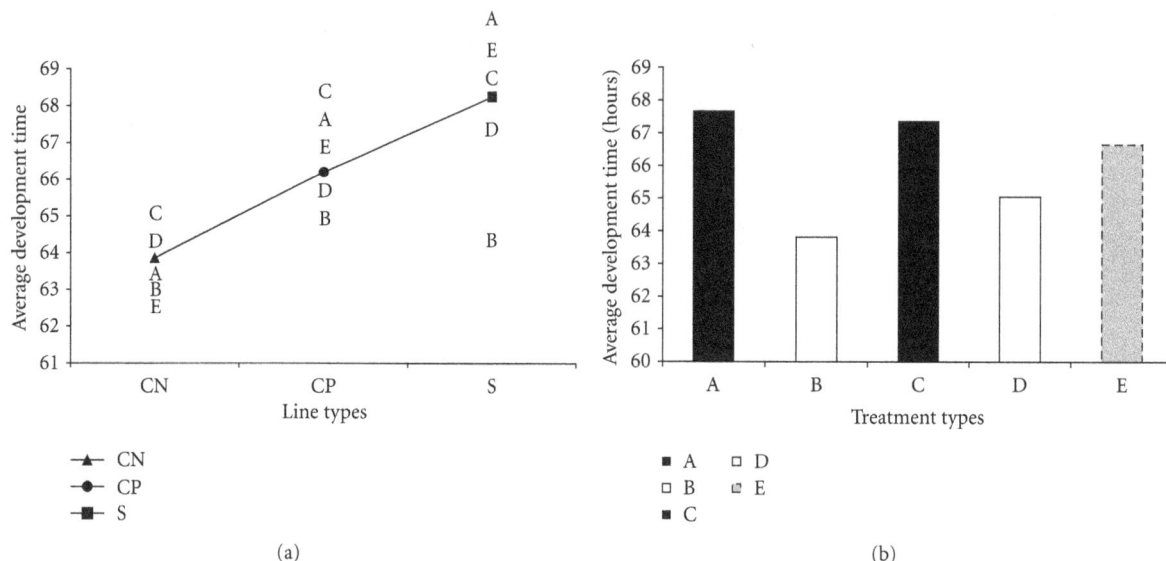

FIGURE 6: (a) Line graph of progeny development time from different F_1 matings shown for each line type. Line types: S—selected lines, CP—lines punctured with H_2O each generation, and CN—no perturbation lines. The F_1 treatments are described in Table 1 and reiterated here: A—only females treated with autoclaved spores, B—only males treated with autoclaved spores, C—only females treated with sterile H_2O, D—only males treated with sterile H_2O, and E—both sexes untreated. (b) Bar graph of overall means of progeny development time for each F_1 mating. The F_1 matings are described in Table 1 and reiterated here: A—only females treated with autoclaved spores, B—only males treated with autoclaved spores, C—only females treated with sterile H_2O, D—only males treated with sterile H_2O, and E—both sexes untreated.

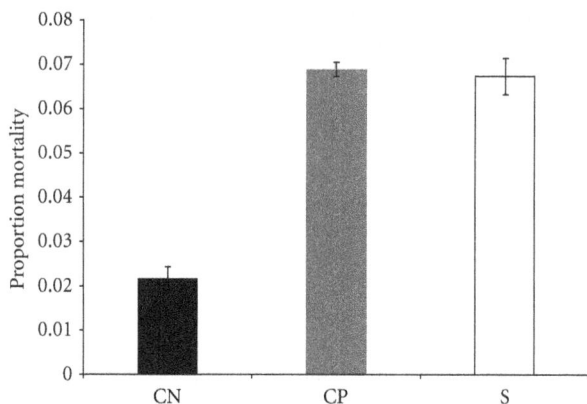

FIGURE 7: Average proportion mortality after treatment with autoclaved spores. The average was obtained by pooling all of the variates for all lines of the same type. S—selected lines, CP—lines punctured with H_2O (wound control), CN—no perturbation lines.

spores, whereas the selected lines (6.74% mortality) and wound-control lines (6.89% mortality) showed a marked increase in relative mortality after introduction of autoclaved spores. This difference was statistically significant (P = 0.0003). The increase in mortality of the selected lines compared to no-perturbation lines was 3.19-fold and 3.27-fold for the selected and wound-control lines, respectively. It is important to emphasize that there was almost no mortality after the flies were only wounded (sterile H_2O treatment). Thus, the effect shown in Figure 7 was not an evolved response to mortality from wounding. In general, there was

an early-age spike of mortality in the selected lines and wound-control lines after autoclaved spore administration. This observation is consistent with the hypothesis that there was cost of inducing an immune response (deployment).

4. Discussion

Our work here has established that D. melanogaster can evolve high levels of resistance to B. cereus spores. Specifically, a 3.3-log change in the number of spores required for 50% mortality over 24 generations was documented. After a substantial direct response to selection, we investigated life history trait indirect responses to selection which was the focus of this study. Extrapolating from McKean and Lazzaro [19], selection responses per se represent the "standing defense" which can exert a cost. Moreover, when an inducer of the immune response was introduced, autoclaved spores in the present study, then "deployment" is the cost of maintaining the immune system in an activated state under conditions where its function is unnecessary and indeed detrimental, and this can also exert a cost [19]. This is in many ways analogous to the detrimental effects of inflammation on multiple systems in humans (e.g., metabolic disorders, IBD, and arthritis) as illustrated by the inflammatory bowel diseases [20]. In the present selection experiment, we could compare the cost of selection under noninducing conditions and differentiate this cost from that observed when a putative inducer (autoclaved spores) was introduced into flies.

There were six principle indirect responses in the present study. The first was that life span was reduced after introduction of autoclaved spores into males in the selected lines, and females and males in the wound-control lines. This observation is consistent with the hypothesis of a cost associated with induction of an elevated immune response. Second, egg production was markedly elevated in untreated selected lines and to a lesser degree in untreated wound-control lines. There might have been increased titers of one or more hormones that could have a mutually stimulatory effect on reproduction and the immune response. Or, selection for increased adult fat body could underlie this indirect response as described below in the paragraph starting with "An alternative perspective…." Third, the relatively high egg production in the untreated selected lines, and to lesser degree in the wound-control lines, was abolished after introduction of autoclaved spores. The observation is consistent with the hypothesis of a cost of induced immunity in these lines. Fourth, untreated selected lines exhibited relatively slow development time. This observation is consistent with the hypothesis of an evolved cost associated with a high degree of resistance to live B. cereus spores (standing defense). However, there is an alternate adaptive hypothesis that could explain delayed development (see the paragraph below "An alternative perspective…"). Fifth, the progeny of selected line males exposed to autoclaved spores, then mated to untreated females, developed relatively rapidly. The biological interpretation of this observation is not obvious, but it might reveal a novel effect of male accessory gland secretions. Sixth, a pathological outcome (decreased early-age survival) was observed in the selected and wound-control lines after treatment with autoclaved spores. This observation is consistent with the hypothesis of a deleterious effect associated with induction of the immune system (deployment). This was an intriguing evolutionary observation as it suggests how the selection response for elevated immunity could eventually be constrained. Overall, there were multiple instances in which deployment was associated with a cost, and one case in which standing defense was potentially associated with a cost.

Evolution of resistance to bacterial spores has been observed in Aedes aegypti populations when Bacillus thuringiensis subspecies israelensis was the pathogen [21]. In this instance, resistance likely evolved due to changes in toxin-gut receptor interactions as observed in the diamond back moth [22]. In the present study, the level of resistance to B. cereus spores was many-fold greater than previously observed for selection on A. aegypti using B. thuringiensis spores. The direct response to selection in the present study was unprecedented for insect resistance to Bacillus spores.

The absence of a negative effect on life span in the selected lines in the present study differs from the selection experiment on D. melanogaster for resistance to P. aeruginosa [6], in which life span was negatively affected. There were appreciable differences between the selection process in Ye et al. [6] and the present study that could account for the different outcomes. This includes the use of vegetative cells versus spores, use of a gram-negative versus gram-positive pathogen, the size of the selected and control populations, and using approximately 50% mortality level for every generation of selection in the present study. Another possible explanation might be that P. aeruginosa is a more virulent pathogen and could exert a greater effect on life span than did B. cereus spores in the present study.

Generally, a negative relationship between reproduction and immunity has been observed [19, 23]. For example, decreased reproductive activity is associated with increased immunity in the cricket and in Drosophila [24, 25]. Thus, the major increase in egg production in the untreated selected lines in the present study was an unexpected outcome. The endocrine system of D. melanogaster might provide insight into understanding this phenomenon. In females, the evolution of elevated juvenile hormone (JH) could stimulate egg production, but elevated JH would also be expected to suppress immunity [26]. It is possible that male accessory gland proteins and peptides could have evolved to stimulate female egg production as there are a number of peptides in the male ejaculate that have this effect [27]. Almost all of the male ejaculate effect on female egg production is a result of the action of the sex-peptide [28]. However, the sex-peptide also stimulates JH production [28], and this would tend to suppress immunity. Importantly, the elevated egg production response in the selected lines was manifest for most of the reproductive life time. Thus, it does not seem likely that elevated egg production was a male effect resulting from mating for 24 hours relatively early in life. Another endocrine candidate is insulin, insulin-like signaling (ISS). ISS mutations in Drosophila result in sterility or extremely low levels of oocyte production [29, 30]. ISS in D. melanogaster is known to be the key hormone in the endocrine control of vitellogenic oocyte development [31], and ISS mediates the signals from nutrients to upregulate egg production [32]. However, the effect of elevated insulin signaling on D. melanogaster is to suppress innate immunity in this species [33]. At present, it is not clear which hormone in adult D. melanogaster could have caused a positive correlation between evolved elevated survival after B. cereus infection and high levels of egg production in the untreated selected lines.

An alternative perspective, perhaps involving hormones, might provide an explanation for the strong positive correlation between the evolution of a high degree of resistance to spores and elevated egg production. Specifically, part of the evolved response to spore infection might have been a delayed development rate which was observed in the selected lines in the present study (Figure 5, Table 4). If this delay resulted in an increase in adult fat body tissue, then there would be more of the tissue that principally secretes antimicrobial peptides. It is known that an acceleration of development time results in lower levels of fat in D. melanogaster larvae and adults [34, 35], and, conversely, a delay in development is expected to increase adult fat content. Hormones could mediate delayed development if juvenile hormone titers were relatively high in larval and pupal stages, and relatively low in the adult stage to avoid suppression of the immune response in the life stage at which selection occurred. The hypothesis of delayed development time as an adaptation could explain the positive correlation between egg production and the response to selection. Moreover, the interpretation of the delay in development as a standing defense cost of selection would be replaced by delay of development as an adaptive response.

Introduction of autoclaved spores abolished the relatively high egg production in the selected and wound-control lines; after the autoclaved spore treatment, these lines did not produce more eggs than the no-perturbation lines. This result is consistent with the hypothesis of a cost of inducing (deployment) of the immune response.

There was a striking difference between the results for egg production in the present study versus the results for progeny production. Progeny production is a much different trait. For example, the total number of progeny is limited by the number of sperm stored after a mating. Lifetime egg production does not have this kind of constraint and the numbers can be much greater.

Two examples of changes in progeny development time were observed in the present study. The selected lines exhibited delayed development when the F_1 generation was untreated. This observation is consistent with the hypothesis of an evolved (standing) cost manifest in progeny development. In the context of a species that develops rapidly, delayed development could be interpreted as a cost. However, delayed development time could be an adaptation as described in a paragraph above ("An alternative perspective…"). The implications of observing accelerated development after treatment of adult males with autoclaved spores or sterile H_2O (Figures 6(a) and 6(b)) are less clear. It might be the case that the effect of male seminal fluids is normally to delay the development time of progeny. In this scenario, when males are impacted by wounding or spores, then the normal male effect might be blocked. However, an effect of male seminal fluids on progeny development time is not established; the present study may suggest a novel function for male seminal fluids.

An interesting evolutionary observation in the present study was that introduction of autoclaved spores into the selected and wound-control lines resulted in elevated early-age mortality. This effect was observed during the three-day period when mortality was monitored after introduction of autoclaved spores, puncture with H_2O, or no treatment before life-history assays were conducted (Figure 1). In two line types, selected and wound-control, exposure to dead spores resulted in relatively high mortality. One hypothesis is that the wound-control lines evolved a similar response to the selected lines because wounding each generation activated immune responses in these lines. If these responses were costly each generation, then the wound-control lines might have evolved indirect responses that were similar to the selected lines. Another hypothesis is that the short-term mortality response of the selected lines was entirely due to wounding. However, there was no increase in mortality in either line type after treatment with sterile H_2O which is the wounding alone treatment. In general, the evolved higher mortality in selected and wound-control lines in response to autoclaved spores suggests one way in which selection for immunity can be constrained which is through counter-tending negative effects that oppose the direct response to selection.

In this study, the effect of exposure to autoclaved spores is consistent with the hypothesis of a cost of induced immunity (deployment) on life history traits. This was observed for life span, egg production, and early-age mortality. Overall, the observation of an inducible cost resulting from the introduction of autoclaved spores is consistent with the hypothesis that the selected and wound-control lines have evolved to become hyper-inducible in response to autoclaved spores. The evolution of inducible responses apparently is a general response to selection in experimental evolution studies [36, 37] and perhaps in natural populations.

An extension of the studies described here will provide insight into the evolution of *B. cereus* spore infection resistance and/or tolerance in *D. melanogaster*. Whole-genome mapping of the responses to selection is underway as is a whole-genome transcriptome study. Through these approaches, and by other means to investigate resistance/tolerance, there is potential to increase our understanding of mechanisms underlying the dramatic response of *D. melanogaster* to selection by *Bacillus cereus* spores. Novel mechanisms of resistance/tolerance may emerge from these studies.

Acknowledgments

This study was supported by a grant from the Army Research Office (AR45071-LS-DPS). Comments on the paper by two reviewers were consistently valuable and are appreciated.

References

[1] R. B. Huey and F. Rosenzweig, "Laboratory evolution meets catch-22, balancing simplicity and realism," in *Experimental Evolution*, T. Garland and M. R. Rose, Eds., University of California Press, 2009.

[2] T. Garland Jr. and S. A. Kelly, "Phenotypic plasticity and experimental evolution," *Journal of Experimental Biology*, vol. 209, no. 12, pp. 2344–2361, 2006.

[3] M. R. Rose, "Laboratory evolution of postponed senescence in *Drosophila melanogaster*," *Evolution*, vol. 38, pp. 1004–1010, 1984.

[4] L. S. Luckinbill, R. Arking, M. J. Clare, W. C. Cirocco, and S. A. Buck, "Selection for delayed senescence in *Drosophila melanogaster*," *Evolution*, vol. 38, pp. 996–1003, 1984.

[5] A. J. Zera and L. G. Harshman, "The physiology of life history trade-offs in animals," *Annual Review of Ecology and Systematics*, vol. 32, pp. 95–126, 2001.

[6] Y. H. Ye, S. F. Chenoweth, and E. A. McGraw, "Effective but costly, evolved mechanisms of defense against a virulent opportunistic pathogen in *Drosophila melanogaster*," *PLoS Pathogens*, vol. 5, no. 4, Article ID e1000385, 2009.

[7] A. R. Kraaijeveld and H. C. J. Godfrey, "Trade-off between parasitoid resistance and larval competitive ability in *Drosophila melanogaster*," *Nature*, vol. 389, no. 6648, pp. 278–280, 1997.

[8] M. D. E. Fellowes, A. R. Kraaijeveld, and H. C. J. Godfray, "Trade-off associated with selection for increased ability to resist parasitoid attack in *Drosophila melanogaster*," *Proceedings of the Royal Society B*, vol. 265, no. 1405, pp. 1553–1558, 1998.

[9] A. R. Kraaijeveld, E. C. Limentani, and H. C. J. Godfray, "Basis of the trade-off between parasitoid resistance and larval

competitive ability in *Drosophila melanogaster*," *Proceedings of the Royal Society B*, vol. 268, no. 1464, pp. 259–261, 2001.

[10] A. R. Kraaijeveld, J. Ferrari, and H. C. J. Godfray, "Costs of resistance in insect-parasite and insect-parasitoid interactions," *Parasitology*, vol. 125, pp. S71–S82, 2002.

[11] M. Lipsitch and A. O. Sousa, "Historical intensity of natural selection for resistance to tuberculosis," *Genetics*, vol. 161, no. 4, pp. 1599–1607, 2002.

[12] D. P. Kwiatkowski, "How malaria has affected the human genome and what human genetics can teach us about malaria," *American Journal of Human Genetics*, vol. 77, no. 2, pp. 171–192, 2005.

[13] M. E. Westhusin, T. Shin, J. W. Templeton, R. C. Burghardt, and L. G. Adams, "Rescuing valuable genomes by animal cloning: a case for natural disease resistance in cattle," *Journal of Animal Science*, vol. 85, no. 1, pp. 138–142, 2007.

[14] E. M. Ibeagha-Awemu, P. Kgwatalala, A. E. Ibeagha, and X. Zhao, "A critical analysis of disease-associated DNA polymorphisms in the genes of cattle, goat, sheep, and pig," *Mammalian Genome*, vol. 19, no. 4, pp. 226–245, 2008.

[15] D. C. Ko, K. P. Shukla, C. Fong et al., "A genome-wide in vitro bacterial-infection screen reveals human variation in the host response associated with inflammatory disease," *American Journal of Human Genetics*, vol. 85, no. 2, pp. 214–227, 2009.

[16] F. Calenge, P. Kaiser, A. Vignal, and C. Beaumont, "Genetic control of resistance to salmonellosis and to *Salmonella* carrier-state in fowl: a review," *Genetics, Selection, Evolution*, vol. 42, article 11, 2010.

[17] T. E. Schwasinger-Schmidt, S. D. Kachman, and L. G. Harshman, "Evolution of starvation resistance in *Drosophila melanogaster*: measurement of direct and correlated responses to artificial selection," *Journal of Evolutionary Biology*, vol. 25, pp. 378–387, 2012.

[18] W. Nicholson and P. Setlow, "Sporulation, germination and outgrowth," in *Molecular Biological Methods for Bacillus*, C. R. Harwood and S. M. Cutting, Eds., John Wiley and Sons, 1990.

[19] K. A. McKean and B. P. Lazzaro, "The costs of immunity and the evolution of immunological defense mechanisms," in *Mechanisms of Life History Evolution*, T. Flatt and A. Heyland, Eds., Oxford University Press, 2011.

[20] C. A. Anderson, G. Boucher, C. W. Lees et al., "Meta-analysis identifies 29 additional ulcerative colitis risk loci, increasing the number of confirmed associations to 47," *Nature Genetics*, vol. 43, pp. 246–252, 2011.

[21] I. F. Goldman, J. Arnold, and B. C. Carlton, "Selection for resistance to *Bacillus thuringiensis* subspecies israelensis in field and laboratory populations of the mosquito *Aedes aegypti*," *Journal of Invertebrate Pathology*, vol. 47, no. 3, pp. 317–324, 1986.

[22] B. Tabashnik, "Evolution of resistance to *Bacillus thuringiensis*," *Annual Review of Entomology*, vol. 39, pp. 47–79, 1994.

[23] L. G. Harshman and A. J. Zera, "The cost of reproduction: the devil in the details," *Trends in Ecology and Evolution*, vol. 22, no. 2, pp. 80–86, 2007.

[24] K. M. Fedorka, M. Zuk, and T. A. Mousseau, "Immune suppression and the cost of reproduction in the ground cricket, *Allonemobius socius*," *Evolution*, vol. 58, no. 11, pp. 2478–2485, 2004.

[25] K. A. McKean and L. Nunney, "Bateman's principle and immunity: phenotypically plastic reproductive strategies predict changes in immunological sex differences," *Evolution*, vol. 59, no. 7, pp. 1510–1517, 2005.

[26] T. Flatt, K.-J. Min, C. D'Alterio et al., "*Drosophila* germline modulation of insulin signaling and lifespan," *Proceedings of the National Academy of Sciences of the United States of America*, vol. 105, no. 17, pp. 6368–6373, 2008.

[27] F. W. Avila, K. R. Ram, M. C. Bloch Qazi, and M. F. Wolfner, "Sex peptide is required for the efficient release of stored sperm in mated *Drosophila* females," *Genetics*, vol. 186, no. 2, pp. 595–600, 2010.

[28] E. Kubli, "Sex-peptides: seminal peptides of the *Drosophila* male," *Cellular and Molecular Life Sciences*, vol. 60, no. 8, pp. 1689–1704, 2003.

[29] D. J. Clancy, D. Gems, L. G. Harshman et al., "Extension of life-span by loss of CHICO, a *Drosophila* insulin receptor substrate protein," *Science*, vol. 292, no. 5514, pp. 104–106, 2001.

[30] M. Tatar, A. Kopelman, D. Epstein, M. P. Tu, C. M. Yin, and R. S. Garofalo, "A mutant *Drosophila* insulin receptor homolog that extends life-span and impairs neuroendocrine function," *Science*, vol. 292, no. 5514, pp. 107–110, 2001.

[31] D. S. Richard, R. Rybczynski, T. G. Wilson et al., "Insulin signaling is necessary for vitellogenesis in *Drosophila melanogaster* independent of the roles of juvenile hormone and ecdysteroids: female sterility of the chico1 insulin signaling mutation is autonomous to the ovary," *Journal of Insect Physiology*, vol. 51, no. 4, pp. 455–464, 2005.

[32] D. Drummond-Barbosa and A. C. Spradling, "Stem cells and their progeny respond to nutritional changes during *Drosophila* oogenesis," *Developmental Biology*, vol. 231, no. 1, pp. 265–278, 2001.

[33] J. R. DiAngelo, M. L. Bland, S. Bambina, S. Cherry, and M. J. Birnbaum, "The immune response attenuates growth and nutrient storage in *Drosophila* by reducing insulin signaling," *Proceedings of the National Academy of Sciences of the United States of America*, vol. 106, no. 49, pp. 20853–20858, 2009.

[34] A. K. Chippindale, J. A. Alipaz, H.-W. Chen, and M. R. Rose, "Experimental evolution of accelerated development in *Drosophila*. 1. Developmental speed and larval survival," *Evolution*, vol. 51, no. 5, pp. 1536–1551, 1997.

[35] N. G. Prasad, M. Shakarad, D. Anitha, M. Rajamani, and A. Joshi, "Correlated responses to selection for faster development and early reproduction in *Drosophila*: the evolution of larval traits," *Evolution*, vol. 55, no. 7, pp. 1363–1372, 2001.

[36] L. G. Harshman, J. A. Ottea, and B. D. Hammock, "Evolved environment-dependent expression of detoxication enzyme activity in *Drosophila melanogaster*," *Evolution*, vol. 45, pp. 791–795, 1991.

[37] T. Garland Jr. and S. A. Kelly, "Phenotypic plasticity and experimental evolution," *Journal of Experimental Biology*, vol. 209, no. 12, pp. 2344–2361, 2006.

Permissions

The contributors of this book come from diverse backgrounds, making this book a truly international effort. This book will bring forth new frontiers with its revolutionizing research information and detailed analysis of the nascent developments around the world.

We would like to thank all the contributing authors for lending their expertise to make the book truly unique. They have played a crucial role in the development of this book. Without their invaluable contributions this book wouldn't have been possible. They have made vital efforts to compile up to date information on the varied aspects of this subject to make this book a valuable addition to the collection of many professionals and students.

This book was conceptualized with the vision of imparting up-to-date information and advanced data in this field. To ensure the same, a matchless editorial board was set up. Every individual on the board went through rigorous rounds of assessment to prove their worth. After which they invested a large part of their time researching and compiling the most relevant data for our readers. Conferences and sessions were held from time to time between the editorial board and the contributing authors to present the data in the most comprehensible form. The editorial team has worked tirelessly to provide valuable and valid information to help people across the globe.

Every chapter published in this book has been scrutinized by our experts. Their significance has been extensively debated. The topics covered herein carry significant findings which will fuel the growth of the discipline. They may even be implemented as practical applications or may be referred to as a beginning point for another development. Chapters in this book were first published by Hindawi Publishing Corporation; hereby published with permission under the Creative Commons Attribution License or equivalent.

The editorial board has been involved in producing this book since its inception. They have spent rigorous hours researching and exploring the diverse topics which have resulted in the successful publishing of this book. They have passed on their knowledge of decades through this book. To expedite this challenging task, the publisher supported the team at every step. A small team of assistant editors was also appointed to further simplify the editing procedure and attain best results for the readers.

Our editorial team has been hand-picked from every corner of the world. Their multi-ethnicity adds dynamic inputs to the discussions which result in innovative outcomes. These outcomes are then further discussed with the researchers and contributors who give their valuable feedback and opinion regarding the same. The feedback is then collaborated with the researches and they are edited in a comprehensive manner to aid the understanding of the subject.

Apart from the editorial board, the designing team has also invested a significant amount of their time in understanding the subject and creating the most relevant covers. They scrutinized every image to scout for the most suitable representation of the subject and create an appropriate cover for the book.

The publishing team has been involved in this book since its early stages. They were actively engaged in every process, be it collecting the data, connecting with the contributors or procuring relevant information. The team has been an ardent support to the editorial, designing and production team. Their endless efforts to recruit the best for this project, has resulted in the accomplishment of this book. They are a veteran in the field of academics and their pool of knowledge is as vast as their experience in printing. Their expertise and guidance has proved useful at every step. Their uncompromising quality standards have made this book an exceptional effort. Their encouragement from time to time has been an inspiration for everyone.

The publisher and the editorial board hope that this book will prove to be a valuable piece of knowledge for researchers, students, practitioners and scholars across the globe.

List of Contributors

He-ping Jiang and Qing-song Zheng
College of Natural Resources and Environmental Science, Key Laboratory of Marine Biology, Nanjing Agricultural University, Nanjing, Jiangsu 210095, China

Ming Zhu
College of Natural Resources and Environmental Science, Key Laboratory of Marine Biology, Nanjing Agricultural University, Nanjing, Jiangsu 210095, China
College of Oceanography, Huaihai Institute of Technology, Lianyungang, Jiangsu 222005, China

Bing-bing Gao, Wen-hui Li and Chang-hai Wang
College of Natural Resources and Environmental Science, Key Laboratory of Marine Biology, Nanjing Agricultural University, Nanjing, Jiangsu 210095, China

Chun-fang Zheng
Zhejiang Mariculture Research Institute, Zhejiang Key Laboratory of Exploitation and Preservation of Coastal Bio-Resource, Wenzhou, Zhejiang 325005, China

Stuart J. Lucas and Hikmet Budak
Faculty of Engineering and Natural Sciences, Sabanci University, Orhanlı, Tuzla, 34956 Istanbul, Turkey
Sabanci University Nanotechnology Research and Application Centre (SUNUM), Sabanci University, Orhanlı, Tuzla, 34956 Istanbul, Turkey

Bala Anı Akpınar
Faculty of Engineering and Natural Sciences, Sabanci University, Orhanlı, Tuzla, 34956 Istanbul, Turkey

Masami Hasegawa
School of Life Sciences, Fudan University, Shanghai 200433, China
Department of Statistical Modeling, Institute of Statistical Mathematics, Tokyo 190-8562, Japan

Takahiro Yonezawa
School of Life Sciences, Fudan University, Shanghai 200433, China

M. Mosquera and N. Geribàs
Universitat Rovira i Virgili (URV), Campus Catalunya, Avinguda de Catalunya 35, 43002 Tarragona, Spain
Institut Catala de Paleoecologia Humana i Evolució Social (IPHES), Campus Catalunya, Avinguda de Catalunya 35, 43002 Tarragona, Spain

A. Bargallo
Institut Catala de Paleoecologia Humana i Evoluci´o Social (IPHES), Campus Catalunya, Avinguda de Catalunya 35, 43002 Tarragona, Spain

M. Llorente and D. Riba
Institut Catala de Paleoecologia Humana i Evolució Social (IPHES), Campus Catalunya, Avinguda de Catalunya 35, 43002 Tarragona, Spain
Unitat de Recerca i Laboratori d'Etologia, Fundacio Mona, Carretera de Cassa 1 km, Riudellots de la Selva, 17457 Girona, Spain

Hikmet Budak, Melda Kantar and Kuaybe Yucebilgili Kurtoglu
Biological Sciences and Bioengineering Program, Faculty of Engineering and Natural Sciences, Sabanci University, 34956 Tuzla, Istanbul, Turkey

P. K. Rout, A. Mandal and R. Roy
Central Institute for Research on Goats, Makhdoom, Farah, Mathura 281122, India

K. Thangraj
Centre for Cellular and Molecular Biology, Uppal Road, Hyderabad 500007, India

Salvatore Bordonaro, Anna Maria Guastella, Andrea Criscione, Antonio Zuccaro and Donata Marletta
DISPA, Sezione di Scienze delle Produzioni Animali, Universita degli studi di Catania, Via Valdisavoia 5, 95123 Catania, Italy

Masahiko Muraji
Division of Insect Sciences, National Institute of Agro biological Sciences, Ibaraki 305-8634, Japan

Norio Arakaki
Plant Disease and Insect Pest Management Section, Okinawa Prefectural Agricultural Research Center, Okinawa 901-0336, Japan

Shigeo Tanizaki
Arakawa 2357-11, Ishigaki, Okinawa 907-0024, Japan

Wenying Liu, Kenming Yu, Tengfei He, Dongxu Zhang and Jianxia Liu
School of Life Science, Shanxi Datong University, Datong 037009, China

Feifei Li
School of Agriculture and Food Science, Zhejiang Agriculture and Forestry University, Hangzhou 311300, China

Hangxia Jin, Guangli Xu, Qingchang Meng, Fang Huang and Deyue Yu
State Key Laboratory of Crop Genetics and Germplasm Enhancement, National Center for Soybean Improvement, Nanjing Agricultural University, Nanjing 210095, China

A. J. Durston
Sylvius Laboratory, Institute of Biology, University of Leiden, Wassenaarseweg 72, 2333 BE Leiden, The Netherlands

José Luiz Rybarczy k-Filho
Departamento de Fısica e Biofısica, Universidade Estadual Paulista, Distrito de Rubiao Junior, S/N, 18618-970 Botucatu, SP, Brazil

Ricardo D'Oliveira Albanus, Rodrigo Juliani Siqueira Dalmolin, Mauro Antônio Alves Castro and José Cláudio Fonseca Moreira
Departamento de Bioquımica, Universidade Federal do Rio Grande do Sul, Rua Ramiro Barcelos 2600, 90040-180 Porto Alegre, RS, Brazil

Zhiwen Ding
School of Life Sciences and Biotechnology, Shanghai Jiao Tong University, Shanghai 200240, China
Department of Physics, Shanghai Jiao Tong University, Shanghai 200240, China
Institutes of Biomedical Sciences, Fudan University, Shanghai 200032, China

Gaibin Lian, Dabing Zhang and Jie Xu
School of Life Sciences and Biotechnology, Shanghai Jiao Tong University, Shanghai 200240, China

Qin Wang
Department of Physics, Shanghai Jiao Tong University, Shanghai 200240, China

Ufuk Tatli, Mehmet Kürkçü and Mehmet Emre Benlidayi
Department of Oral and Maxillofacial Surgery, Faculty of Dentistry, Cukurova University, Saricam-Balcali, 01330 Adana, Turkey

Yakup Üstün
Private Practice in Oral and Maxillofacial Surgery, 01120 Adana, Turkey

Dieter Anseeuw
Interdisciplinary Research Centre, K. U. Leuven Campus Kortrijk, Etienne Sabbelaan 53, 8500 Kortrijk, Belgium
KATHO, Wilgenstraat 32, 8800 Roeselare, Belgium

Paul Busselen
Interdisciplinary Research Centre, K. U. Leuven Campus Kortrijk, Etienne Sabbelaan 53, 8500 Kortrijk, Belgium

Bruno Nevado and Erik Verheyen
Vertebrate Department, Royal Belgian Institute of Natural Sciences, Vautierstraat 29, 1000 Brussels, Belgium
Evolutionary Ecology Group, University of Antwerp, Middelheimcampus G.V. 332, Groenenborgerlaan 171, 2020 Antwerp, Belgium

Jos Snoeks
Zoology Department, Royal Museum for Central Africa, Leuvensesteenweg 13, 3080 Tervuren, Belgium
Laboratory of Biodiversity and Evolutionary Genomics, K. U. Leuven, Charles Deberiotstraat 32, 3000 Leuven, Belgium

Etienne Bezault
UMR 110, Cirad-Ifremer INTREPID, 34398 Montpellier, France
INRA, UMR 1313 Genetique Animale et Biologie Integrative, 78352 Jouy-en-Josas, France
Department of Biology, Reed College, Portland, OR 97202, USA

Bernard Chevassus
INRA, UMR 1313 Genetique Animale et Biologie Integrative, 78352 Jouy-en-Josas, France

Karim Gharbi
INRA, UMR 1313 Genetique Animale et Biologie Intégrative, 78352 Jouy-en-Josas, France
Institute of Evolutionary Biology, School of Biological Sciences, University of Edinburgh, Edinburgh EH9 3JT, UK

Jean-Francois Baroiller
UMR 110, Cirad-Ifremer INTREPID, 34398 Montpellier, France

Xavier Rognon
INRA, UMR 1313 Genetique Animale et Biologie Integrative, 78352 Jouy-en-Josas, France
Agro Paris Tech, UMR 1313, Genetique Animale et Biologie Integrative, 75231 Paris, France

Matti Jalasvuori
Department of Biological and Environmental Science, Center of Excellence in Biological Interactions, University of Jyvaskyla, 40014 Jyvaskyla, Finland
Division of Evolution, Ecology and Genetics, Research School of Biology, Australian National University, Canberra, ACT 0200, Australia

German Avila-Sakar and Cora Anne Romanow
Department of Biology, The University of Winnipeg, Winnipeg, MB, Canada

Olavo Pinhatti Colatreli, Natasha Verdasca Meliciano, Izeni Pires Farias and Tomas Hrbek
Laboratorio de Evolucao e Genetica Animal (LEGAL), Universidade Federal do Amazonas (UFAM), 69077-000 Manaus, AM, Brazil

Daniel Toffoli
Laboratorio de Evolucao e Genetica Animal (LEGAL), Universidade Federal do Amazonas (UFAM), 69077-000 Manaus, AM, Brazil
Departamento de Genetica e Evolucao, Universidade Federal de Sao Carlos (UFSCar), 18052-780 Sao Carlos, SP, Brazil

Hie Lim Kim
Center for the Promotion of Integrated Sciences, The Graduate University for Advanced Studies (SOKENDAI), Hayama, Kanagawa 240-0193, Japan
Department of Biochemistry and Molecular Biology, Pennsylvania State University, 312 Wartik Laboratory, University Park, PA 16802, USA

Mineyo Iwase and Tasuku Nishioka
Center for the Promotion of Integrated Sciences, The Graduate University for Advanced Studies (SOKENDAI), Hayama, Kanagawa 240-0193, Japan

Takeshi Igawa
Institute for Amphibian Biology, Graduate School of Science, Hiroshima University, Higashihiroshima, Hiroshima 739-8526, Japan

Satoko Kaneko
Laboratory of Plant Genetics, Graduate School of Agriculture, Kyoto University, Kyoto 606-8502, Japan

Yukako Katsura and Yoko Satta
Department of Evolutionary Studies of Bio systems, The Graduate University for Advanced Studies (SOKENDAI), Hayama, Kanagawa 240-0193, Japan

Naoyuki Takahata
The Graduate University for Advanced Studies (SOKENDAI), Hayama, Kanagawa 240-0193, Japan

Junjie Ma and Andrew K. Benson
Department of Food Science and Technology, University of Nebraska Lincoln, Lincoln, NE 68583, USA

Stephen D. Kachman
Department of Statistics, University of Nebraska Lincoln, Lincoln, NE 68583, USA

Zhen Hu and Lawrence G. Harshman
School of Biological Sciences, University of Nebraska Lincoln, Lincoln, NE 68588, USA

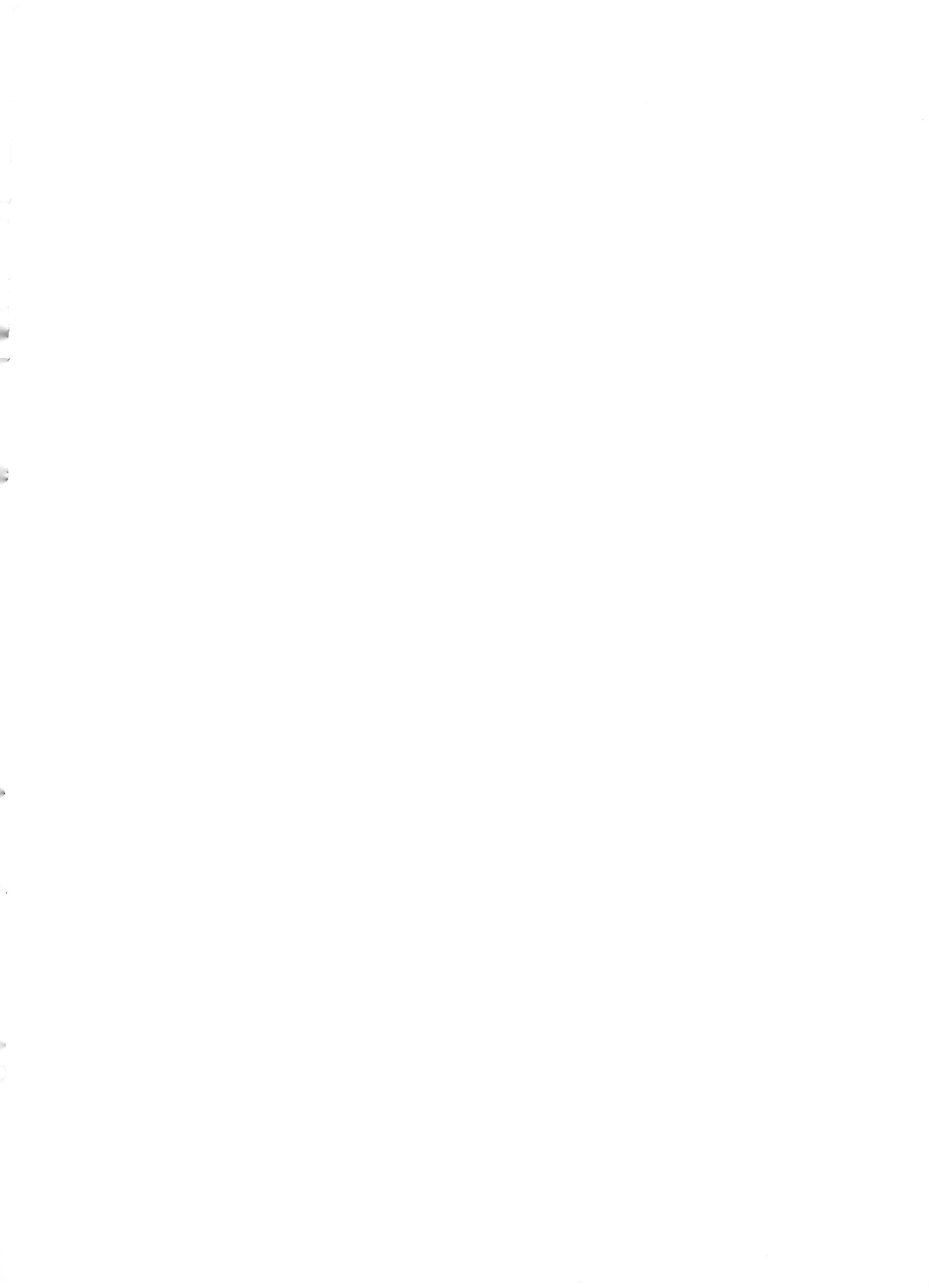

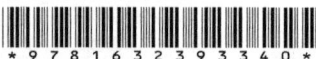